McTaggart's
RED KEG

1867 1868

McTaggart's RED KEG

LOGGING *from* A–Z

on the TITTABAWASSEE in Michigan

ILLUSTRATED

IRENE M. HARGREAVES HAROLD M. FOEHL

International Standard Book Number 0-911230-01-7
Library of Congress Catalog Card Number 88-92030

Red Keg
PRESS
Bay City, Michigan

Printed in the United States of America

Book Design and Illustrations by the Authors

Printed by BookCrafters

Henry Foehl
1875–1960

An Artisan In Metals

The woodsman stands and gazes upon the massive pine, nor stops to think of the ages of time it represents. He only thinks how many saw logs it will make.

—John W. Fitzmaurice
The Shanty Boy

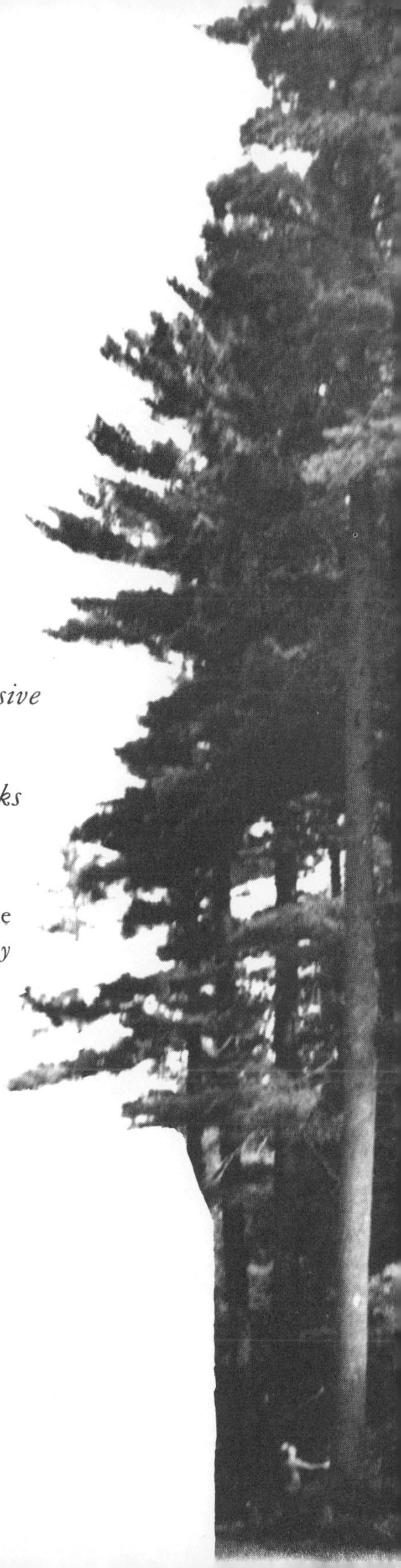

CONTENTS

CHAPTER

BOOK TWO
LOG DRIVE

Chapter titles are from the old loggers' "Alphabet Song," modified by the authors.

ILLUSTRATIONS

The pictures and maps listed below were created and executed by the authors—unless otherwise noted.

MAPS

BOOK ONE
LUMBER CAMP

BOOK TWO
LOG DRIVE

BOOK ONE
LUMBER CAMP

Once upon a time there was a place in Michigan called Red Keg*. Whether it started out as a logging camp, or a saloon (as the legend goes), is lost to the mists of the past.

The camp called McTaggart's Red Keg and the characters associated with it are figments of the authors' imaginations. Everything else—descriptions of the times, places and methods in the woods and on the Tittabawassee and Saginaw rivers are true and based upon a background of facts.

*Now called Averill

Introduction

Here they come! The season is autumn. The year is 1867. And the choppers, the sawyers, the teamsters, the loaders, the swampers and the road monkeys are on their way back to work another winter at Camp 3—the lumber camp that's called "McTaggart's Red Keg" by the crew.

The camp is owned by the lumberman John Shannon, and it is the third one he has built on the Tittabawassee[1]—the log-driving river that will have floated, by the time the great pine days are gone, over thirteen billion, five hundred

million board feet of Northern Pine[2] to the sawmills that dot the mainstream of the industry—the historic old Saginaw.

You can hear the boys coming down the tote road—full of beans (and a little Saginaw red-eye). Some of the woodsmen are native Michiganians, but for the most part they are Scots from Ontario, French Canucks from Quebec, Bangor Tigers from Maine and Blue Noses from New Brunswick. The accessible pine has all but disappeared from the Northeast, and the men who cut it down have come west to work in Michigan's forests, bringing with them their know-how as well as their tools, their customs and their songs.

All summer long the men have been laboring as rivermen on the log drives—as mill hands, dock wallopers and shipyard workers in the sawmill towns on the Saginaw. Some have been wresting crops from stump-farm homesteads. Others have been sailing the Lakes on ships with holds loaded to the plank sheer and decks piled high above the rail with "Saginaw Pine"—pine lumber bound for the docks in Buffalo and Albany.

Trees once grew to the water's edge the length of the Tittabawassee's five-mile-wide, eighty-mile-long pine belt. But now, you can see openings where the men in the lumber camps pile their crops of pine and where in the springtime they send the logs rolling from the banks into the cold water below to float them to the sorting booms at the river mouth.

There at a bend in the river, some twenty-five miles up from the mouth of the Tittabawassee, you can see the lumber camp called McTaggart's Red Keg, a cluster of log buildings built close to the ground, sitting snug against a background of Norways. The camp looks much like the lumber camps the Bangor men knew on Maine rivers.

One of the buildings is where the men sleep (and until this year, ate their meals as well). You can distinguish the bunk-camp from the other buildings by its one dominating feature—a great stick and clay chimney built over an eight-foot-square fireplace which occupies the center of the building. A new log structure, only recently erected, has been built at one end of the bunk-camp—the two buildings connected by a roofed-over storage area that is called a "dingle." A metal smokestack rises from the new building. It shows that the cook at Red Keg will cook his fat pork on an iron wood-burner and not over a fire in the camboose, as has been his practice until this season. A pole roof covers the buildings, and on top of the roof are laid boughs of spruce and hemlock to insulate the inside against frigid Michigan winters. A little apart from the bunk and cook camps is the van where the foreman and the scaler sleep. And behind the "push's" quarters is the stable which is handy to another log building that is the blacksmith shop.

The camp buildings sit back from scarred river banks in the midst of tree stumps, tangles of trampled yellowing vines and ferns. Rutted logging roads, leading to and from the woods, skirt clumps of discarded pine tops and stumps that show where the woodsmen toiled last season. The roads wind through scatterings of hardwood trees, sumac and underbrush (now blazing in their autumn colors)—through cedar thickets and Norway pines, hemlocks and spruce trees—and through clumps of white pine trees that have not yet felt the keen edge of the chopper's ax.

A sandy ridge follows the winding course of the river. On its slopes grow splendid pines that have reached their full maturity. They stand head and shoulders above the other trees in the forest—their billowy tops seeming to sweep the sky overhead. These and the lesser noble of the

Pinus strobus are what the lumbermen seek in the Saginaw Valley.

What is it about Northern White Pine that moves men to refer to it as "green gold"—that motivates them to travel hundreds of miles on foot, by rail, canal boat, sailing ship and steamboat to search it out? What is it about the cone-bearing tree that compels fortune seekers to go into debt or to risk their capital to invest money in lands on which the trees grow?

This is a time when most everything in town and countryside is built of wood. Houses, out-houses, barns and places of business—sidewalks, cisterns, ships and railroad cars—all are built of wood. And as the nation grows—so grows the hue and cry for more and more lumber.

The wood that best accommodates both builder and lumberman is that of the Northern White Pine tree. Wood workers like the clear creamy wood of the pine tree, for it is easily worked by saw and plane. It holds nails and paint well and does not split like the wood of the Norway.

The lumberman is partial to pine because its purchasing price is so cheap. There is a lot of it on government lands for just the taking. And unlike the hardwoods, pine floats like cork, requiring nothing more than a river and rivermen to run it to the sawmill.

In 1867, lumber means *pine* (as it will for twenty years more). It is the backbone of the lumber industry.

The great stands of virgin white pine in the Saginaws withstood the assault of the fortune seekers—until 1847. That was the year when a canal boat loaded with clear pine lumber hauled up to a wharf at Albany. The lumber had come from the Saginaw Valley in Michigan—having traveled via the Lakes and the length of the Erie Canal to reach the great lumber center. Nothing like Saginaw Pine had been seen in Albany in years.

When the eastern lumbermen learned where the pine came from, it wasn't long before there appeared in the forests of the Saginaw Valley a hardy breed of men who were equipped with packs on their backs and Government land maps in their pockets. The men were timber cruisers and they came to seek out pine for their eastern backers. Theirs was an occupation that was called in the early days of logging "the great guessing game of Maine."[3]

John Shannon was only one lumberman who was lured to the Saginaws by the pine lumber he had seen on the wharves in Albany. After investigating the pine situation in the Valley for himself, he engaged a fellow "York-stater" to go west to Michigan to look for pine lands to purchase. The name of the "looker" was Alex McTaggart, a Scotsman whose reputation as a master woodsman in the Chemung Valley pineries was well known to the lumberman.

On his cruising expeditions, McTaggart disappeared into the woods for months at a time, enduring great privation to seek out the green gold. In the winter he paced on snowshoes through the frozen wilderness in his search. In the close heat of summer he continued his survey, tramping through bogs, climbing over windfalls and struggling through dense brush with clouds of black flies and mosquitoes hovering around him wherever he went.

When the looker discovered a stand of white pine worthy of investigation, he carefully examined the terrain where the trees grew, noting their proximity to river or stream. Satisfied that the pine was close enough to water for floating to the mill, he would search for old government blaze marks on trees to locate corners of land sections, so he could note them on his maps. He paced off the sections, following old and many times poorly defined boundary lines—all the while jotting down descriptions of his findings in his note book—notations that included the number of pine trees to the acre—their quality—and an estimate of the number of board feet of lumber that could be produced from the trees.

At nightfall, the cruiser built a fire to boil his tea and cook his rasher of salt pork. Before retiring he would take out his Minutes, and by the wavering light of his camp fire, check over the figures and descriptions he had entered in his notebook during the day. When he was satisfied that his descriptions were in order, he would put them in a safe place, roll up in a blanket and fall asleep with his feet to the fire.

McTaggart located section after section of government land for John Shannon. He entered them at the government land office in Flint, where they were purchased for the boss-lumberman in New York for the going price of $1.25 an acre.[4]

Most of the tracts of pineland that the cruiser McTaggart entered for his employer were located on the watersheds of the Saginaw River—once said to be the richest pine tracts in the world. No other logging stream in the nation would float such enormous quantities of pine logs to its mills.[5]

In the month of October, the year 1864, while Sheridan was sweeping through the Shenandoah Valley, the lumberman John Shannon was building his first set of camps on the Tittabawassee in Michigan. By the time the war between the States was over, he was well established in the Saginaw Valley with two more lumber camps on the log-driving river and a sizable sawmill on the Saginaw.

Now, the year is 1867. The summer is over. Fall is in the air and the woodsmen have answered the call. They are on their way back to work another winter in the pine. You can see them coming—straggling down the tote road with their turkeys slung over their shoulders. There's Jimmy-the-Chopper, Crosscut Bill, Frenchy LaRiviere, Pat Toohey, Jimmy MacDonald and all the others who are rearin' to have another go at falling the pine.

The reader is introduced to McTaggart's crew and learns about the jobs they do at Red Keg — and about how Camp 3 got its name.

1

A is for the AX-MAN that you all know

It didn't take long for the boys to settle in at Red Keg. They threw their turkeys into their bunks and were out in the woods in no time, working at the job they had come to do. Alex McTaggart, foreman, wanted one million board feet of pine cut and piled at the skidways by the time the sleighs could be gotten into operation.

And while everyone, from the head push on down to the lowly chickadee, kept their fingers crossed for freezing weather and an early snowfall, they went about their business six days a week working in the woods from dawn until it was too dark to see.

No matter the season, the men never saw the bunkcamp in daylight, except on Sundays. It was strictly against lumber camp policy to leave for the woods after sunrise, or to return to camp before it was dark. But, like people who take pride in their hard life, the woodsmen at Camp 3 could joke about it—especially about the long hours they worked.

"Day begins in the lumber camp anytime after midnight," they would say. Or—"the hours per day we work in the bush all depends on how little sleep the foreman can worry along with." That was a favorite saying of the Canadians.[1]

There was one private joke the boys at Red Keg never tired of repeating. It started early one morning last winter when they were waiting outside the bunk-camp for their work instructions from McTaggart. The stars were still bright in the sky when Frank Cole remarked, to nobody in particular, that it was "goin' to be a real nice day." And Whispering Bill, in his usual droll way answered him, saying, "I don't know what kind of a day it's goin' to be, Frank, but it sure is a nice night." That was a real knee-slapper at Red Keg.

After supper, when the men relaxed around the camboose trying to see who could tell the biggest lies about their experiences in the woods, Jim McHugh, the chopper from Maine, would tell (before anyone could stop him), about early rising in a lumber camp he worked in up on the West Branch of the Penobscot. There not being a clock or a pocket watch in camp, the cookee often mistook the moonlight for dawn, and called out the wretched crew for breakfast in the middle of the night. On those "mornings," the men would eat their beans and tramp to the cutting area, only to build a fire to keep from freezing to death and to keep the howling wolves at bay, while they waited around for daylight.[2]

That old camp was where Jimmy McHugh learned the trick of chopping—where he learned to swing an ax with long muscular arms that never ceased motion until the pine tree he was cutting toppled to the ground.

The chopper was born in pine country in Arrostook County, going to work in a lumber camp as a choreboy before he could hold an ax, let alone swing one. When he was older, he "graduated" to a swamper's job. But being a swamper was to Jim no better than being the camp choreboy. In his mind, there was no difference between the jobs. Instead of using a broom to clean up around camp, it was the swamper's job to wield an ax to make clearings—to hack out trails for the ox teams so they could skid or drag the big logs from the skidway to the stream. To Jim, the swamper's job was something like being a choreboy out in the woods.

In those days, if there was any class distinction among the workers in the lumber camp, the choppers were rated at the top of the heap. It was Jim's one ambition to join the elite corps. It began the day he saw one of the chosen fall a tree.

Once the boy got it into his head that he wanted to be an ax-man, he began pestering one of the head choppers to teach him his art. Finally, to get some peace, the older man gave in and showed Jimmy how to plant his feet firmly on the ground and how to hold and swing the heavy tool, so he wouldn't lose his balance (or jolt his arms out of their sockets) when the ax bit into the tree.

Jim's first day at chopping nearly put an end to his career. That night, and for a week afterward, he ached in every muscle. He was so sore all over, if it hadn't been for his kind-hearted chopper friend rubbing his aching arms and shoulder muscles with bear grease before the blazing heat of the fire, he might have been lame for life. At least—that's the way Jim told it.

That first week's venture with an ax didn't discourage the would-be chopper, and after practicing for months on end the proper stance and rhythm, he at last became an expert ax-man. That's when the woodsmen in camp began calling him "Jimmy-the-Chopper." And the name stuck.

Jim's tool was a single-bit ax (the double-bit ax had not yet been introduced to McTaggart's choppers). It weighed about four pounds—the single blade tapering to a keen cutting edge with rounded corners (after the Michigan ax pattern). The ax was designed to cut deep and not bind in the timber.[3]

The camp blacksmith knew how to forge ax heads—but handforged axes couldn't be depended upon for a good uniform blade every time—so, it was the policy of the lumberman Shannon to order axes for his camps from the Collins Company, a firm that had been manufacturing woodsmens' axes in Canton, Connecticut since 1826.[4]

To make a clean cut and not waste his strength, Jim kept the cutting edge of his ax very sharp. Every night he sharpened the steel bit by hand, using a hand-held whetstone.

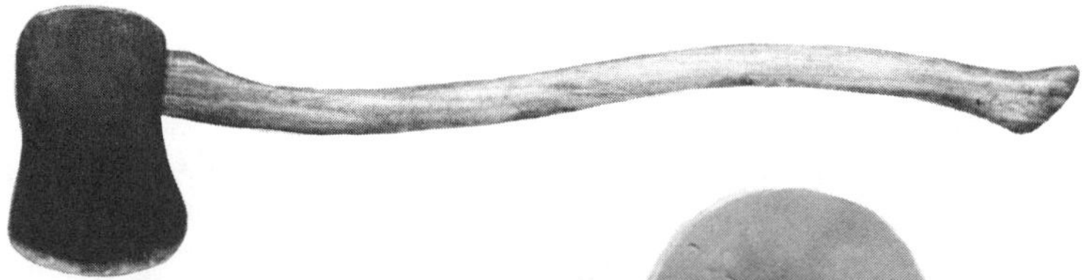

Jim's ax was made after the Michigan ax pattern, with rounded corners, designed to cut deep and not bind in the timber.

The grindstone.

After his nightly chore, he could have used the blade to whittle or to shave with—the ax's cutting edge was so keen. To keep the blade behind the edge thinned down, Jim sometimes took his turn at the grindstone—but he always finished the edge by hand with a whetstone.

The satin-smooth, gracefully-curved helve of Jim's ax was made from the wood of the hickory tree—the wood having a natural elasticity or "spring" that choppers liked. But even the best made handles sometimes broke and had to be replaced, so the foreman purchased them by the box from a lumberman's supply house in Saginaw City.

Unlike the other choppers in camp, Jim made his own ax handles, and on any Sunday, he might be found in the woods searching for a hickory tree that would make good helves. On nights when he wasn't sharpening his ax, he sat on the deacon's bench, ankle deep in hickory shavings, shaping the wood into new handles.

The blacksmith "hung" most of the axes in McTaggart's camp—that is, he fitted and made fast the handle to the "eye" of the ax head—a job that required considerable skill. But again, Jimmy-the-Chopper preferred to do his own work. And after he tried out a new ax for its "heft" (weight and balance), and found it to his liking, he would allow no one else to use it.[5] At night, the ax lay at his side, wrapped in a piece of an old horse blanket.

The men had been in camp for a week, and although it was not unusual in Michigan to see snow flurries in October, there had been no such welcome signs of approaching winter. Each day, while the hardwoods slowly shed their bright mantles of crimson and yellow, the men labored in the smokey heat of a late Indian summer. When night came, they tramped, dog-tired, back to the bunk-camp feeling weighed down, depressed by the thought of working another day in such unseasonable heat.

"It can't last forever—the weather's bound to change," the foreman kept telling himself.

And change it did—the crew waking that morning to find that a cold snap had hit the camp sometime in the night, making the trees and undergrowth ghostly white with hoarfrost. The air crackled with cold and the woodsmen were heartened by the change. For the first time since they came into camp, they could feel winter in the air, and they went off to their work with new found energy.

As was his usual practice, McTaggart had given his instructions to the choppers as to which trees were to be felled that day. Only the choicer pines were cut at Red Keg—the more common grades were ignored.[6] (The foreman acted upon instructions from the lumberman Shannon). When it was light enough to see to work, Jim and his partner, Frank Cole, were hard at their business.

Northern White Pine grows in a mixed forest—that is, the trees are usually found growing in clumps, or scattered among hardwoods or other cone-bearing trees such as the Norways. Pure stands of pine were not uncommon, but when timber cruisers discovered virgin trees such as the ones the crew were working in that wintery fall morning, they must have felt like Cortés, the Spanish explorer, when he first saw the dazzling gold treasures of the Aztec prince, Montezuma.

The pines grew on a broad sandy ridge that followed the winding Tittabawassee. They were prize "cork" pines[7]—mature trees that soared high above the earth, their small crowns intertwining nearly two hundred feet above the floor of the forest. Because they had grown up among trees standing thickly around them, their lower branches had long since disappeared. Standing among these great pillars, it was difficult to imagine them as ever being delicate seedlings topped by clusters of slender blue-green needles.

When the pines first poked their heads above the moist earth is not known—but certainly it was close to the time when the Jesuit Marquette founded Michigan's first mission at Sault Sainte Marie in 1668. The seeds from which the pines were issued were scattered from trees that grew not too far distant—the wind catching up the tiny-winged seeds and blowing them through the air to drop them on the sandy loam in the shade of other trees.

The pines survived attacks by forest fires, grubs, rust, winter frosts and the fury of summer storms. They grew slowly at first, protected by the mature hardwoods around them. Then, the pine saplings began to reach above the forest canopy, sometimes one foot—two, three and more feet a year, in their search for light and air. As the years passed, the older trees, like old friends, began to die, while the pines continued to grow in girth and height. By the time the pine trees reached their one hundredth birthday, the old hardwoods lay mouldering at their feet. In their own advancing years, the pines grew but inches until their growth suspended altogether.

Many of the great trees measured from four to five feet in diameter. But there was one tree—the largest of the giants—that measured six feet, eight inches in diameter, four feet above the ground. When the pine was felled, the stump was so broad a team of oxen could have turned around on top of it.

The sawyers cut the gray, deeply fissured tree trunk into eight logs—each one over twenty feet long. When the scaler measured them, he estimated they would yield over 28,000 ft. (board measure)[8] of ivory-clear pine boards without sap, rot or knot.

Because of the trees' proximity to the river, the logs were skidded or hauled by teams of oxen directly to the stream where the logs were rolled down the embankment to splash into the water. A chain or boom was stretched across the stream so the valuable property wouldn't drift off before the river froze over.

The first pine that Jimmy-the-Chopper selected to fall that morning was one of the smaller trees in the lot. It was about three and one-half feet in diameter, running one hundred and fifty feet into the air. Looking up, his eyes following the trunk to the top, the chopper could see the tree sway in the light breeze. The green boughs, stirred by the gentle wind, looked peaceful, murmuring softly to themselves. But Jim took no notice of them. He was observing which way the tree trunk was moving.

He hit the side of the tree with the flat side of his ax.

"Sounds good," he said to Frank. There was not a hint of rot or insect damage.

"Cully!" he hollered to one of the swampers. "We'll throw her over to the foot of them old windfalls yonder. Break away those sapplings in the way there—then lay some pine tops down for her to fall on. No use in breakin' her back."

The trees that had protected the big pines in their infancy lay rotting beneath a spongy carpet of moss, pine needles and cones. Some spindly evergreens and hardwood saplings had struggled up in the shade amongst the pines. Where daylight had been let into the woods by the choppers' axes, the small trees, for the first time, were feeling the

Looking up, his eyes following the trunk to the top, the chopper could see the tree sway in the light breeze.

warmth of the sun on their thin branches. But before they had time to benefit from the sun's rays, the saplings were laid to rest. The woodsmen would leave nothing standing in the path of the pine tree which was about to fall. They knew the awful consequences of a falling tree trunk glancing off another object, throwing it off its course and into another tree—or worse, on the men and beasts of burden working near by.

Satisfied at last that the way was made clear, Jim set to work without further delay. He stood next to the pine tree (on the side where he intended it to fall) and planted his feet as firmly on the ground as he was able. He swung his ax from his shoulder, letting it fly into the pine, notching the trunk about three feet from the ground. When he finished, he came back to the side where his partner was standing. Frank automatically took his position to the right of Jim.

"Ready?" asked the chopper. And when Cole nodded his head, the two men began to chop a little higher than the notches that Jim had made on the opposite side of the tree trunk. Alternately, each man threw his ax blade into the tree, holding and swinging their axes much like a baseball player holds and swings his bat—letting the bits on the axes do the work.

Chip-chop. Chip-chop. Working with perfect grace and harmony, one man cut into the tree with powerful upward strokes—the other, cut downwards, forcing out big wedge-shaped chips that flew out of the gash with a steady rhythm.

The men worked on until they felt the tree shudder. Without a word, both choppers straightened. And while Frank backed away from the doomed tree, Jim cupped one big hand to his mouth and shouted, "TIMBER-R-R-r-r-r!"

The warning had an immediate effect on the men working nearby. They scrambled for safety, out of the way of falling trunk and branches that could kill at a stroke.

Jim swung his ax into the deep gash, once—twice. The faint cracking sound made by the tree became louder. Then, gathering momentum as it fell, the great pine rushed toward the earth in a shower of tree boughs, pine needles and hoarfrost—the feathery green top swinging in a blurred arc in the direction the chopper intended it to fall—until it thundered to the ground.

When the monarch lay prostrate on the floor of the forest, a fleeting silence hung suspended in the frosty air—a silence more poignant to those among the crew who might have felt a twinge of conscience when they saw the giant fall, or felt the earth quake when it hit the ground. But, it was "saw logs the men were after" and such feelings were hastily put aside. Almost at once, the air was again filled by the sounds of the woodsmen at work.

While Jimmy and his partner were off sizing up another tree to cut down, a team of swampers with sharp axes began to top and de-limb the fallen monster pine—shaving off its boughs close to the trunk to make it ready for the sawyers who would come with their crosscut saws and cut the tree into logs.

It had taken the pine tree over one hundred years to grow. Jimmy-the-Chopper and his partner had felled it in less than one hour.

The blacksmith's tool carrier.

2

B is for the BLACKSMITH who makes the forge glow

While the crew tramped out to the woods to labor another day in the pine, Angus McCloud, the camp blacksmith, was in his shop replenishing the fire in the forge.

It was the middle of November—the crew had been at work in the woods almost a month, and the foreman was pleased to write in his weekly report to the superintendent down at the mill that he would have his million feet of pine on skids by the first snowfall. But in spite of repeated frosts and a few scattered snow flurries, there was still no snow on the ground—and while logs piled up along the sleigh roads awaiting transportation to the river banks, McTaggart was beginning to worry about the winter being one of those snowless seasons that plagued lumbermen every so often.

This morning there might have been a hint of a storm brewing in the air. Ben Sweeney, one of the sawyers, had been complaining for two days about his old war wound "actin' up." (His injury was a very reliable weather forecaster). But Angus McCloud had other things on his mind besides the weather. He was more concerned about the work that he had to do that day.

The gloom of the shadowy room was only a little dispelled by the light of a barn lantern that he hung on a hook near the forge. Outside, a rosy glow in the east suggested that a new day had begun.

If you asked the smith how many times in his life he had begun a new day by hanging up a lantern so he could see to work, Angus wouldn't be able to tell you right off. He would have to stop and do a little "figgerin' " to recall all the lumber camps where he had worked as a blacksmith. It seemed only yesterday when he had left Oxford County in Ontario to go north into the lumber woods to seek his fortune.

Unlike his father, who was the village blacksmith, Angus, Jr. had some schooling—just enough to make him wonder what the world outside Oxford County was like. Because of his interest, the schoolmaster had given him permission to read the only geography book in the schoolhouse, and when Angus examined the pale patchwork of colors that made up the northeastern provinces of Canada and read the names "Quebec," "New Brunswick," "Nova Scotia," "Newfoundland," "Prince Edward Island," printed across the maps, he dreamed of adventures he might have in those faraway places.

One day, his dreaming over the old geography book came to an end when Angus's father told him it was high time he began his apprenticeship in the forge. There was no need for more book learning, explained the smith. "You don't have to know how to cipher in order to shoe an ox or sharpen a plowshare—and what accounts we have at the forge," he said, "I keep in the back of my head."

Young Angus reluctantly put away his books to learn how to turn out axes, steel traps and kitchen utensils—to learn how to make and repair plows, farm wagons, sleighs and carriages—and how to shoe horses and oxen for the people who lived in and around the settlement.

By the time the apprentice reached his seventeenth year, he had served out his training period. Angus knew a trade. But instead of being happy working at his father's side, he felt only a nagging discontent. Nothing would relieve his feeling of unrest, he told himself, but to seek out those places he had read about in the old geography book. Finding work would be no problem. Hadn't he often heard the old men, who loafed around the forge, tell about the need for good blacksmiths in the lumber camps out East?

So one day, while a tearful mother, a grim father and a parcel of envious brothers and sisters stood at the gate of the rustic homestead, he solemnly said goodby to each one and then walked down the lane and out the gate. Over his shoulder he carried a meal sack that contained all of his worldly goods—two home-spun shirts, hand-knitted wool socks, a leather apron and a collection of his grandfather's hand-forged tools (tightly wrapped in a grain sack). His apprenticeship papers (signed and bearing the seal of the Clerk of Oxford County), stating that he, Angus McCloud, Jr., had served as an apprentice to Angus McCloud, Sr., Master, and was fully qualified "this day of April 1850 to serve as a Journeyman Smith who forged Iron," was his passport to the outside world. The document was carefully folded and wrapped in a piece of oiled paper and tucked into a small leather bag which he hung around his neck. In his hand he carried a bundle which contained some cold salt pork and a loaf of bread—food to sustain him for several days—or until he could find work on his travels.

Since that day, when the young blacksmith started out to seek his fortune in Northeast Canada, he worked in lumber camps on the Salmon River above the Grand Falls in New Brunswick, and on the St. Croix, on both sides of that river in Canada and in Maine. And as the pine dwindled under the choppers' axes, he moved westward with the woodsmen on their migrations, until one day in late summer 1866, he found himself in a place called East Saginaw—a mill town in Michigan that was just gathering up steam to become the roaringest of the rip-snorting pine towns in the Lakes states. The town was twenty-two miles up the Saginaw River from Saginaw Bay.

It was on a Friday when he landed in East Saginaw—a sultry hot day. He had just disembarked from the crowded shore-line steamer *Susan Ward*, that had brought him the two hundred-odd miles up from Detroit,[1] and he was cooling off with a glass of lager in the dim recesses of a roistering water-front saloon. He remembered thinking as he stood at the bar drinking his beer, jostled on one side and then the other by rivermen who were still coming in from the log drives, that St. Nick's in East Saginaw could have been any one of a dozen saloons he had been in, up in Canada and in Maine—when he felt a light touch on his shoulder.

The blacksmith turned around, half expecting to see an acquaintance. He saw, instead, a stranger—a tall man with flaming red hair, who was clean-shaven except for a big moustache that matched the color on his head. The man stuck out a hard calloused hand and introduced himself—speaking loudly so he could be heard over the noise in the barroom.

"The name's McTaggart—one of John Shannon's foremen," he said.

Like most men in the logging game, the blacksmith was acquainted with the name John Shannon—a name well known in pine lumber circles in Buffalo and Albany—and now in Saginaw. He shook the proffered hand and told the "push" his name.

Introductions out of the way, McTaggart said, "The barkeep tells me you're a blacksmith looking for a bunk. We're starting up a new camp up on the Tittabawassee and I need a good iron-burner. Job pays $45* and found. If you're willing to come, we'll be leaving as soon as you can get your stuff together."

Angus reached down to where his turkey was resting at his feet. "Tools are right here," he said. "I'm ready when you are."

And that is how the Scotsman from Oxford County in Ontario met up with another Scot by the name of McTaggart and how he became one of the crew who helped the foreman hack out a new camp for the lumberman Shannon in a pine forest in Michigan.

Now, if the reader is disappointed that Angus McCloud did not realize the adventurous life he dreamed about in the old log schoolhouse—or feels bad because he did not become a millionaire-lumberman with an elegant mansion in Toronto and a seat in Parliament, please don't bother your head about it—because Angus (now a master blacksmith) was a happy man who worked at a job he liked. He was a true artisan who could fashion tools, conveyances and other things of both iron and wood. Like the blacksmith-inventor, Joseph Peavey of Stillwater, Maine, he was a master of improvisation, often improving upon the original design of the article he was working on. He was able to

*a month

visualize an item and make it. He could forge by hand anything from a hook to a length of chain and he could repair anything in the lumber camp from a broken ax helve to a busted sleigh runner. He could shoe a horse or an ox—or "hang" an ax or sharpen a saw. In short, Angus McCloud

Angus MacCloud could forge by hand anything from a hook to a length of chain, and he could repair anything in the lumber camp from a broken ax helve to a busted sleigh runner.

was a man who took pride in his work—a man who worked at a job that was as indispensable to the camp operations as was the cook's.

The building in which Angus spent his days (and many nights as well) was built near the camp stables—and like the other buildings in Camp 3, it was made of logs with the cracks between the logs chinked with moss and clay. The building was low (its eaves almost touched the ground). In front of the big open doors, a scattering of wood shavings and chips covered the well-trodden yard. Several oak logs (for making sleigh runners) and a half-finished pung lay in their midst. Empty horse-shoe kegs were laying higgledy-piggledy to one side of the building where the smith had unceremoniously tossed them. And parked in the clearing between the tree stumps were sleigh bobs, a light two-seater cutter, the tote wagon and the road-icer. Only one small window graced the log building under which the smith sharpened saws on days when the weather was too inclement to work outside.

Inside the shop there was a not unpleasant ammonic smell of horses (peculiar to stables and blacksmith shops)—of wood shavings, hot iron, burning coal and tar. In the murky light furnished by the lantern, one could see the forge was equipped with a big wood and leather bellows. The anvil, on which the blacksmith shaped hot metal to make and repair camp equipment, rested on a knee-high block of hardwood fashioned from a rock-maple stump.

When the sky became brighter, the light that came in the open doorway revealed the hard-packed dirt floor of the shop. It was littered with metal flakes, wood chips and shavings, hoof parings, bent nails and discarded iron shoes (worn out by the camp horses and oxen). A clutter of tools—sledges, rasps, files, chisels and hammers, farriers' knives, tongs, nippers and hoof parers—was strewn over a crude work-bench.

As the light advanced across the room, one could see more distinctly an odd-looking contraption standing near the forge. It had wooden rails on each side of it—and between the rails, a wide leather sling, which was suspended from ropes and pulleys. The oddity was called an ox-sling or ox-stocks, and it was used by the smith to hoist (with the aid of a hand windlass) the heavy oxen off the ground so he could fit their small cloven hooves with the two iron half-moons that were their shoes.[2] Kegs of horse and ox shoes, boxes of shoe nails and toe calks, cans of pungent-smelling pine tar and coils of oakum (used for cooling and healing the feet of the animals) were shoved out of the way against the wall while peavies, cant hooks, saws and other tools (in different stages of repair) leaned in dusty corners. Over the forge, new ax handles were curing in a rack, and logging chains (in a variety of sizes and lengths) dangled from wooden pegs and from the massive log beam that ran overhead the length of the building. Propped in assorted bunches against the back wall of the shop were lengths of various-sized iron bar stock from which sleigh runners, chains, hooks, fittings and repair parts were made.

The iron-burner, wearing a long leather apron (split to the knee and black and shiny with wear), busied himself at the forge. With a small hand-shovel he sprinkled fine blacksmith's coal on what was left of the fire he had banked only a few hours before. With his left hand he gently pulled on the frayed rope that operated the bellows, encouraging and coaxing the coals to a cherry-bright hue with the life-giving oxygen. A puff of smoke and a few flames rewarded his efforts as they leaped from the coals.

As Angus worked, he made a mental note of tools and

equipment he would repair or make that day. There were a couple of cant hooks to be fixed, two chopping axes with broken handles to be re-hung. Yesterday, one of the teams parted a 16-foot length of skidding chain that had to be mended—and he had to make up a couple of dump hooks. Last night the barn-boss told him he'd bring in Buck, one of the camp oxen, who had a loose shoe. The two roans needed new iron—and Angus had been promising the cookee for two weeks that he would finish making a new pung for him. "I'll clean up all the little stuff today and then get back on the snowplow," he thought to himself. This was the 17th day of November, and the snows would surely be upon them soon.

By the time dawn arrived, the sounds of squeaking bellows, the rhythmic ring of hammer on anvil and the hiss of hot metal dropped into water for cooling, filled the shop and bounded out through the double doors and over the trampled yard to echo through the woods. When it was daylight, Angus doused the light in the lantern and worked on by the light that streamed in through the open doorway.

About noon, the smith put down his tools, not having to consult his pocket watch to see what time it was. The growling in his stomach had been accompanying the tune his hammer was making on iron for the last hour. He banked his fire and then plunged his hands into the water butt, splashing cold water over his face and neck. Without drying off or even removing his apron, he went out the door to tramp up to the cook-camp to see what the cook had to offer in the way of eats.

Today, Kelly dished out warmed-up stew and beans left over from breakfast which McCloud shared with Billy Flynn, the barn-boss and the barn-boss's helper, big George Pepper. They were the only men besides himself, the cook and the cook's helper, who worked around camp during the day. The crew ate out "on the job," not taking time from their work to come back to camp for their noon meal. (The cookee was presently hauling their dinner out to them on a hand sled).

Kelly was in a talkative mood, but Angus didn't linger over his beans to chew the fat with him or the other men. He was back at the forge in short order to work on through the long afternoon—not paying attention to the leaden sky and gathering snow clouds. When it became too dark to see, he lit the barn lantern and kept on working.

It was dark when the blacksmith heard the crew coming in from the woods. He put down his hammer and began to bank the forge fire for the night—but hearing sounds of someone approaching the shop, he looked up from his work to see the foreman, holding a lantern over his head, peering into the shadowy room. The light revealed a sprinkling of snow on McTaggert's cap, on his shoulders and even on his moustache. And Angus was surprised to see the foreman kick away a small snowdrift that had piled up inside the open doorway.

"Is the new plow ready, Angus?" he asked. "If this drifting keeps up, we'll be needing it in the morning."

Angus looked back into the shadows where several big oak beams lay—wooden pieces hewn and assembled for the still unfinished snowplow.

The foreman's eyes followed those of the blacksmith's, but he made no comment when Angus said flatly, "The plow'll be ready if you want it tomorrow."

Satisfied, McTaggart turned to go, knowing that the smith was as good as his word.

After supper and all through the night, an occasional

flare of the forge fire could be seen through the little window in the blacksmith shop—when it wasn't blotted from view by gusts of snow swirling around the building. And even in the bunk-camp, the noise of the wind howling and the snores of the sleeping assembly could not drown out the faint sounds of hammer on iron.

The snow fell all night long—the wind blowing down the forge chimney so that the coals sometimes glowed white without the benefit of the bellows. It was after dawn before the storm blew itself out.

Angus finally put down his tools and wearily hung up his apron on a peg behind the door. He made his way up to the cook-camp on the path the cookee had shoveled for him. He felt dead beat—but he didn't care. He had finished making the big V-plow and now he could hear the profane shouts of the teamsters as they urged six plunging horses to maneuver the big machine into position. Tinkling Swiss bells that hung from horse collars filled the cold morning air with their cheerful sounds—the bells seeming to herald the new-fallen snow and give it welcome as the plow slowly moved off into the woods behind the hard-working teams. Two great wings of snow fanned out on either side of the sleigh road, and a high-spirited party of choppers, sawyers, swampers and road monkeys, buoyed up by the occasion of the first snowfall, followed behind, walking down the wide path left in the wake of the big plow.

The cook looked up from his bake board when the iron-burner stepped into the cook-camp and watched him drop, exhausted, onto a bench near the stove.

Kelly, who wasn't known for his subtlety or his timing—chose that moment to say, "Angus—I been meanin' to ask—why ain't you turned out that pie-fork I wanted? I dang near burned off my hand again this mornin' trying to get a batch of bread out of the oven."

A wooden snow shovel that the cook's helper had been using to dig out paths to the barn and woodpile was handy but the blacksmith was too tired to pick it up—let alone hit Duff Kelly over the head with it. But he'd get even. After his untimely remark, the dough-pounder could burn his fingers down to his knuckles before Angus McCloud would make a pie-fork for him.

3

C is for CROSSCUT not only for Cant Hook

"Ben!" exclaimed Crosscut Bill in exasperation, "I don't mind you ridin' on your end of the saw—only quit draggin' your feet!"[1]

Bill's sarcasm wasn't original. The saying was one that belonged to the sawing trade, going back to those ancient times when two sawyers first operated a saw together.

Ben Sweeney, the man at the other end of the crosscut, smiled good-naturedly and without upsetting the rhythm of the saw, waved a mittened hand to acknowledge that he had gotten the message.

"It's hard enough," said Bill (not willing to let Sweeney off so easily), "trying to cut up a pine standing up to your neck in snow—but dang it all Ben, when you keep pushin' instead of pullin'—"

No one in camp, not even his partner, could tell you much about Crosscut Bill. They didn't even know his last name. All they knew about him was his nickname, which indicated he had an uncommon talent in the use of the crosscut saw.

It was the custom in lumber camps to call men by nicknames. If there was something distinctive about a man—whether it was his occupation (Jimmy-the-Chopper), his inclination (Billy-the-Bum), or habit (Whiskey Jack), he sooner or later ended up with a handle that was tacked onto him by one of the men in camp—sometimes out of necessity.

Like the season when four Canadians blew into a lumber camp up on the Muskegon—all with the name Dan MacDonald. The situation almost stumped the clerk when he began writing their names on the camp books, until he got the idea of providing each man with a nickname. After giving the woodsmen a quick study, he wrote, "Big Dan," "Curly Dan," "Dirty Dan" and "Dan-with-the-gold-watch."[2] It was a decision that settled the question which Dan it was among the MacDonald clan who owed $3.95 for tobacco on the books in the van, and who had what was coming to him in wages at the end of the season.

Whoever gave Bill his moniker recognized the sawyer's skill at handling the "ornery" saw. As Crosscut Bill himself put it, he knew all the "ins and outs of the old briar."

If Bill seemed immodest about his talent, no one blamed him. Even the improved saws—the flexible ribbons of steel that were employed in the Great Lakes region a decade later—took a certain knack and a lot of muscle to operate. But the old-style crosscut saws that Bill and other sawyers used up until 1876 were even more difficult to work. Called "man killers," the tools didn't hold their cutting edges and "set"—which made them hard to manage. Moreover, the wooden handles of the old saws were riveted or screwed in place.

Bill was often heard to complain that he had to "almost chop off the dang log to get the saw out." After the mid-seventies, saw handles were fastened on in such a way a sawyer could slip them off and draw the saw blade out of the cut with one easy motion. This was the tool that Bill dreamed about but wouldn't see until the "brass" introduced the saws into their camps a few years hence.

It was several days after the big snow storm; and Crosscut Bill and Sweeney were standing hip deep in pine boughs and snow sawing up one of the big pines that had just been felled. When it hit the ground, the crew of swampers had swarmed over the tree to cut off the limbs—cutting them close to the trunk with sharp axes and throwing the tops and other waste aside. Using a long measuring pole, Sweeney "laid off" the tree at twenty-foot intervals while Bill followed him, marking with clips of his ax the places where the sawyers would cut the trunk into sections.

Crosscut Bill and his "ornery saw."

Pine trees were usually reduced to sixteen-foot lengths, but at Red Keg, trees like those from the prize clump were being cut into twenty-foot logs. Pines with rot or insect damage were sawed into fourteen and twelve-foot lengths to eliminate the "conky" parts.

The pine that the sawyers were sawing was five feet in diameter at the butt end, which meant that Bill and his partner had to use a saw that was at least eight and one-half feet long to have room to draw it through the tree trunk. The longest crosscut saws purchased for Red Keg were six and one-half feet—longer in length than any man in camp was tall—but these saws weren't long enough to cut through the trunks of the bigger trees. They had to be lengthened for the job by McCloud the blacksmith.

To do this, the iron-burner removed the handles from a six-foot crosscut and then sliced off the ends where the handles had been fastened. He cut two lengths from an old saw (the smith never threw anything away), trying the pieces for size before fitting them on the ends of the altered saw. When he was satisfied that they were the right size and shape, he scarfed, or tapered the parts being joined (so there would be no lumps where the metals overlapped) and fused the pieces together (using borax as a flux)—first heating the edges to be joined until white hot. Then he pounded the joints until flush and flat. When the saw was completed—with parts welded together and handles screwed back on—Crosscut Bill and Sweeney were provided with an eight and one-half foot saw that was like new—the splices at each end hardly noticeable and as strong as if the saw were one piece of metal.

A verse from an old woodsman's song tells about how sawyers felt when their saws were too dull to pull through the pine.

Filing saws is a ticklish job;
It has to be done just right.
If the saw pinches and drags,
the sawyers cuss
And get mad enough to fight.[3]

Keeping crosscut saws in shape so they cut smoothly was a never-ending occupation in the lumber camp. Some larger outfits hired a man who did nothing but keep the saws sharpened, gummed and set. But at McTaggart's Red Keg, the filer's job fell to Angus McCloud the blacksmith. He could "fit a saw" with the best of them.

Saws were sharpened by filing—the saw being held in a vise or saw-clamp while the cutting teeth and "rakers" (the teeth that cleared the kerf of sawdust), were worked upon with hand files. When a saw's teeth were repeatedly sharpened over a long period of time, they became too short to cut into the pine effectively. When that happened, Angus used a tool called a saw "gummer" to cut away the plate of the saw in order to deepen the gullets (the spaces between the teeth)—thus making the teeth the same height as they were when the saw was new. When the blade became worn away by repeated filing and gumming, the saw was discarded by the sawyers (but not by the blacksmith).

There was one more important thing that had to be done to the saw before it could be used. The cutting teeth were "set"—that is, offset—one tooth bent a fraction of an inch one way and its neighbor bent the opposite way—and so on down the length of the saw. It was a technique that prevented the saw from sticking in the saw cut. To do this Angus might have used another hand tool called a saw-set—but he preferred using a small steel block as an anvil, bending the saw teeth to one side and the other by a light blow of a hammer.

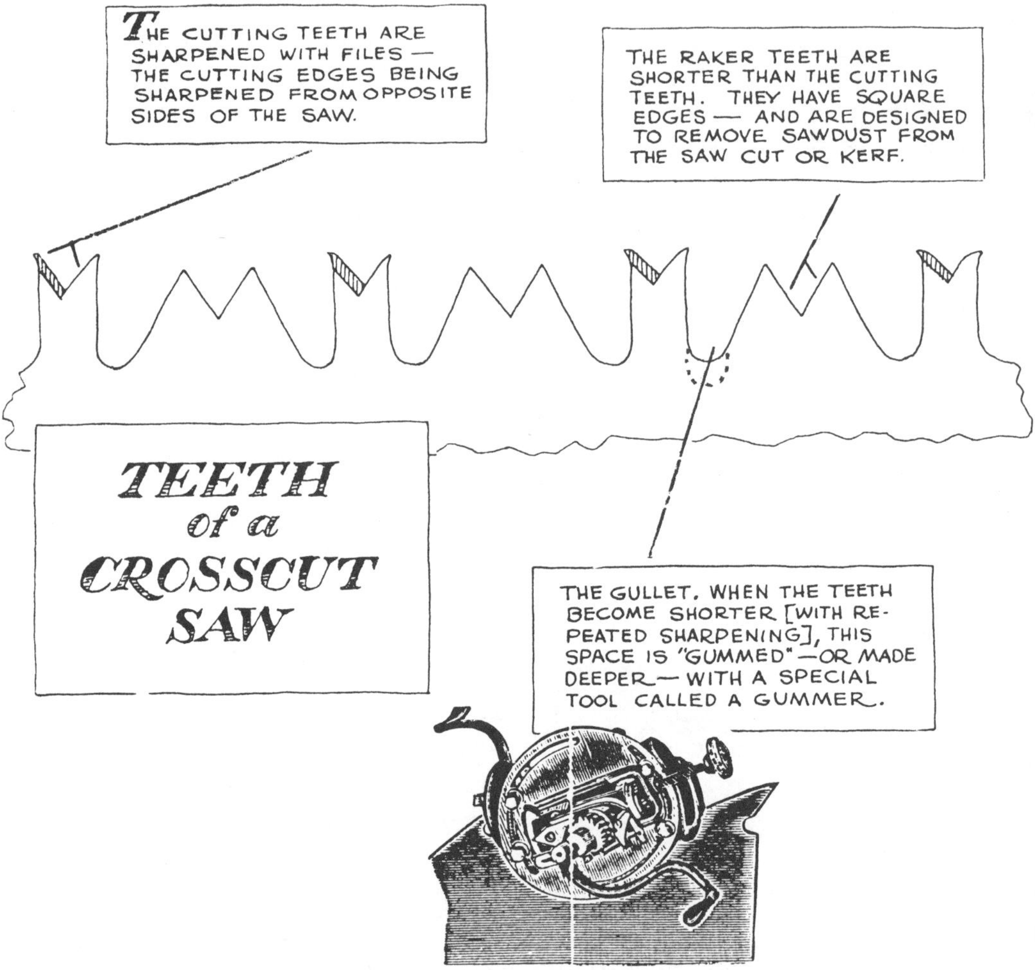
THE CUTTING TEETH ARE SHARPENED WITH FILES — THE CUTTING EDGES BEING SHARPENED FROM OPPOSITE SIDES OF THE SAW.
THE RAKER TEETH ARE SHORTER THAN THE CUTTING TEETH. THEY HAVE SQUARE EDGES — AND ARE DESIGNED TO REMOVE SAWDUST FROM THE SAW CUT OR KERF.
TEETH of a CROSSCUT SAW
THE GULLET. WHEN THE TEETH BECOME SHORTER [WITH REPEATED SHARPENING], THIS SPACE IS "GUMMED" —OR MADE DEEPER— WITH A SPECIAL TOOL CALLED A GUMMER.

Saw gummer

Crosscut Bill and his partner, Sweeney, had begun to saw up the felled tree by first "squaring" or sawing off the butt end. They then proceeded to cut the trunk into measured sections.

The tree lay in the snow on uneven ground, which made it difficult for the sawyers to find a good place to stand in order to operate the saw effectively. To add to the difficulty, the men could scarcely get a grip on the upturned handles of the saw because of the height of the tree trunk from the ground. If the trunk had been any bigger in girth, they would have had to unscrew the handles of the saw and turn them downward (like they had done when they sawed up the big six-foot, eight-inch diameter pine), changing the handles back to their original position when the saw went far enough into the soft wood. This morning, the saw was newly sharpened and set, and after a few awkward strokes, it sank out of sight. The men worked on as they had done so many times—one man on one side of the tree trunk pulling the saw through the kerf and the man on the opposite side pulling it back. The partners worked in close harmony, keeping up a steady rhythmic pace—the great crosscut swishing back and forth, its cutting teeth severing the great tree trunk while the raker teeth cleared the cut of sawdust.

The blade wobbled—and stopped.

"Whuups—she's stickin'," said Bill. "Cussed dang thing. Wait'll I squirt a little oil on the blade."

Sweeney took the moment to relight his pipe while his partner reached for a whiskey bottle that leaned against a stump near by. The bottle contained coal oil. The sawyer was about to flick a few drops of the oil on the saw when he looked up to see his partner calmly sucking on his stub of corncob, enjoying the brief rest and looking dreamily off into the woods.

Bill frowned, then said peevishly, "Since you're only standin' there, Sweeney, you could jest as well be drivin' in some wedges." (Tapered pieces of hardwood to keep the saw cut open and the pine from pinching the saw).[4]

Sweeney, accustomed to his partner's grumpiness, good-naturedly picked up a mallet and wedge-shaped pieces of wood from where they lay near the big stump, and quickly drove several of them into the kerf as close behind the saw as was possible.

"There," he said, "that ought to do 'er."

The sawyers continued sawing, stopping only to drive in more wedges, and from time to time, to oil the saw blade—until at last the crosscut went through the trunk of the pine.

The saw was worked up and out of the kerf, and the men, without further ceremony (or rest), went immediately to the next mark on the tree trunk and began sawing all over again.

Bill and Sweeney continued their work until noon when they heard Cookee blow the nooning horn. It was a welcome sound. They straightened aching backs and flexed cramped muscles and then slipped on their coats (which they had removed when they became over-heated from their exertions) and went to join their comrades at the dinner hole.

Flaggins, or the noon meal, was brought out to the woods on a pung by Cap Wilkes (the cook's helper), and when Bill and his partner reached the place where the food was set out, they jostled for a position with the others to pick up eating utensils and food. They ate hurriedly while balancing tin plate and teapan in one hand and knife in the other—gulping down fast congealing food and cooling tea—at the same time trying to warm themselves at the fire.

While they ate their food, Bill and the others felt the wind whip around to the northeast. But in spite of the darkening sky and the raw wind blowing down the back of their necks, they trudged back to work after only a few minutes of rest, leaving the warmth of the fire—and the dirty dishes to the cookee.

When the last empty plate was handed back to him, Cap dumped it with the others into the pine soap box that was fastened on top of the pole runners of the pung. After carefully smothering the fire, he put on his snowshoes, picked up his dinner horn and headed back to camp—pulling the sleigh behind him through the woods and into the swamp in a direction that was considered by the cookee to be a short-cut back to camp.

Since the freeze-up, Cap had shortened his daily noon-time hike to and from the cutting area by walking through, instead of around, a place that was a dismal sea of muck and water from about April until November. Moss-covered pines and spiky tamaracks grew there amongst a confusion of vines, dead trees, broken tree limbs and windfalls. Even in the wintertime, the swamp was a forlorn place and was avoided by everyone but the cookee.

As Cap tramped on, the dirty tin plates, pannikins, empty kettles, tins and a miscellany of lids and serving spoons rattled together in the box as the pung bumped over and against stumps and fallen limbs of trees.

"Makes enough noise to scare off every hodag[5] in the woods," grumbled the cookee, who liked peace and quiet. He would have enjoyed the walk through the woods every day if it wasn't for the infernal racket made by the sled and its contents. But this wasn't a day for lingering. The wind was rising, and above his head he could see the tops of the trees bending this way and that.

Cap stopped to turn up his coat collar and was about to continue on his way when, without warning, a freakish gust of wind sent a smashing blow through the trees to hit, broadside, a shaky pine, uprooting it from the partially frozen earth as easily as if it had no roots at all.

The cookee heard the pine tearing from the half frozen muck—he heard the cracking, snapping sounds the tree made as it began its descent—knocking down branches and limbs from trees near by. He saw it coming. He was walking directly into the path of destruction. Without letting go of the rope on the pung, Cap quickly moved out of the way of the tree and into a thicket of cedars, pulling the sled in after him. But as luck would have it, the obstinate pine, intent upon a victim, glanced off a tamarack that stood in its path. Again, the cookee saw that he was in direct line of the falling tree. He stood rigid, fixed to the ground. He watched helplessly as the pine swung in its deadly arc toward him only recovering his senses in time to throw his dinner horn to one side before falling flat. The tree dropped with a WHUMP into the cedars, bending them almost to the ground with its great weight. Dead or alive, the cookee was trapped.

The foreman was one of the crew who was working close to the swamp that day and when he heard the cry of distress, he immediately sounded an alarm. Instantly, woodsmen came wallowing through the snow from all directions.

McTaggart was the first to reach the swamp. He followed for a short distance the path made by the cookee in his daily route to and from the camp. It was empty. No one was in sight. There were other footprints in the snow—but they seemed to lead nowhere. The foreman shouted, "Hey there! Are you in trouble—?" No answer. Had the woods claimed another life? McTaggart impatiently shrugged aside the morbid thought. Then he heard a faint noise. It was one that could only be made by a man—muffled sounds of rage coming from the depths of a cedar thicket. A conky pine tree was cradled in its midst. The foreman began issuing orders.

"Bill!" he hollered, "Get over here with that saw of yours." He didn't know who was trapped in the middle of the cedars but whoever it was, he was lucky to be alive—but not for long if the pine decided to shift its position.

He heard someone say, "It must be Wilkes—there's his horn." The crewman was pointing to a half-buried object glinting in a snowbank.

McTaggart shouted for the sawyer again—only to see him coming with his big saw bobbing awkwardly over one shoulder. Sweeney was right behind him.

The foreman pointed into the cedar clump and said, "He's right about there, Bill. The cedars are protecting him—holding the buckwheat[6] off the ground. If you and Ben can cut through," McTaggart parted a tangle of branches, using an ax as a pointer, "one end of the trunk will drop down at the edge of the clump, and the branches on the pine will make the short end of the trunk cannon up high enough so we can get the man out."

Without a word, Bill waded around the cedars to find a spot for his partner to work. After directing Sweeney where to stand, he went back to the other side of the tree trunk and carefully inserted the long wobbling blade of the crosscut saw through the tree branches. At the same time, he prayed the pine wouldn't settle. It seemed as secure as a baby in a cradle—but you never could tell—.

Sweeney, who couldn't see his partner from where he stood, waited for the handle of the crosscut to appear from the other side of the pine and when it did, he heard Bill shout across to him, "Catch aholt, Sweeney—and when I say 'go'—start sawin'!"

Luckily, the fallen tree wasn't much over two feet in diameter. Its trunk was about three feet off the ground, held up by the cedars and its own branches. At the signal, both men began sawing with measured strokes, their muscles straining in their endeavor.

While the saw blade slowly sank its teeth into the soft wood of the tree, the sawyers could hear Cap cussing in the midst of the greenery.

"Shut up, Cookee—it ain't nobody's fault you're under there," hollered Bill. Didn't he have enough to worry about without listening to old Wilkes kickin' to himself? If the briar should bind and then stick in the kerf, they'd have to chop out what was left of the old geezer.

"She's startin' to pinch," shouted Ben, "what'll we do?"

Even if they could reach the saw cut, pounding a wedge into the kerf might make the pine settle farther into the cedars. So Bill reached for the whiskey bottle and flitted a long thin stream of oil in the direction of the saw blade, hoping to hit it somewhere near the center. He gave a signal to Sweeney to begin sawing again.

Working steadily—carefully—the partners made good progress and before long Bill called to Cap, "We're nearly through—hold on now!"

At last, the saw blade cut the tree in two. One end of the heavy trunk dropped with a sickening thud into the snow at the edge of the cedars—exactly where McTaggart said it would fall. The short end of the tree trunk swung up to reveal the cookee who was already struggling to get out from under the crushed cedars. He was half smothered in snow—he was mortified and mad—but alive.

"Where's my horn?" Cap said as he finally gained the use of his feet.

"Your old horn's all right," growled Bill, "but take a look at the pung."

The cookee's eyes followed those of the sawyer's. The sled and its contents were squashed flat under the trunk of the pine tree.

Cap heaved a great sigh and turned to Bill and Sweeney. He said, "Thanks—I appreciate—." His voice trailed off in embarrassment. Without another word, he turned and walked off toward the camp, holding his dinner horn at his side.

The spell was broken, and McTaggart used the moment to say, " It's time we got back to work. We've got pine to cut."

Kelly's 600-pound, six-hole, cast-iron wood-burner.

4

D is for DOUGH-POUNDER both baker and camp cook

While the cookee trudged disconsolately back to camp, speculating about what the cook would say about the smashed pans and kettles, the man whose displeasure he was anticipating was at work in the cook-camp not knowing how close he had come to losing his flunkie.

It has been said that an army travels on its belly. That old saying couldn't be more true than when speaking of the army of woodsmen who battled daily with the saw log.

The man at Red Keg, whose task it was to furnish the battlers with the energy and stamina to do their work, was a dour autocrat whose name was Duff Kelly. He did the cooking honors at Camp 3. And as a cook he was pretty good at his job if you judged his skill in the cookery department by his ability to keep the men's complaints about the food down to a minimum.

Kelly was proud of two things. One, a walrus mustache that almost covered his chin—the other, his cooking. It wasn't hard to see why the cook was proud of the mustache—it being a marvelous creation that only an autocrat would dare to wear. And his pride in his occupation could be clearly understood if one were acquainted with the thankless job of cooking for a crew of hungry woodsmen three times a day, seven days in the week. It wasn't easy. It was a job that everybody took for granted—unless the cook forgot to put salt in the fried mush or fed the crew fat pork twenty-one times a week.

Duff Kelly was a man of medium height who stooped out of habit (after cooking over a camboose fire for so many years). He wore a long apron of dubious ancestry and a black derby hat that had dabs of flour on it where he repeatedly pushed it back on his head before it toppled into his batter, the fry pan or into the fire.

This day, like all days, the cook was going about his work in a business-like manner. Because he could have performed the ritual with his eyes shut, he would sometimes let his thoughts wander while he went through his exercises with meat cleaver and rolling pin.

Kelly learned how to cook for woodsmen up on the Kennebec, and today he was thinking about that old camp. The cook wasn't one to indulge himself in anything so frivolous as laughter, but recalling an incident that happened long ago, he unconsciously let a dry chuckle slip out from behind the mustache—a sound that encouraged a feline observer who had been sitting quietly near the water barrel watching the cook's every movement. The cat let out a squall, making Kelly aware that he was not alone.

The pussy was a big yellow tom the boys had named Jack. His bailiwick was the barn, but every time Cap went out for water or stove wood, the cat was there waiting outside the cook-camp door ready to slip in for a handout that Kelly or the cookee might choose to toss to him.

Jack made another plea. This time the cook threw a piece of pickled beef to him and absent-mindedly watched the cat make short work of the scrap.

Kelly's job was one that everybody took for granted, unless the cook forgot to put salt in the fried mush or fed the crew fat pork twenty-one times a week.

"Think I'm crazy, don't you—laughin' to myself like that—."

The cat made no comment—rather, he expressed his disinterest by licking first one paw and then the other. Upon completing his toilet, he walked to the door. Kelly lifted the latch and let him out, saying under his breath as he watched Jack lope down the path to the barn, "You would of laughed too if you could of seen old Larky Sharky kneadin' his bread dough with his mittens on."[1]

The cook closed the door and went back to his work table thinking to himself, "Maybe I never went as far as old Larky to keep warm in them old camps, but I kin remember the time when I didn't have no table to work on—when the only pieces of furniture in the whole blame lumber camp was the deacon's seat and the grindstone."

Many a time he had kneaded his dough sitting on the deacon's bench, holding a wooden bread trough between his knees, working by the light of the fire in the camboose. He thought for the umpteenth time what a far cry McTaggart's Red Keg was from the old-time camps like the ones he worked in up at Moosehead Lake where the Kennebec took its rise. In those old days the whole crew ate their meals out of a common fry pan, slept on the earthen floor under one big spread,[2] and drank rum, fought and cussed under one roof.

It was up at Lilly Bay where he learned to cook. That season, the woodsmen, out of necessity, had been taking turns doing the cooking. The foreman had not been able to hire on a dough-pounder. The food was bad enough, but when somebody discovered that one of the teamsters was mixing soap in his batter ("to make it raise," he said)[3]—the boys put their heads together and decided they would elect one of the crew to take on the job of cook full time. If one of their outfit worked at it long enough, they reasoned, maybe he'd learn how to handle a sour dough tub and get good results.

Young Duff Kelly was the flunky in camp and the most likely candidate (in fact, he was the *only* candidate) for the job. So, without ceremony, he was elected to be the camp cook by unanimous vote. Kelly swore he would get even, if it took him the rest of his life.

The cook at Camp 3 smiled grimly when he thought about the days that followed the 'election.' "I got even all right," he said to himself, "for that was the winter I learned how to cook."

McTaggart's Red Keg seemed pretty tame in comparison to those old camps up on the Kennebec. But the dough-pounder wouldn't trade times or places now. Maybe it was a sign he was getting soft, he thought to himself, as he pushed his hat to the back of his head for the hundredth time since he got up that morning.

Up to this year Kelly had never seen a cooking stove in a lumber camp. In camps from Maine to Michigan he cooked his fat pork, beans and bread in iron pans and flat-bottomed bake kettles over the open fire, or in the ashes of the fireplace.[5] But that fall when the crew blew into camp, they were surprised to see a new log building perched on the banks of the Tittabawassee. It was built behind the bunk-camp—separated from that building by a roofed-over, open storage area called a dingle. A shiny stovepipe stuck out of the roof of the new addition—a contrast to the big square stick chimney of the camboose which still dominated the older building.

Inside, and occupying almost one end of the new cook-camp, stood the pride of Duff Kelly—a six hole, cast-iron, woodburning range—six hundred pounds of new convenience for the cook. (To Cap Wilkes, the cook's helper, the range was just another botheration to be fed with wood, and on Sundays to be polished with stove blacking to keep it from rusting).

The big cooking stove had been purchased down in East Saginaw at B.B. Buckhout's hardware store, where it was loaded aboard a raft or scow which was towed up the Saginaw and then up the Tittabawassee as far as the company tug could go (it being the fall of the year, the river was low). The scow, bearing its cast-iron burden, was poled the rest of the way to the camp—arriving on a day when the thermometer registered eighty degrees in the shade.

The stove was moved from the scow with great effort (and many expletives) to a stone-boat and then dragged by a team of sweating horses up the steep river bank and finally installed in the new building with help from the cookee, two swampers and several swearing teamsters.

From that moment, Kelly's life was changed. He was one cook who would no longer suffer the indignities that were common to cooks in most lumber camps—a singed mustache, red eyes and a crick in the back from bending over a camboose fire.

The new cooking stove almost dominated one end of the cook-camp. At the other end were rough plank tables and wooden benches where the crew ate their meals. There was a double bunk built up against the log wall near the stove. It was where the cook and his helper slept (Kelly keeping an eye on the larder in case somebody should risk his wrath and try to steal in for a snack). His big work table was built under the only window in the building; and beneath the table, within easy reach, were barrels of flour, tubs of lard and sacks of corn meal, beans and potatoes. Stacked on the floor and on rough-sawn storage shelves were barrels, boxes, bags, cannisters, crocks and wooden tubs containing an assortment of everything from coal oil, Lucifer matches, lamp chimneys and molasses—tea pots, griddles, kettles, pans, tin plates and boxes of soap and "silverware." Iron spiders, a meat cleaver, basting spoons and cake turners dangled near the stove from iron hooks driven into the massive log beam that stretched overhead the length of the building.

The pork barrel (an indispensable part of the cook-camp) stood handy to the wood-burner. It could be identified by the salt-encrusted meat-hook that hung from its side, or by the puddle of brine it sat in (where the pickle ran down the barrel). At least once a day, seven days a week, Kelly fished out chunks of pork that floated in the brine to cook it for the boys' breakfast or supper—replacing one empty barrel for a full one from the stock pile that was kept outside in the dingle.

Stove wood was piled up neatly against the wall (near the door) and close to the big water barrel stood the dry-sink, which was lined with galvanized tin. The sink drained into a wooden bucket, and when the bucket was full, Cap Wilkes simply tossed the contents out the back door where two or three pigs, in a penned clearing near the cook-camp, did away with potato peelings and whatever other "slops" there were from the camp.

All in all, Kelly was proud to say that his cook-camp was the best-equipped and the envy of every dough-pounder on the Tittabawassee. It gave him a sense of power and deep satisfaction to look over the low, slant-roofed room with its familiar, ever-present odors of

The pork barrel.

cooking, burning hardwood, coal oil, tobacco and the yeasty fragrance of fermenting dough.

Everyone knew that having a good cook in camp was the secret of success in the woods, and Kelly was well aware of his importance to the camp. But woe to the cook if he spoiled the simplest camp fare—if he made poor bread and dished out too much pickled meat. He knew from experience that, at the first signs of mutiny within the ranks, the foreman would tell him to "pick up his time," and "hit the trail."

Duff Kelly was one of the "long-hour boys" and if he was short-tempered, he couldn't be blamed much. He worked by the light of oil lamps more often than by the gray winter's daylight that filtered through the little flour-specked window over his bake board. It took muscle and a lot of time to prepare ribsticking "vittles" for a crew of fifty men who made food disappear three times a day faster than water poured on sand.

Each new day began the same way. There was no alarm clock in the cook-camp—but with instinct instilled in his mind from years of habit, the cook's helper would roll out of his bunk at precisely 3:30 a.m. Numb with sleep and shivering with cold, he lighted the lamps and started the fire in the stove. He raked ashes out from under the grates and shoved chunks of rock maple on top of the few live coals still burning there—throwing in a hunk of fat from the pork barrel to make the hardwood burn faster. Satisfied that the fire was roaring up the smoke pipe, he would pull on his boots, light a barn lantern and let himself out the door and into the night to rouse the teamsters—and to light the fires and fill the water pails in the van and in the bunk-camp for the still-sleeping crew.

Kelly was already up and making initial preparations for

cooking breakfast.

By five o'clock the fire in the cook stove had settled down to a steady blaze, and the air in the smoky, lamp-lighted room was thick with the smell of wood burning and meat frying. Steam curled from fry-pans and from the spouts of two-gallon tea pots. Kelly, his bowler almost lost to view in a fog of wood-smoke and cookery, was a perfect example of a man who knew what he was about. He worked mechanically with perfect ambidexterity at the stove. With one hand he deftly moved thick slices of sizzling salt pork around the fry pans with a long iron fork, while with his other hand, he turned over in another pan, browning, chopped-up potatoes with a cake turner—all the while keeping an eye on his helper. Cap had come back from the bunk-camp carrying two iron pots filled with beans which had been cooking all night buried in the bean-hole in the camboose. The cookee put the pots on the back of the stove and then hurried around setting out platters of meat and bowls of potatoes and gravy for the teamsters (who ate breakfast before the others in order to get into the woods with their teams before their crew mates arrived). In the midst of tinware and granite, Cap completed the table setting by planking down, at intervals the length of the shabby oilcloth, plates heaped with thick slices of bread and molasses cookies.

The ox drivers and teamsters, who had been grooming, feeding and harnessing their charges, drifted in one by one from the barn and took their places at the table. They ate hurriedly, silently—then departed for the barn, and thence to the woods.

As soon as they left, Cap quickly cleared off the table and replenished it with clean dishes and pannikins[6] and more platters and bowls of hot food—completing the chore by setting steaming tea pots at the end of each table. When this was accomplished, Kelly stood with hands on his hips and surveyed the tables with a practiced eye. Everything was in place. "Blow the horn," he said. And Cap picked up his tin "bugle" and made tracks for the door to signal the hungry woodsmen who were waiting impatiently in the bunk-camp to "come and get it."

It was just 6:00 a.m. when the men crowded through the doorway of the cook-camp for their breakfast.

By the time the first rays of daylight wavered through the little window over the work table, the crew had eaten their way through everything in sight that was edible and

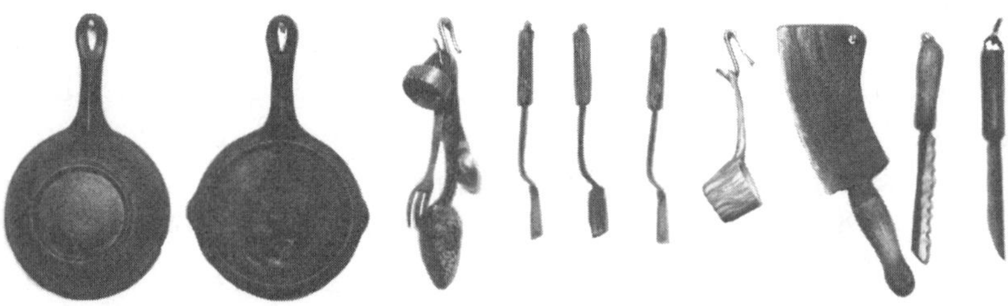

were at work in the pine. Kelly already had a half bushel of cookies and eight dried-apple pies out of the oven and was hurrying around preparing the crew's noon-time meal that the cookee would haul out to the woods.

So it went all winter long. The dough-pounder wrestling with iron kettles, tin pans, flour, salt pork, pickled beef, beans and potatoes. He made mountains of pancakes, kettles of stew and kegs of donuts. Every other day, Kelly mixed dough for bread, setting it in big tubs on the back of the stove to rise. He roasted, toasted, basted and boiled salted meat (fresh meat when the cookee provided it with his flint-lock rifle), emptying a flour barrel every other day throughout the long winter months, (a practice that kept George Pepper and the tote team busy hauling in flour and other supplies).

It took real talent to turn out good-tasting, nourishing food that provided the hard-working crew with the warmth and energy they needed to cut pine. And Kelly did it without fresh vegetables or fruit, and without butter, eggs, sugar or milk—or a single written recipe.[7] With a pinch of this, a dab of that, a handful of something else—the cook managed somehow to make the plainest food palatable.

Kelly's wages were the same as the blacksmith's and he was looked upon as being as indispensable to the camp's operation as was the iron-burner. But only so long as the crew didn't kick about the food would Kelly, the autocrat of the cook-camp, keep his bunk at McTaggart's Red Keg.

TIMBER-R-R-R-R

5

E is for the ECHOES that through the woods rang

By Thanksgiving Day (which was spent in the pine just like any other work day), the woodsmen at Camp 3 had settled into a daily routine that was as familiar to them as was that of the cook or the blacksmith. And the camp account books showed that the cutting, skidding and banking routine was going according to the foreman's schedule.

If there had been a way in those old times to record the every day sounds of camp life in the pine woods, it would have preserved forever the noises that were as common to men laboring on the Tittabawassee as they were to men working on the Saguenay: pine tops sighing in gentle winds or tossing wildly in a winter's storm—the spooky sounds of naked hardwoods snapping with frost, their sharp reports echoing into the night—a snow storm raging around the little cluster of camp buildings—or rain dripping down the camboose chimney to sizzle in the moldering fire below.

The stillness of the November night is broken by the melancholy howling of wolves. The sound frightens the horses—and the teamster LaRiviere mutters a curse as he turns out of the warmth of his bunk to go into the Arctic-cold midnight to help the barn-boss quiet the racket the horses make kicking the sides of their stalls. But Frenchy's going out and his returning does not waken the worn-out men who snore on in a variety of shifting accents that soar and fade in the gloom like surf on an ocean shore.

There is no hint of dawn in the black sky when the sounds of the night give way to those of early morning. No one in the bunk-camp stirs when Cap Wilkes pushes open the door. The feeble light from his smoky lantern reaches into the shadows to reveal bunks lumpy with the forms of sleeping men who are covered with an assortment of buffalo robes and "shoddy" blankets. The room is freezing cold. As the fire died down in the night, the chilling winds found their way down the great chimney over the camboose and through the chinks in the log walls.

The cookee goes about the business of rolling logs into the dying embers, and the sounds of snap-cracking birch and showers of sparks shooting up the chimney do little to diminish the "symphony" rising from the tiers of bunks. The grunts of the teamsters and ox-drivers, in answer to Cap's silent tap on each shoulder, and the sounds of weary men pulling on boots and outer garments and picking up barn lanterns to go to the stables (to curry, feed and harness their animals) make no impression on the men still sleeping.

But the jarring, discordant noise of the cookee's horn

brings thirty-odd heads, with hair and whiskers askew, from the depths of the bunks in an instant. Salty remarks about the cook's helper and where he and his tin horn can go (to a warmer place than the bunk-camp), follow the unconcerned cookee out the door.

Cap leaves a bunk-camp alive with noises of men making ready for a day in the pine. Here and there grumbles about lost socks or a misplaced boot are heard over a fusion of other sounds—men stamping feet, which are bulging with three and four pairs of socks into greased boots—the snapping of suspenders over shoulders—icy water splashing from the water pail into a tin basin and the blowing, snorting sounds of men going through the motions of washing grimy faces—the fumbling for and gathering up of pipes, tobacco, mitts and mackinaws.

Five minutes after the first horn, another blast is heard and the crew starts for the cook-camp. A few quiet comments about the weather (more snow has fallen in the night) are heard here and there—but words are cut short as the men bunch through the low doorway to find their places at the long tables.

There is no sound of men conversing in the cook-camp. The unwritten law of the woodsmen from New Brunswick to Puget Sound is "no talking at the dinner table." But the clatter of knife and fork on tinware, the noise of blowing and drawing in of hot tea from basins, and the sounds of thirty-odd pairs of jaws moving in unison is astonishing.

When stomachs are full, there is a last minute sopping up of gravy with pieces of bread reserved for the purpose, a brushing of sleeves across lips and a resounding belch or two (compliments to Kelly) as the men rise one by one from the benches to go back to the bunk-camp to pick up their tools.

The choppers, sawyers, swampers and the others gather in the dusk outside the log building. The men stand in a circle of yellow light that the foreman's lantern makes on the trampled snow. McTaggert, with icicles already forming on the ends of his mustache, gives the men their orders for the day. As he speaks, the faces of the crew come into view here and there when flaring matches light stubby pipes that are clamped between teeth. The faces are lost again in the twilight and in clouds of tobacco smoke that blend with the vapor created by the warm breath of the men. They listen to the foreman while stamping their feet and flailing their arms around their bodies to keep warm.

There is only a hint of dawn when the crew starts off down the sleigh road, their boots creaking in the thick layer of snow left by the plow.

By the time the bleak, cheerless sun has cleared the horizon, the men are at work in the woods. The sounds they make echo and re-echo through the forest: the biting of axes into pine—the traditional warning cries when trees thunder to the ground—the swish of crosscut saws—teamsters clucking and chirruping to their horses—the hoarse shouts of the bull-whackers cussing at their plodding ox teams, coaxing, commanding them to lean into yokes, "Haw Blaze! Buck! Dang yer ornery hides!"—the noise the swampers and road monkeys make with ax, shovel and broom, building up, smoothing and cleaning off roadways—breathless men standing shoulder to shoulder battling snow drifts, breaking the way for heaving horse teams—men digging in and applying their cant hooks to move stubborn logs into position—the musical tinkling of bells hanging from horse collars—the creaking and jingling of harness and chains—the screeching, creak and crunch of heavy sleigh runners carrying tons of logs over iced logging roads—the sounds of straining snub ropes and horses scrambling for footing as a sleigh-load of logs is eased down a treacherous grade.

At noon, the cook's horn swells and echoes through thicket and swamp—over the noises made by ax and saw. The notes call the men to their flaggins.

Work stops—and the crew can be heard coming from all directions down beaten paths and sleigh roads to the dinner-hole. The men gratefully reach out hands, stiff with cold, to warm them at the fire. They jostle to get into line to pick up their food and then gulp it down before it has a chance to freeze on their plates. Their teams munch nose-bags of oats or chew on their cuds and enjoy their brief rest as much as the men do.

These and the recurrent sounds of man and beast laboring in the pine continue through the day to disturb the silence and tranquility of the forest—until at last, the waning light of the winter's afternoon disappears and darkness

brings relief to the weary, hungry crew—and peace to the woods.

Back at camp, the blacksmith and barn-boss hear the faint sounds of the men and their beasts drifting up to them from the logging road as the crew trudges back to the shelter of the bunk-camp. Angus McCloud lays down his tools in the blacksmith shop, and the barn-boss Billy Flynn and his helper George Pepper open wide the doors of the barn and stand ready with lanterns to light the way in. In the cook-camp, Kelly gives the food on the stove a last minute stir, and Cap Wilkes makes ready to blow the horn for supper.

The bunk-camp door is at last kicked open, and the men crowd into the welcome heat of the fire-lighted room. Tools thump and clank as they are dumped into boxes inside the door. Men haul off their boots. They peel off layers of wet socks and shove chilblained feet into dry stags before crowding around the fire to hang wet clothing to dry on haywire strung up around the camboose. And a question is posed by an impatient woodsman who feels like his belly is touching his backbone—"When is Cookee going to blow his blankety-blankety horn—I'm hungry enough to eat the old so-'n-so." No sooner are the words spoken when, on target, Cap Wilkes blows the supper horn. In no time the men are seated in their customary places in the smoky, flickering light of oil lamps. Without ceremony, they "set to" over the steaming food.

The evening sounds are those of men relaxing in the bunk-camp—men who have earned the precious hours of rest after long hours of strenuous exercise in freezing cold. Smoke rises from the camboose fire and does not always find its way up the wide stick and clay chimney. It drifts in eddies about the room, mingling with a fog of tobacco smoke from the woodsmen's pipes. Some men sit hunched on the deacon's seat staring into the fire—too exhausted to do more than unclasp frost-cracked hands from around knees to remove their pipes from their mouths. Others are occupied with nightly chores—greasing boots or putting an edge on an ax with a whetstone. The sound of ax blade scraping against a whirring grindstone almost drowns out the soft rise and fall of men's voices and utterances made by the card players who continue their non-stop game of High-Low-Jack-and-the-Game[1]. A few warm-up notes are breathed in and out on a mouth organ.

"Give us *Canaday-I-O*,"[2] says a voice from one of the bunks.

The player (Cully Wills) swings into the sprightly old woodsman's tune. The words are taken up by voices here and there around the room (the men substituting the word "Michi-gan-I-O" in place of Canaday-I-O). The baritone voices make up in vigor what they lack in tune.

O say, my jolly fellows,
How would you like to go,
And spend one pleasant winter
Up in Michi-gan-I-O?

All of the verses remembered are sung—ending with . . .

Our food the dogs would laugh at,
Our beds was on the snow,
We suffered worse 'n poisen
Up in Michi-gan-I-O.

Before Cully can swing into another song, the cookee walks into the perspiring room, and with no apologies, says, "Nine o'clock!" and ducks out again.

Immediately, stags are kicked off, pants are removed and burning tobacco is knocked out of pipes. Within minutes the only sounds heard in the smoky, highly-scented room are those of burning birch settling into a bed of red ashes and the discordant chorus of snoring men.

As the fire dies, the air in the bunk-camp grows colder. The woodsmen, each deep in their private dreams, are oblivious to the faint tinkling of sleigh bells rising and fading away in the night—sounds that portend their morrow's labor in the pine.

While the men sleep, George Pepper, muffled up

Big George Pepper bundled up for a night on the icer.

against the bone chilling cold in a great fur coat, is perched high on the seat of the road-sprinkler, slowly driving the icer up and down the logging roads, guided only by flaring torches stuck in poles along the roadways. He is preparing the sleigh roads so a single span of horses will be able to draw a load of logs ten times their weight over the roads the next day. Streams of water splash from two holes in the back of the contraption's big wooden tank into the man-made grooves in which the sleigh runners glide.

The teamster listens to the trickling sounds of the water over the rumble of the runners. When the sound stops, it will be the signal for Pepper to refill the tank at the water hole and continue his circuit until the roads are ready for the coming day.

Pocket compass.

6

F is for the FOREMAN, the head of the gang

Alex Taggart was the boss of the outfit. There was no doubt about that. He stood tall—six feet, three inches in his socks—and his head was topped by hair that was only a little less brilliant in hue than the firey mustache that curled up on either side of his nose. To anyone seeing or meeting the foreman for the first time, he gave the impression that here was a man with whom one should not trifle. In truth, Alex McTaggart had a long fuse and was not easily roused to anger—but the red on the top of his head served as a warning to those in camp who strayed too far from the rules laid down by him.

The crew respected the foreman. They respected him for his honesty and his integrity—but more so, for his know-how and drive—qualities he had acquired after working years in the pine and on log-driving streams. Most of the woodsmen at Red Keg had followed the white pine west to Michigan and in their travels they had worked in all kinds of camps, so they knew how to judge men as well as how to distinguish a good camp operation from a haywire outfit.[1]

McTaggart was law at Camp 3. He was the chief—the leader who acted as peacemaker, judge and jury in the isolated lumber camp. There wasn't a job there he couldn't do. He could fall a tree or scale a log. He could stitch up an ax gash in a chopper's foot or sew together the frost cracks in his hands. He could doctor a lame ox or a sick horse, and he knew as much or more about timber cruising and running logs down log-driving streams than anyone in the Saginaw Valley.

The head push first saw the light of day in the Chemung Valley pineries in the State of New York. His father was a woodsman who believed that children should earn their own keep as soon as they were able. And when the boy reached the age of eleven, he was sent off to work as a cook's helper in a lumber camp.

The dazzling display of the new choreboy's hair immediately attracted the attention of the men in camp. Glad for any diversion from the monotony of their hard life, they teased him morning and night, calling him names like "Rosey" and "Little Red Riding Hood," getting a kick out of the way he flushed scarlet up over his ears when he got mad. But by the time McTaggart was fifteen, there were no more jokes, in his presence, about red hair. You didn't tease a young man who was suddenly taller than you were—who had long muscular arms and cool blue eyes that stared through yours into the back of your head. Maybe it was all the beans, salt pork and sour-dough bisquits he ate—or the hundreds of water pails he carried—or the wood he chopped, to keep camboose fires going day and night, that gave him his height and strength. He had brains, too—and in spite of the harassment from the men the choreboy eagerly took to the woods life and learned all phases of the lumber game. By the time he was eighteen, he could out-chop any chopper in the Chumung Valley—and he had already worked on a log drive on the west branch of the Susquehanna. The year 1851 was when McTaggart grew his big mustache, as was befitting a dashing young riverman.

By 1854, the white pine forests in New York State were almost depleted, and like so many other woodsmen of the times, Alex McTaggart packed his turkey and headed west to seek out the white pine in Michigan. He was drawn, as if by a magnet to the Saginaw region where he worked in the woods as chopper, scaler, riverhog and lastly, as timber cruiser for New York parties.

The lumberman John Shannon was one who benefited from McTaggart's skills—the cruiser having located extensive pinelands for him in the Valley. Impressed by the timber looker's performance in the woods, Shannon hired him to lay out a lumber camp for him up on the Tittabawassee. It was the third camp the boss-lumberman would build on the log-driving river. Upon its completion, it was understood that Alex McTaggart would take charge of the camp.

It was early summer, 1866, when the new "push" went into the woods to lay out Camp 3. The location chosen for the new camp was about four and one-half miles upriver from Shannon's Camp 2.

The importance of a good network of roads to a successful camp operation was well understood by the foreman. To determine the best (and shortest) routes for hauling logs, McTaggart criss-crossed the sections of land on foot, following the contours of the ground, tramping through thicket and brush—over swamp and sand ridge to mark out trails and roads over which logs could be drawn from cutting areas to skidways and thence to the river. For a road to the outside, the foreman investigated an old Indian trail that followed the winding river down to Camp 2. If it were widened, it would make a good tote road to be used in "soft" weather. When the river froze over, the tote teamster would drive his team on the ice, preferring the smoother sleigh ride to the bone-cracking torture of driving over the narrow rutted road through the woods. A road from Camp 2 down to the base camp (where Shannon's supplies were stored in the winter) had been built two years before.

It was mid-summer, just fourteen days after McTaggart went into the woods, when he completed laying out Camp 3. Satisfied at last with his efforts, he rolled up his gear and started on foot back to Saginaw, following the old Chippewa trail to Camp 2 and then by the tote road down to Midland where he took the stage for Saginaw City.

As soon as the foreman reached town he went directly to Shannon's sawmill where he set up temporary quarters. He let out the word, "Shannon's hiring—a dollar a day and found." In answer to the call, woodsmen straggled into the mill office to sign up for a winter in the new camp. For the rest of the crew, McTaggart scoured the saloons, bawdy houses, music halls and every other place where woodsmen were likely to hang out. In between hiring his 50-man crew, the foreman managed to sandwich in the business of making out lists and ordering camp equipment and provisions for his road builders and wood-butchers.

At last, in the final weeks of August, camp equipage and a skeleton crew (the rest of the woodsmen would follow later), were put on board a flat boat. The company tug, the

Satisfied with his efforts, McTaggart rolled up his gear and started out for Saginaw City.

Fannie S., towed the loaded scow up the Saginaw River past a place called Green Point and entered the mouth of the Tittabawassee where it proceeded up river until the water became too shallow for the tug to continue. Lines were cast off the scow. The woodsmen poled the flatboat and camp paraphernalia the remainder of the way to the new camp site.

It being late in the summer, the crew was hampered by hot weather and low water. There were times when the boat stuck on the river bottom. Each time it happened, the men would have to unload most of the supplies and drag the big scow by ropes into deeper water. Finally, after several days of frustrating, exhausting hard labor, they arrived at their destination where they found horse teams and oxen waiting for them. The animals had been driven to the site by the barn-boss and his helper from a farm that was owned and operated by John Shannon near a crossroads on the Tittabawassee called Freeland.

After the barge was unloaded, tarps were thrown over the supplies, and while the cook and blacksmith were setting up under canvas, the wood-butchers were falling trees and erecting the log buildings that would shelter the crew and animals that winter. Another part of the crew was set to work on the tote road, widening and leveling the old trail down to Camp 2. And during the weeks that followed, the road-builders worked on the logging roads, following the trails that the foreman had blazed only weeks before. They cut down trees, cleared out stumps, roots and brush, and leveled and filled in the low areas and holes—all under the close supervision of McTaggart.

In October, the rest of the crew and more teams and supplies came in by the new tote road; and by the time the river was frozen solid from shore to shore, camp activity was being pushed ahead by the foreman at full speed.

To the unitiated, it may sound like a simple thing to lumber off a "piece of pine." But the vexations that seemed designed solely to bedevil the head of a lumber camp, usually began the instant he took on the job. Accidents were one kind of trouble—men hurt or killed by falling trees or tree limbs, or by a glancing blow of an ax. The scaler getting behind in his work, was another. An epidemic of the epizooty in the stables, finding a jug of red-eye in the barn, mutiny in the bunk-camp because of too much pickled meat on the menu—all of these things were headaches. But the weather—sleighing operations held up or suspended altogether because of the lack of snow or unseasonable thaws—was the biggest headache of all. McTaggart and other lumbermen like him who worked the pine before the days of "steam and highball," were always at its mercy. If anything turned a foreman's hair white, it was high temperatures and spring-like showers—the fear of which made his battle with time almost constant.

The foreman considered himself fortunate to have sixty days of good sleighing weather. If he had less than thirty, he was in real trouble. Logs that were left lying in the woods or on the river banks were usually ruined[2]—either by timber borers or by fire. Timber borers perforated the bark of pine logs to such an extent as to make them unmerchantable. Failure to get logs out promptly could mean a bankrupt owner and unpaid woodsmen.

The 1866–1867 season, the first at Red Keg, had not been successful in that the weather had not been favorable for hauling logs or log driving. If the lumberman Shannon had not been on solid financial footing with his creditors, he would have felt the loss more than he did.

The foreman seldom blew out the light from his lamp

until his crew had been in their bunks for an hour or more—and he was often up and going when Cap opened the door of the van in the morning to light the fire. By day he directed camp operations in the woods—at night, he wrestled with accounts in camp ledgers and tally books. For six or seven months in the year, McTaggart thought "pine"—dreamed "pine"—and until that day in spring when the last log was hustled down the Tittabawassee on its way to the sorting booms at Green Point, the foreman would live "pine" around the clock.

Cap's gabriel, behind the water barrel.

7

G is for the GABRIEL, the cookee's dinner horn

Unlike the foreman, the cook and the blacksmith, the cook's helper was expendable in the lumber camp operation. If Cap Wilkes were to pick up his turkey and leave without notice, the camp routine would go on without him. Everyone in camp might have felt inconvenience at the cookee's quitting (especially Kelly), but the routine would have gone on as usual. And, there was always someone ready to step in as a replacement for the cook's helper in spite of the low pay and long hours that were the only rewards of the job.

In most lumber camps, the cookee was either a boy, not yet strong enough to work in the woods, or was a woodsman grown old who had no where else to go. Cap was neither of these. Until recently, he had been a hunter and guide in the woods and on the streams of the Valley—his tools of trade being traps, fish lines and an old flintlock rifle. But, because of injuries suffered when a "widow-maker" put him out of commission, Cap had taken up the stiff broom, the ax, chopping block, sawbuck and bucksaw—as well as the dinner horn in John Shannon's camps up on the Tittabawassee.

The "gabriel" was an instrument that was well named—for surely the message the cookee blew on the dinner horn at chew time was as eagerly listened for as was the archangel Gabriel's message in ancient times. Although it must be said, when the jarring notes of Cap's horn probed the depths of the bunk-camp before the crack of dawn seven days a week, it was greeted with less enthusiasm.

Not all lumber camps employed the use of horns to signal the men. It depended upon the originality of the cook's helper what kind of a noise-maker he used. Some preferred such alarming ear deafeners as the gut-hammer—an iron triangle or a length of iron rail on which the cookee beat a tattoo with another piece of iron or an old horseshoe. Others had a liking for bells or the melodious sound of a conch shell. Then there was the satanic flunkie who shot off a double-barreled shotgun to wake the crew—a practice that never caught on in most camps.

Cap's horn was homemade of tin. It was five feet long—slender and conical in shape and was fitted with a brass mouthpiece that had been salvaged from an old canaller's horn. The brass piece was all that remained of the horn, a memento of former times and happier days.

Like the other men in camp, this was Cap's second season at Red Keg—and like the work he did in seasons past at Shannon's Camp 1, his tasks were menial and varied. Other than the usual fetch and carry jobs, he rustled firewood, cleaned the stoves, kindled fires, "slopped" the pigs, kept the water barrel filled, cleaned lamp chimneys, filled lamp reservoirs, scrubbed floors, peeled potatoes, set tables, dished out grub, washed stacks of dishes and pans, took "sass" from the boys and orders from the overworked autocrat Kelly, from early morning until he climbed into his

The menial job wasn't much to Cap's liking, but he was thankful for it just the same.

bunk at night. So, it can be understood why Cap's dinner horn meant so much to him. Like the camp flunkie who shot off the shotgun every morning—if there ever was a time when the cookee at McTaggart's Red Keg had any feelings of satisfaction in his work, it was in the morning when he routed out the "cussed crew" from their bunks by the "diabolical toot" he blew on the gabriel.[1] The "music" that flew from Cap's horn was enough to wake the dead—and that was about the state the crew was in when they were brought back to consciousness by its din every morning.

Like Jimmy-the-Chopper, who at night kept his ax at his side, Cap, who slept in the bunk over Kelly's, propped the gabriel on a couple of hooks placed handy under the eaves where he could reach it first thing every morning. During the day, the horn rested behind the water barrel where nobody could step on it.

This morning, for some reason, the cook's helper had wakened earlier than usual and knowing he had a few more precious minutes of comfort under his blankets, he let his mind steal back to the "old days" in the Valley. It was a comfort the cookee rarely had time for.

As he lay there, Cap thought of the old Saginaw as he remembered the broad river when he was a boy—the shining, mirror-like surface broken only by resting water fowl and the dugouts and birch canoes of the Indians. The sweet recollection called him back, and another picture of the river, as it used to be, welled up in his mind—the sloop *Savage* and how she looked each spring when she made her way down stream laden with furs for John Astor's fur company at Mackinac Island. Cap felt his eyes moisten with the rememberance.

Was it true what they were saying, he wondered, that every time a new sawmill was built on the Saginaw more of the river's surface was covered with sawdust?

"What will the old river look like when the lumbermen get through with it," the cookee thought sadly. How the place had changed since he was a boy!

Cap sighed aloud and turned over to lie on his back. Below him, Kelly's rhythmic snoring went on undisturbed by the movement overhead. His helper continued to stare into the darkness. "I'm caught up in the lumbermen's business like most everybody else in the Saginaw Valley," he lamented to himself.

Again, the cookee's mind drifted back in time, his thoughts carried along by irresistible currents of longing for the past. He recalled the first time he saw the Saginaw. Another picture came into focus, and this time Cap smiled in remembrance.

It was so cold in the cook-camp there was a skim of ice on top of the water barrel—but the cookee only felt the warmth of the summer's day in 1834 when he, Thaddeus Wilkes, his father and mother and little sisters stood huddled together on a crowded wharf in the frontier town that was Detroit. The Wilkes family had just debarked from a sailing vessel that had brought them up from Buffalo. They stood there, an island of homesickness and misery in the middle of a crazy crowd of land buyers,[2] speculators and other emigrants like themselves who were swarming over the docks in search of luggage and other personal belongings. Cap's mother, thinking of the small white homestead they had left in New York, with its shade trees and neatly fenced-in fields, looked at her husband with eyes that said, "Jonas, what business do we have here?" And for one fleeting moment, the head of the Wilkes family would have given anything to be back home.

Like so many others in the 1830's, Jonas Wilkes had

gotten caught up in the fever of the great land rush days, when enormous sales of public lands were bringing speculators, land buyers and settlers into Michigan at the rate of two thousand people a day.[3] Sitting by his fireside on long winter evenings, he had dreamed over gazetteers and pamphlets which told about the unlimited opportunities that awaited those who dared to take a chance and settle in the "west." Sensing opportunity knocking, Jonas sold his farm. A place called Saginaw in the Michigan Territory was where he would take his family to live.

At Utica, Jonas loaded tools, household goods, an ox cart and a brace of oxen aboard a freight-carrying canal boat and then boarded the line boat, *May-Fly* to travel with his family, via the Erie Canal, across State to Buffalo.

For Thaddeus and his sisters, the trip across New York on a boat that resembled an ark with green-shuttered windows, was the highlight of the adventure. Cap remembered sitting or strolling on the cabin roof—and ducking when the man at the helm cried "low bridge!" He recalled the two mules that towed the boat walking in single file on the tow path at the side of the canal—and the driver (a boy no older than himself) who rode on the back mule trying to keep the beasts moving at four miles an hour. And Cap remembered the many locks they passed through and how the steersman signalled each approaching town by blowing his brass horn.[4]

At Buffalo, the Wilkeses boarded the schooner *Red Jacket*. The cookee smiled wryly in the dark when he thought about the length of time it took the vessel to reach Detroit—nine days, counting the delays on account of stormy weather. Barn yard animals, poultry, furniture and freight, not lashed down, had pitched about on deck at the height of one storm—and he remembered longing fervently for the journey to end, bemoaning the fact the family had not taken passage on a steamboat, which was faster, but as his father explained, more expensive.[5]

As the *Jacket* bore up the Detroit River, the sight of the American shore on one side of the ship, and the Canadian shore on the other, made up for all the discomfort the Wilkes family had suffered on the Lake. Lying in his bunk at Camp 3, nearly thirty-five years later, Cap fancied he could still hear the skirling of bagpipes and see the Highlanders at Fort Malden parading in plumed bonnet and kilt—"the dress that Waterloo made famous."[6]

While the *Red Jacket* maneuvered to the dock, Cap saw the masts of other sailing vessels at anchor and heard the whistles of steamboats on the river. He saw Indians—their faces and bodies painted in vermillion and white—lounging about with blankets thrown over their shoulders. Others mingled on the streets with frontiersmen dressed in leggings and buckskins, and with French ladies who wore straw hats on their heads and short gowns that flounced over ankle-length petticoats.[7]

Cap and his father had to shove their way through the crowds to sort out their belongings from the mountain of goods piled on the wharves and he remembered seeing frantic parents running around looking for lost children—and hearing people shout at their usually docile animals, trying to move them off the docks. In his mind, Cap could still hear the sounds of that day—the bawling cattle, bleating sheep, wailing children, church bells ringing and the singsong cries of the street venders—stage coaches lumbering up to the wharves to take people away or to bring passengers to ships setting sail—covered wagons jolting over narrow rutted roads and low, two-wheeled carts lurching around street corners, splashing mud on pedestrians who walked on sidewalks that were nothing more than loose planks thrown down to keep the townspeoples' feet dry.[8]

Standing in the midst of the noise and confusion, Cap vowed then and there that he would never become a city dweller if he could help it.

The taverns and rooming houses were filled and overflowing—and after being turned away from several lodging places, Jonas and Cap loaded the family belongings on the ox cart, hitched the oxen to the cart and started north, on foot, up the Territorial Road to Saginaw. The family would "camp out" along the way.

The hundred mile overland trip to the Valley settlement was torturous, traveling over a road that wasn't much more than a wide path cleared of trees—the stumps cut low so wagons could pass over them.[9] Finally, after five days of crossing rivers and streams and walking where it was almost impossible to distinguish the main "road" from the trail—the family, worn out, half starved (having eaten nothing but bacon and hardtack along the way), their faces swollen with mosquito bites—arrived at their destination, completing the final leg of their journey from Flint to Saginaw over a trail that had been only recently cleared of fallen timber by Valley settlers to admit sleighs in the wintertime.[10]

Their first sight of the place called Saginaw—hemmed in by forest trees and marshes—made the Wilkeses think they had come to the end of the earth. Fort Saginaw, which afforded a splendid view of the Saginaw from its vantage point high on the west bank of the river, still dominated the place. Its pickets partially surrounded an old barracks, hospital and other log buildings that had been erected by the military in 1822.[11] A dozen or so other buildings—rude structures that included a trading post, a warehouse for Indian goods, two shops and several dwellings—lay scattered below the fort along the river front.

The buildings inside the stockade were still in use—being occupied by settlers, as well as by the fur traders and surveyors stationed at Saginaw. It was into one of the blockhouses (now a tavern) that the Wilkes family moved, until Jonas could erect a cabin of their own.

Cap remembered happily how he took to the pioneer life—like he was born to it. While his mother was sorrowfully writing back home about the "millions of musquitoes that pour over us,"[12] Cap, now dressed in buckskins with an eagle feather stuck in his hair, was exploring every corner of the old fort and making friends up and down the river. His memories of those happy old days were of dugouts lining the sandy shore of the river—of Indians trading at the post, their cooking fires at night winking like fireflies in the dark—ducks and other wild fowl winging in morning mists, and dark clouds of passenger pigeons in their endless flights over the settlement.[13] He could hear yet the hubbub and hullabaloo, and see the tablecloths and handkerchiefs waving when someone caught sight of a ship sailing up river bringing the settlers provisions—and news from the outside.

The cookee recalled going to school in the barracks inside the fort, and he couldn't help but smile when he recalled the day of the logging bee, when their neighbors came to help his father put up their new home. They brought along Uncle Jimmy Cronk and his fiddle box[14] and when he played that night, Jimmy almost raised the new roof (or perhaps it was the contents in the demijohn that old Major Mosley brought to the "bee" that raised the roof!).

Cap shook his head when he thought about the clatter-bang and whine of the steam sawmill that the fur traders Gard and Eph Williams put into operation in 1835—the settlers never dreaming that the noise the big sash made, when it fell with each revolution of the twelve-foot fly wheel, was a foretelling of what was to come on the river.[15]

Growing up, Cap lived close to nature and like his Indian companions, he needed no compass in the woods. One of his tasks at home was to help his father provide for the family by hunting and fishing in the hunter and fisherman's paradise that was then the Saginaw Valley.

There was time for fun, too. Cap earned his nickname playing "soldiers" in the old fort when he and the other boys and girls marched around the flagstaff and up and down the hard-packed earth keeping time to the battered brass canaller's horn that had been the parting gift of the steersman on the Erie Canal. When they weren't imitating soldiers, they were playing a rough game of ball the Indians taught them — a game that was played with sticks fitted with leather sacks at one end. In the winter time they skated on the ice-covered river and bayous wearing skates they fashioned from the jawbones of wild hogs.[16]

The cookee stiffened — visions of the old times vanishing like spirits in the wind. Kelly had stopped snoring.

Cap held his breath — and waited (it wasn't time to get up), and when he heard the cook resume his breathing exercises, he relaxed — his thoughts returning to the past — when he and the Indians gathered hickory nuts, wild plums and grapes and searched in the marshes for wild celery and flagroots — remembering how the cranberries were so plentiful the settlers ballasted outgoing vessels with them.[17] And sadly he recalled the day when he awoke to find the forest had melted from around the settlement. With the appearance of more and more settlers, the lands were cleared of stumps and fenced in. The clatter of several more sawmills and the noise of caulking hammers at the shipyard could be heard over the sounds of carpenters' saws and hammers. And after the steamboat *Governor Marcy* showed up from Buffalo, more and more vessels started prowling up and down the river.

Cap decided to move on. There was nothing to keep him in Saginaw. Cholera, which had wiped out most of the Indian population, had taken a young wife, a baby son and both of Cap's parents. So he packed a few belongings in an old meal sack and sought employment as guide and hunter for the timber cruisers who had come to the Valley forests to look for white pine.

In 1858, Cap began working for the lumberman John Shannon—still managing to keep away from city life—until—. Even now the cookee winced when he thought about the day when he was struck down by a tree limb while out on an excursion with Alex McTaggart. As he waited, half frozen and almost unconscious with pain, for the cruiser to come and remove the limb from across his back—he thought his days in the woods were over, never dreaming he would live to see the time when he would be working as a lowly cook's helper in a lumber camp. But that's what happened after they patched him up so he was able to get around on his own feet again. The menial job wasn't much to his liking—but he was thankful for it just the same—thankful he could still earn a living away from the hustle of Saginaw City.

The vibrations below the cookee had stopped. This time he knew it was time to get up, and he gritted his teeth, anticipating the discomfort of stepping onto the cold floor boards with his bare feet. But before he swung his legs over the side of the bunk, Cap reached up with his hand to where the long tin horn was under the eaves. He touched the smooth, cold metal and smiled to himself.

"Today," he said under his breath, "I think I'll just blow her extry loud and long for them jolly old days on the Saginaw."

8

H is for HOW the name Red Keg was "born"

The summer of '66 when McTaggart was laying out Shannon's third camp, he had to select a place to build a set of camp buildings. The foreman's decision where to build was greatly influenced by a mark he had drawn a decade before on a map while cruising the area for the lumberman. The mark was an "X"—and next to the X-mark was a penciled note that read, "Trapper's hovel." Later on, the notation was crossed out and another written beneath it. It read, "Robb's place since '65."

The cruiser's mark covered a "forty"—forty acres that was surrounded by land owned by John Shannon. Virgin timber stretched across the forty—pines that might well have been the lumberman's if the trapper, who had homesteaded the land, had been willing to sell his property to him. The old man chose, instead, to sell it to one Ross Robb, then a soldier in the Army of the Potomac—a man whose main interest in life centered around a popular drinking establishment in Bay City, Michigan called the Domino Saloon.

When McTaggart first saw Robb's place, sitting back from the river bank, surrounded by a scattering of stumps and encircled by pine trees—there was no doubt in his mind the kind of influence it would have on his men. So the foreman built his camps two miles down river from Robb's—as far from the river-front property as he could—and later on, in no uncertain terms, he made it plain to his crew that the place up stream was strictly off-limits to them. If they didn't abide by the rules, they knew the consequences.

Camp 3 was in operation only a few weeks when late one moonless night McTaggart went outside for a breather before turning in. He was about to re-enter the van when he saw a flicker of lantern-light through the trees. The foreman thought he heard shreds of a woodsman's song. The sounds seemed to be keeping time to the swaying light. He waited—watching the lantern weave past him—winking up the path leading to the bunk-camp. The next thing he heard was the door being kicked open and muffled protests within, "What are you tryin' to do—wake up McTaggart! Shut your yap—an' while you're at it—shut the blankety door!"

The plank door slammed shut. And within minutes, all was silent.

The foreman heaved a sigh of relief—at the same time he made a mental note to find out who had waltzed in at that hour.

The next day, as might be expected, the guilty party was seen going down the tote road on his way out of the lumber camp.

The foreman's old anxiety about the place up-stream had proved him correct—and again, he warned his men about staying clear of it.

Ross Robb wasn't bothered by McTaggart's off-limits rule. There were other lumber camps operating on the Tittabawassee—and tramping three or four miles through the woods to his place, meant nothing to woodsmen who were weary of early rising, hard work and living in close quarters. They could be counted on to break camp rules for an hour or so of pleasure that was offered them at his establishment.

In spite of those men who risked unemployment for a brief smile or two, Robb had to admit that business was punk in the wintertime. But in the spring and summer—now, that was a different story. From the time the ice melted until well into the fall of the year, the Tittabawassee, being a log-driving river, was covered from shore to shore with a shiny, dark-brown mass of floating pine logs. And, balanced on top of the logs were thirsty rivermen, working with peavey and pike pole directing the logs down stream to the sorting booms below. There were men in wanguns, as well, men who followed the rivermen—and others in dugouts and in bateaux—and hundreds more thirsty woodsmen coming down from the lumber camps who had in their pockets wages earned for their winter's work.

This was the picture Ross Robb dreamed about while he was serving with Meade's crack VI Corps far off in the Shenandoah Valley in Virginia.

It was on a bleak snowy morning in December 1864 when an idea came to him—an idea that was so simple Robb didn't know why he hadn't thought of it before. He rummaged through his knapsack for paper and pencil—and before his squadron received orders to move on, he had scribbled off a hasty letter, addressing it to "Miss Mattie Morgan, The Domino Saloon, Bay City, Michigan."

This is what he wrote:

Somewhere in Va.

Dear Mattie,

In spite of the war it sounds like things are going great at the Domino. I'm doing pretty good myself. Card winnings from the boys in the outfits just hit a little over $500 ! Hope you been getting it regular! Which brings up an idea that come to me out of the blue. Why not invest some money in another place, lets say up on the Tittabawass' after the war is over? We could just as well be getting the boys wads before they blow it down to Saginaw or some place else. Dont know of any land what the lumbermen haven't already got tho. Ask around Matt honey and see if there is some place on the Tittabawassee or on some other log driving river we can get cheap. Old Jolly Timble, who used to cruise some for Daddy Wright, might have some idea on property we could get. If he is still coming in to the saloon, ask him on the sly—but *don't* tell him the reason you want to know about the land. Write me same address (it'll find me sometime, I hope).

I surely miss you. But there's hope in the air that the next campaign will be the last—with the north winning out. Maybe it'll lay to rest the croakers who keep on prophising a rebel victory! I hope so. In spite of the easy pickings, I sure am sick of this old war.

Yours,

Ross

P.S. This idea cant hurt the Domino none, since Bay City is closer to the lake traffic and to the boys coming down from the camps farther north. Don't let *nobody* in on this!!

As the reader might have deduced, after reading Robb's letter, the writer was something of a card sharp. He was also Miss Morgan's partner in the entertainment business, being half owner of the river-front saloon in Bay City.

The Domino was located at the foot of Third Street, where the steam ferry crossed the Saginaw to the west side of the river. Sailors, raftsmen, rivermen, log dealers, woodsmen, forwarders, dock wallopers, mill hands and other imbibers had only to step from the ferry, the steamboat dock or the nearby lumber wharves to swing through its doors and into the welcoming arms of Mattie Morgan—who was the main attraction at the water-front saloon.

Robb's joining the army had come as a surprise to Mattie. Her paramour had always shown more interest in the welfare of Ross Robb than in the well being of his country. The fact was, Ross had not volunteered for service—nor had he been drafted. Rather, he had been offered an opportunity and, like the gambler he was, took it when a man of some means tendered him a $500 commutation fee to take his place in the army. So, when Robb went off to fight Johnny Reb (thinking the war between the States would be over in a month), Mattie stayed behind to run the place in Bay City that had become a handy subject for church sermons on Sundays and for newspaper editorials on any day in the week—publicity which did much to spread the fame of the Domino Saloon.

Once a month, to keep her partner up to date on the business, Mattie would sit down in the little office behind the bar and in her bold, free-wheeling script, write a letter to Robb. She always enclosed the latest trial balance sheet, as well as a report on the goings on around town. Answers to her letters came only intermittently—depending upon troop movements.

It was almost springtime in 1865 before Mattie received the letter from Robb in which he disclosed his plan to put into operation another establishment for the pleasure of woodsmen and rivermen—and incidentally, for the profit of Robb & Morgan.

Mattie was impressed by Robb's idea. Draining the boys' pockets before they roared into the mill towns down river, after the lumber camps broke up, appealed very much to her business sense. So after making a few discreet inquiries in regard to the property mentioned in her partner's letter, she sat down at her neat rolltop desk and wrote:

Bay City, Mich.

Dear Ross,

I like your idea. Talked to Jolly like you wanted—and he told me about a old trapper who homesteaded a small place up on the Tittabawass and has been living on the river since about '32. The lumbermen have been pestering him for his land but he is so mad at them for spoiling the woods he wont sell it to them. But I think he will sell to *me*. Its a real good location. About 30 mi. up from the booms at Greenpoint or about 35 miles up from Saginaw. After thinking it over maybe I better go ahead and make a deal with the old boy, since he aint getting any younger—and you know what the mails are like between me and wherever you are! We might miss out on this if we dont take it now.

Things are still good here. The new toll swing bridge that took the place of the ferry is finally open for bizness. Its so close to the saloon I can stand on our roof and almost reach out and touch it. You can bet it will be great for bisness. Also a big new saw mill thats supposed to be the biggest one in the world (can you believe it?) has just started up almost across from us. That makes about 70 mills operating on the river so far. A couple of New York state lumbermen by the name of Sage and McGraw is building the new mill and they have laid out a village over there too, calling it Wenona (after a Indian I think) and selling building lots which also will be very good for us—which I dont have to tell *you*. What with the new bridge and all we should be able to retire on easy street. Also theres a rumer that another shipyard is in the works over there south of the new mill.

I sure will be glad when you get back home. The place is boomin up and down the river what with the river traffic, and the like—and the shipyards and mills all going full tilt. Last summer it was nothing but *logs and lumber* around here and am looking forward to a even bigger season this summer.

I hope, like you do, the war will be soon over and you'll be on your way back to B.C. all in one piece. I miss you a lot!

Love,

Mattie

March 31, 1865

The old trapper did indeed sell his piece of property to Miss Morgan and when Robb showed up at the Domino one day the following June, the first thing he did (in the way of business), was to go with Mattie on an inspection tour to look over their property on the Tittabawassee.

Just as Mattie had said, the place was perfect for their plans—and while the summer log jams lazily floated by, Robb, with the help of a local wood-butcher, built a wing and another story on the trapper's cabin. After the additions were put up, he covered the whole building with siding, laying it over nailing strips on the old log exterior. When it was completed, the saloon keeper couldn't help but think that the building, sitting against the pines, had a very snug appearance. The riverhogs on the river below thought so, too, and weren't very long in coming up to sample Robb's hospitality.

When Alex McTaggart first saw the place, he thought it looked harmless enough (but he knew better). The building looked like any other two story farm dwelling of the period—with unpainted exterior, pine shingle roof and a long shallow porch built across the front. What made it different from other houses in the area were the swinging doors in front and the large portrait of Mattie (painted in oils by a sign painter down in Saginaw City), that hung in a heavy gilt frame in the barroom. Other distinctions were the whiskey barrels that Robb placed in a row on each side of the loose plank walk that led from the building to the wooden steps that ran down to the river's edge.[1] On the riverbank, at the top of the steps, a ten-foot high pole supported, topside, another barrel. The barrel was painted red—and on its sides were capital letters that spelled out RED KEG SALOON. The unique sign was placed so it could be seen by passersby traveling up or down the Tittabawassee.

Ross Robb and his partner knew what they were doing when they opened their El Dorado in the middle of a wilderness. The logging season of 1865 wasn't over before the saloon had become an oasis for thirsty men.

Robb named the saloon "Red Keg"—for obvious reasons—and because of its proximity to the infamous drinking establishment up river, Camp 3 was dubbed "McTaggart's Red Keg"—to the annoyance of John Shannon who was a teetotaler and a staunch supporter of the temperance movement.

All winter long there was feverish activity in the pine woods. The ring of ax, the swish of saws and the hoarse commands of men urging on their teams kept the woods lively with their sounds.

At the saloon, two miles up river from Camp 3—all was quiet. The only sounds to break the silence of a winter's afternoon were the snapping of burning beechwood in the big iron stove that stood in the center of the barroom—the sounds of creaking chairs and the idle gossip of a woodsman or two (too old to work in the woods but not too elderly to play the fiddle or to do odd jobs around the place)—and the noise that Griswold, the bartender, made when he replenished the stove with wood, opening and closing the iron door with a rattle and clank. Perhaps that evening, or on Sunday, a handful of woodsmen would risk the wrath of the bull-of-the-woods and steal away from camps not isolated by distance—for an hour or two or pleasure. The shelf clock behind the bar ticked off the hours—while Ross Robb waited impatiently for spring when the logs, piled high along the winding river, would tumble into the spring freshets; and the rivermen and woodsmen on leave would tumble through the swinging doors of the Red Keg Saloon to relieve their mighty thirsts.

John Shannon, the boss lumberman, puts in an appearance at camp. He stirs things up with an announcement. In the meantime, log rustlers are discovered operating on his logs.

LEDGER

9

I is for the INK-SLINGER who keeps books in the van

In some lumber camps, the scaler acted as the ink-slinger, but at Camp 3, McTaggart, like other lumbermen, didn't completely trust his scaler. So he kept the books himself.[1]

Every night after supper, when the crew ambled back to the bunk-camp for an hour or so of leisure around the fire, the foreman pushed away his supper plate and headed for the van where more work awaited him—work that was far more tiring than the physical labor he knew in the woods.

Whether a logging contract called for the cutting and hauling of one million feet of logs, with a crew of fifteen men and three teams of oxen in a dray-haul camp in the 1850's—or for the cutting and hauling of one hundred million feet in a whole string of camps with a force of 450 men, seven steam locomotives and thirty-two miles of iron railroad track in the late 1880's—exact account had to be kept of camp operations as they progressed. In the lumberman's game, no sentiment was ever mixed with the cold, hard facts of logging. Every detail of camp activity was written down in daybooks, tally books and ledgers—accounts of daily records of logs cut and scaled; records of the crew's time and wages paid; "van accounts" showing items like tobacco, mitts, socks, and patent medicines that were issued to and charged against the men who purchased them; as well as lists of supplies and inventories of camp property. The lumberman-owner (or whoever was the main party in the lumbering operation) was interested in *facts*—and the man who kept the books in camp had to keep them in order and up to date, ready to show to the owner or his agent who was known to drop in any time, unannounced, to look over the situation in camp.

Almost every night after supper, McTaggart went through the same routine. Upon reaching the door of the van, he kicked the snow off his boots, let himself into the office and went straight to his rough board desk where he was soon up to his elbows in papers and daybooks—all bristling with numbers and jottings. He totted up long columns of figures and made numerous entries in the assortment of books piled on his desk—only interrupted by members of the crew who came in to report on their day's work.

One of the satisfactions the choppers, sawyers and loaders got out of their work was watching McTaggart carefully write down the number of trees they had felled or the number of logs they had "bucked" or piled up along the logging roads or at the river bank. And as each man, in his turn, read off the figures that he had scribbled on small pine-chip tablets (each man carried these little "note books" in his pocket for the purpose), they did it with pride and were not a little put out when they heard rivals call out larger numbers than they had to report.

Although rivalry was a cause of many injuries in the woods, the foreman encouraged competition among the men and as he jotted down the day's work of a chopper, he might pause to gravely look over the figures and then say casually, "What's the matter Jim — the rheumatics bothering you? Looks to me like Elias and Jack beat you and Frank today."

You can be certain Jim-the-Chopper and his partner Frank Cole would put more muscle into their work the next day.

It was almost the end of December, when one night, after the usual ritual of men coming in to give their daily reports, the door slammed shut on what McTaggart hoped would be the last interruption before he could get down to some serious book work. He glanced at the ticking shelf clock. It was almost eight o'clock. Across the dusky room, Uriah Bell, the scaler, with whom the foreman shared quarters, was at a deal table working on his own reports. The light from two oil lamps cast long shadows of the men over the bare log wall behind them.

The office was small and sparsely furnished. Besides the clock, the only other thing that could be called a decoration in the room was a large calendar that hung from a nail in the wall near the foreman's desk. Although the light was dim, one could read the advertising message printed in black letters across the top of the sheet:

1867 JOHN SHANNON CO. 1867
Dealer in Pine Lumber, Logs, Laths,
Shingles & Salt
Mill — S. Water St., Saginaw City
Yards at Albany, N.Y.

All of the pages on the calendar had been torn off, except the last one—for December—and on that page someone had crossed off, with a pencil, the first sixteen days of the month.

The room was colder now, caused by the frequent opening of the door. The foreman got up from his stool to replenish the fire in the box stove. That chore taken care of, he relighted his pipe, adjusted the lamp wick, and then

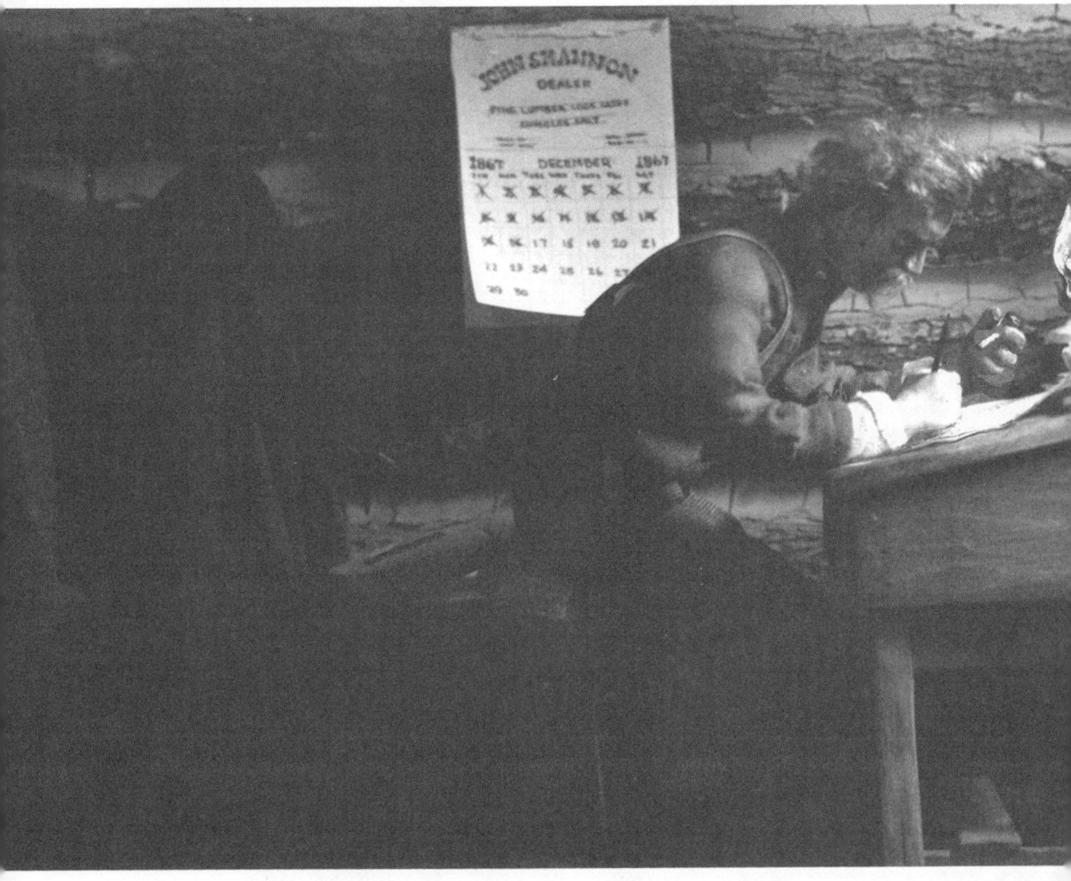

With a sigh, McTaggart settled down to his papers and books.

with a sigh settled down again to his papers and books. The only sounds in the room were the ticking clock and the scratching of pen on paper.

Again, the foreman was interrupted. Cully Wills, one of the swampers, came in to buy a plug of tobacco. McTaggart got up to rummage through several boxes on the store shelves built against one corner of the room. Finding the right brand, he handed the tobacco to Cully, and as soon as the swamper left the room, he went back to his desk and wrote down in the van book, "Wills—1 plug Gold Spray—10¢." The transaction would be added to others made by the man and deducted from his pay at the end of the season.

"When do you expect the Old man to come up?"

The foreman looked up from his books. Bell had asked the question that was foremost in his own mind.

"I have a feeling he'll be up sometime this week. He hasn't been here since before Thanksgiving," answered McTaggart.

John Shannon's lumber camps were strung out along the banks of the Tittabawassee. They were about four or five miles apart. Camp 1 was across the river from the Indian reservation just below the village of Midland. It was the base camp where supplies for all three camps were dropped off from Saginaw City. The lumberman's other two camps were north of Midland—Red Keg being the most northerly—about thirty-five miles from his sawmill in Saginaw. If the weather stayed at freezing, Shannon would travel up from the mill town, riding in his cutter on the winding course of the frozen river. If the weather were mild, he would come in over the tote road.

"By the end of the week we should have a pretty good report on the number of logs we got banked and scaled," the scaler said. "If this good hauling weather keeps up, you should have your four million banked by the middle of March." (The foreman's contract called for cutting and putting into the river four million feet of pine).

McTaggart couldn't help thinking, "Yes, we'll bank our four million *if* the weather doesn't go soft—*if* nobody gets hurt, *if, if*—."

The only thing that was certain about the lumber business was its uncertainty.

The foreman shook his head, as if to shake away his worries. He scraped off the dried ink on his pen, dipped it once again into the ink well and went back to his scribbling—oblivious to the cold draft that had found its way under the door.

At ten o'clock, Bell doused his light and made preparations to turn in—first dropping his report on McTaggart's desk. The foreman hardly noticed.

The clock was striking midnight when at last the nocturnal pen-pusher added up a final column of figures. Then he blew out the light, pulled off his boots and fell into his bunk—falling asleep almost before his head hit the pillow. The crew had been "sawing wood" in the bunk-camp for over three hours.

Morning came only too soon. Cap stumbled in with the water pail and while he was building up the fire, McTaggart got up and began looking over the entries he had made in his journals the night before.

"If the Old Man drops in out of the blue today, I'll be ready for him," he thought to himself.

The cookee was letting himself out the door when his boss looked up from his papers to say, "Cap, I want you to keep your eye peeled for Mr. Shannon this week. I'm expecting him up on an inspection tour. If the river is still hard, he'll drive up on the ice, so watch for his sleigh. The minute you see him, give me a signal on your horn so I can get back to camp. Make it two short and one long toot. And Cap," he added, "fix up the extra bunk in here and make sure the place is swept out and the water pail full. And if you can find a clean towel around the place, put it on the roller. That one," he glanced at the back of the door where the jack towel hung, "hasn't been changed since he was up here the last time."

Cap bobbed his head to show he had heard all the instructions before he closed the door behind him.

After breakfast, McTaggart spoke briefly to the cook about the expected visit of the boss lumberman. "Kelly, Mr. Shannon favors your mincemeat pie. Why don't you make up some while he's here?"

Kelly got the message. He grunted his acknowledgment but continued to fish doughnuts from a wash boiler full of hot grease that was simmering briskly on top of the stove.

Shannon didn't show up that day—or the next, but every evening after supper McTaggart burned the midnight oil to keep the camp books up to date—and during the day, he stayed closer to camp than he was wont to do—half expecting to hear Cap's horn. He wasn't let down. One afternoon late in the week the men looked up from their work to hear Cap's signal ringing through the woods. Immediately, the foreman made tracks for the camp, arriving in time to see the barn-boss come out to unhitch the owner's little chestnut mare and see John Shannon toss aside the buffalo robe that covered his knees before alighting from the cutter.

The temperature was below zero, but the ride up river from Camp 2 in the open sleigh hadn't bothered the lumberman. He was accustomed to the cold of snow-bound camps. Besides, he was muffled up in an ankle length sealskin coat, its great ulster collar turned up over his ears. On his head he wore a fur-lined Scotch Windsor cap. Fur-lined gauntlet gloves covered his hands, and German socks and heavy boots, his legs and feet. Only his black moustache and eyes could be seen between his great coat collar and the brim of his cap.

"Well, McTaggart, how are things getting on here?" (John Shannon didn't waste time getting down to brass tacks). The two men shook hands and followed Cap over to the van—the cookee carrying the boss's carpet bag and satchel of business papers.

Cap deposited the lumberman's things on the foreman's desk and then hastily ran an eye over the room. All was in order. The water pail was full—but, oh lordy, he'd forgotten to change the towel. So, while the two bosses were shedding their outer garments, he quickly slipped the limp, begrimed towel off the roller and replaced it with another of a lighter shade of gray—shoved more chunks of wood on the fire, adjusted the stove damper and then departed, leaving the lumbermen to their business.

Shannon stood for a moment or two before the stove with hands out-stretched to the radiating heat while the foreman began to spread out the camp books on the table. The lumberman consulted his pocket watch and said, "Your crew'll be in for supper in just an hour. Why don't we save the books until later and sit down and talk before we go over to the cook-camp?"

There was only one comfortable chair in the room, an old swing rocker upholstered in a shabby piece of red "Turkey" carpeting. The chair had been discarded by the landlady of a rooming house in East Saginaw where McTaggart had once lodged. It was used but infrequently, usually when visiting brass came to camp (the foreman rarely had time to sit in it). The chair was offered to Mr. Shannon, and McTaggart sat down on the stool next to his desk, crossing one long leg over the other. Shannon drew out of his breast pocket a telescoping leather case from which he offered a cigar to his foreman (who respectfully declined, preferring, he said, the stub of corncob he was smoking). The boss-lumberman lighted his cigar and in a cloud of blue smoke he said, "I don't have to tell you, Alex, the bad time we had last year and what it cost me. The season was the very deuce from start to finish. From February to March the bottom was out of the roads, and in the woods was an ocean of water—and when it did freeze up, there wasn't any snow. Camp 1 finally got some snow, but they had to haul it two miles or more to put it on the roads—and at Camp 2 and here at 3 we had to halt operations altogether."

McTaggart didn't need his employer to remind him of the disaster—he remembered only too well having to tell some of his men to pack their turkeys and sending them down the tote road before the season was over.

The foreman shifted his thoughts back to the van and to his visitor who continued his monologue.

"We left almost two million feet of logs in the woods for the timber borers. I had to buy logs from others for the mill—and when we shut down in September, long before the boom company suspended operations for the season, there wasn't a plank left on the docks. Now that we've got some good weather, we've got to get cracking and make up for the time and money lost."

The lumberman's speech was almost the same as the one he gave when he visited the camp at Thanksgiving time. It was the same old story of hustle, hustle while the going was good.

It was common knowledge that most lumbermen operated on credit—and keeping their credit high at the banks was the only way they could stay in business. The lumber business was a risky one at best—a complex business in which most lumbermen had their capital tied up in nonliquid assets for a full twelve months. Vast fortunes were made by a few so-called lumber barons, but the business proved to be a profitless one for hundreds of other investors.

The pine lumber game was one of chance—not a business for a novice. Even the best equipped gamblers in pine—men and women with superior ability and understanding of the business—or those with substantial financial backing could lose their assets—money, property and other resources while playing the game.

Any number of disasters could break a lumberman—bad weather, fire, errors in evaluation of timberlands, dishonest employes, a glutted timber market, a national recession. These were a few of the hazards of the pine lumber business.

McTaggart knew that John Shannon was into another season as deep as ever. The man owned timberlands, lumber camps and lumber vessels, as well as a big mill, salt works and docks at Saginaw City. In Albany, he owned and operated a big lumber yard, and he maintained a growing family and servants in a commodious house that was surrounded by park-like lawns, green houses and carriage barns. To keep everything going, he needed pine logs for his mill—logs and more logs to keep the wheels turning to pay the banks and to keep a host of creditors happy.

All of this was left unsaid. McTaggart was aware of what faced the older man. But it wasn't his job to worry about his employer's problems. His task was to furnish logs from Section 34 and that was what the lumberman was here for—to see how operations at Red Keg were progressing.

The foreman looked at the van clock and said, "I think we've got time to check the tally book before Cookee blows the horn." He reached for the accounts resting on his desk, handing the book to Shannon. If Kelly's mince pie was a little heavy tonight, the figures in the books might make it rest a little easier in the Old Man's stomach. He added, "I've been keeping Bell humping and he's right up to date on his scaling." (Bell was not in the van to hear this. He would spend the evening in the bunk-camp). "The figures show we've already got one million scaled and on the banks—and over five hundred thousand decked at the skidways—which leaves about two and a half million feet still to cut."

McTaggart waited for Shannon to make some comment and when he did not, he continued speaking.

"As I reported some time ago, we've been cutting and skidding with a full crew since late September. By the way—my original estimates on the kind of lumber our trees would yield are proving not far off. The logs are running about ninety percent uppers."[2]

Shannon looked pleased to hear this. He waved his hand, indicating with the motion for the foreman to continue speaking.

"So far, we've been lucky to have freezing weather since before Thanksgiving—so I've been keeping the icer going at night, and in the daytime, hauling as many logs to the banks as we can—."

The lumberman interrupted the report. "I wouldn't mind if you had at least one and a half million on the banks," he said. "This weather is just too good to last. You'd best start hauling some at night."

The foreman didn't expect any praise for work already

accomplished. He, like his men, would have to work that much harder. Still—he could see that John Shannon was pleased with the way things were progressing. After supper he'd show him the rest of the figures and do a little "beefing" of his own about the poor supplies that had come up from one of the supply houses that the lumberman liked to deal with—and about a few other things that he had on his mind.

The cookee's dinner horn sounded, which put an end to the brief report.

After supper, the two men were back in the van and hard at it within minutes.

Although every week the camp foreman prepared statements and forwarded the reports down to the mill office (to be forwarded by the mill superintendent to Albany), the lumberman believed in seeing for himself how operations were moving and made his trips to Saginaw City and to his camps unannounced just to keep things humming. He enjoyed these excursions, and this evening he seemed to be in a good mood (he didn't once mention the despised saloon up river). Whether his good humor was due to the brandy in the mincemeat pie (McTaggart wondered where Kelly was keeping the bottle), or to the fact that the books showed some progress—whatever the reason, the foreman was thankful for the expansive mood his employer was in.

Just before eleven o'clock the scaler came in, and with his appearance, John Shannon closed the last account book. He lighted another cigar. Blowing out a long stream of smoke he said, "Everything looks pretty good here." Then, as if to counter the praise he added, "The other camps are doing well too—but I've got to have at least twelve million feet of pine to go through the mill this season. I'm going to need every log I can get—."[3] But before he could expound further, Cap Wilkes took the moment to bring in from the cook-camp the remains of the mincemeat pie and a big pot of scalding hot tea.

While the three men were putting away the refreshments, McTaggart noticed the lumberman seemed to be mulling over something in his mind.

Shannon drank three cups of Kelly's tea, and when he drained the last cup he absent mindedly tamped out his cigar in the dregs. The foreman wondered what was going on in the Old Man's head.

It came out as the lumberman was climbing into his bunk. "Alex," he said, "tell the boys there'll be a bonus for the camp that gets out the most logs over what contracts call for."

The foreman, trying not to sound too surprised, replied by saying that everybody at Camp 3 would do their best to get the logs out.

Later, as he lay in the dark listening to the gentle snoring of his roommates, McTaggart, unable to sleep, was thinking about how he could put more logs on the banks. He had an idea in that direction that he had been meaning to try. In the meantime, to spur competition within the ranks, he'd post in the bunk-camp a daily report of each team's work. The "bucking-board" would go up tomorrow. The other idea would take a little more figgering.

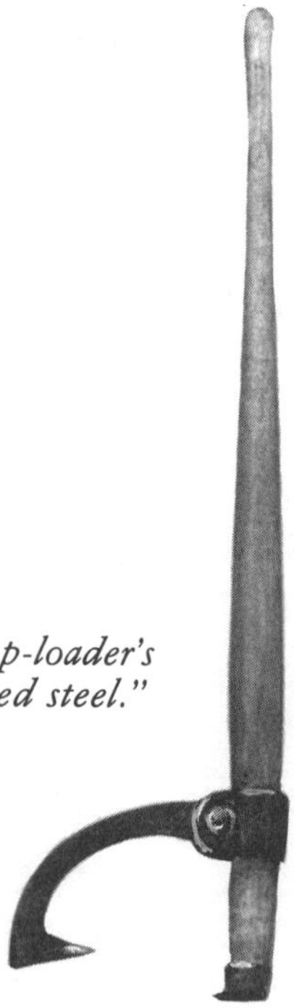

The top-loader's "crooked steel."

10

J is for JACK DRISCOLL, a legendary man

"You don't have to look for trouble in a logging camp, it's always there," was a favorite saying of the foreman. But it was the scaler, Uriah Bell, who voiced the truism on an occasion one night soon after Shannon's visit.

He and McTaggart were in the van waiting for the supper horn—getting out of wet clothes and into something dry when they heard a knock on the door.

"Come on in," shouted the foreman.

When the door opened, the cold wind blew into the room scattering papers over the foreman's desk—making the flame in the lamps bobble and flare. A man dressed in a shabby blanket coat, black felt cold-weather cap, "Soo" wool pants and heavy boots stepped into the room. He dropped a meal sack (his turkey) to the floor, propped a big cant hook against the wall, took off his cap and waited for the foreman to look up.

McTaggart, expecting Cap Wilkes on some errand or other, was surprised to see a tall slender man of about thirty years standing in front of the door. In the dusk, the stranger's hair, his mustache and eyebrows looked the pale color of flax. Or was the color silver?

With a start, it crossed the foreman's mind that standing before him was the notorious Silver Jack Driscoll, who was alleged to be a perfect terror in the woods and in sawmill towns throughout Michigan and Ontario. He noticed in the lamp light that the man's shoulders were broad and his back was straight without the slightest inclination to stoop. The high forehead, blue eyes, prominent cheek bones, and the nose like an eagle's beak—were certainly the physical characteristics of Silver Jack—acknowledged by woodsmen to be the "best man" in all the lumber regions of Michigan.[1]

If his coloring and physique fitted the description of the notorious woodsman, the quiet manner of the stranger did not. He waited patiently for the foreman to speak and finally stuck out his hand to shake with McTaggart, who at last recovered from his surprise.

"The name's Driscoll. Heard in East Saginaw you're lookin' for a top-loader."

McTaggart nodded.

Encouraged, the man continued. "Just come over from Ontario where I been top loadin' in a camp up at Belle Ewart. I got knocked off a sleigh load of logs and broke a couple of my ribs. The boys took me down home to Lindsay where I been mendin' until about three weeks ago. Went back up to Lake Simcoe, but found out they replaced me with another man—so I've been lookin' for work ever since.

The foreman thought with a start, if he wasn't mistaken, standing before him was the notorious Silver Jack Driscoll.

The camps I went to in Canada and the States have full crews."

By this time, the top-loader was standing in a small puddle. Snow had melted from his boots and he was beginning to look uncomfortable in the over-heated room. McTaggart motioned to the man to sit on the desk stool and as Driscoll loosened his coat, the foreman noticed, now that he could see him better, that the man did look a little peaked—but saw too, that he didn't sag on the stool. His back was as straight as an arrow.

"Walk up from Saginaw, did you?" asked McTaggart.

"Came over through Sarnia to Port Huron and walked north from there," answered the man. "In Saginaw, heard about the contest between your camp and Shannon's other camps—so I come up thinkin' maybe somebody could use a good top-loader. But the brass at the camps down river says they don't need—."

The man's voice trailed off when he heard a muffled laugh. The sound was mocking and came from behind the scaler's hand. It was loud enough so both the foreman and Driscoll heard it. The top-loader didn't turn to look at Bell, but kept his eyes trained on McTaggart. The gaze was steady. It seemed to be saying, "I'm no trouble-maker. Any stories you have heard about Silver Jack aren't half the truth. If you take a chance on hiring me, I won't let you down."

Whatever the reason—whether it was the top-loader's look, or the scaler's bad manners—McTaggart made a quick decision.

"Had your supper?" the foreman asked. And when Driscoll replied in the negative, McTaggart said, "Throw your turkey in at the bunk-camp and tell Cookee to set another place on the table. I'll give you a chance to show how good you are with that stick you got there."

Driscoll stood up, shouldered his turkey and then picked up his cant hook. He thanked the foreman and was almost out the door when McTaggart said in a tone that was unmistakable, "By the way, there's a place up river that's strictly off limits to the men in this camp—and contest or no contest, the man who sets foot in the place will be picking up his time and hitting the road as soon as I hear about it."

The loader turned around and again met the steady look of the push. He had gotten the message. He nodded his head and went out into the cold darkness of early evening.

As soon as the door closed behind the woodsman, Bell slid down in his chair, crossing one leg over the other. He looked over at the foreman and said with an amused smile, "Alex, I know you'll do most anything to beat out Shannon's other camps for the prize, but don't you think you just bit off more 'n you can chew?"

McTaggart didn't give Bell any satisfaction. He only said, "I think I hear the horn for supper," and walked out the door, leaving the scaler to wonder what the foreman saw in Driscoll besides trouble.

The new man in camp soon proved his worth with the "crooked steel," as well as with his ability to work in harmony with the other men. And McTaggart was pleased at the improved showing they were making at the banking grounds since Jack put in his appearance. Uriah Bell's constant forecasts of doom didn't let up, however. He, for some unexplained reason, didn't like the top-loader and made no bones about it.

But Bell was the only one in camp who didn't like the new man. Stories about Silver Jack in the role of a fearless fighter had been enlarged upon around countless camboose fires in logging camps on evenings when the older men talked about their heroes and reminisced about the "good old days" when they felled trees of enormous girth and drank rum to keep warm in their log hovels. At Red Keg, as well as in other camps in Michigan, Jack Driscoll was as much a hero to his fellow woodsmen as Jigger Jones was to Maine men.

In truth, Driscoll was a reluctant hero who was not a little embarrassed at being looked upon as a champion by men in the lumber trade. It all started one spring day a decade or more ago when, upon the joyous occasion of being set free after almost eight months of hard labor in a camp up on the Spanish River, he and his winter's companions had sought out a drinking establishment to quench their thirsts. It was in a small Ontario lumber town where he lifted one glass too many. After a word or two in disagreement over some trivial matter with another liberated woodsman, Driscoll became engaged in a classic barroom battle which was soon legend. Ever since that day he had been trying to live down—or up—to that brief moment of glory. It was, perhaps, for this reason he kept so much to himself.

Then he met up with Cap Wilkes.

It was on a Sunday afternoon, soon after he appeared in camp, when the top-loader accidentally ran into the cookee outside the cook-camp. Cap was binding on snowshoes, preparing to go on a tramp in the woods, and he was as surprised to look up and see the new man standing near by as Driscoll was to see him.

Cap said, "Howdy," but kept on lacing his rawhide bindings. He had been the butt of the boys' jokes too many times to rush into a friendship with any of them.

Driscoll noticed, resting against the door step, an old flintlock rifle, a couple of shot pouches and what looked

Cap's 1812 flint-lock muzzle loader.

like a fluted copper powder flask of another day. He said, "My Pa had a rifle something like that one. That a Springfield?" The top-loader leaned over to get a closer look at the rifle, noticing the polished wooden stock of the old firearm and the smooth barrel that indicated a thousand cleanings and rubbings.

Without looking up, Cap nodded his head and then said, "It's a 1812 model. Pa brought it out from York State in '34. Been usin' it ever since he died in 1841." He added (encouraged by the new man's interest in his gun), "You a hunter?"

The top-loader smiled. "Used to hunt some in the woods up near where I was born in Peterborough County in Ontario. Shot with my Grandad's old 75 caliber "Brown Bess" musket. It had a 42-inch barrel and when he shot it, the old man used to say the ball was so big it was just like slingin' a punkin' at a bear."

It was the older man's turn to smile—and then, after a pause, he said, off-handedly, "I'm goin' to see if I can stir up a deer for the stew pot. There's an extry pair of snowshoes there in the dingle, if you want to come along."

After that first encounter, any Sunday afternoon (when the cookee was free), the two men, one tall and straight, and the other of a stocky build with a slight stoop, could be seen going off into the woods on some mission of their own. Neither man spoke over a dozen words while out on their forays, but their close camaraderie was evident—a friendship that had grown out of a mutual respect that one man had for the other.

Christmas came and was celebrated without much ceremony—except for an ox roast dinner prepared by Kelly. New Year's Day, which was on a Wednesday, was just like any other day in the week. The only thing to distinguish it from other work days was the arrival of a new calendar for 1868 which George Pepper brought into camp with the provisions.

McTaggart took down the old calendar and hung up the new one in its place—leafing through its pages, counting the days remaining before the spring thaws would break up the ice on the river.

Since the snow storm in the middle of November, there had followed a succession of cold days and freezing nights, and by Thanksgiving Day, what with the strenuous efforts of the road crew and the almost nightly exercise of the sprinkler, the sleigh roads were put into such top-notch condition that load after load of logs had gone lumbering down the iced roads to the banking grounds on the river.

But in spite of the unusually good sleighing weather, McTaggart couldn't rid himself of the nagging suspicion that it couldn't last. In the back of his mind there persisted an old nonsense verse that his mother used to recite. It was from Elton's Comic All-my-nack:

"First it snew,
Then it blew,
Then it thew,
And then it friz."[2]

And as the first day of January approached, the foreman waited on edge for the weather to "thew." He wasn't disappointed. On the first of the month he was awakened, not by church bells and horns ushering in the new year, but by the ominous sounds of thunder rolling across the heavens. To the foreman, it was like hearing the sound of doom. He lay tense in his bunk, waiting for the rain to fall. And down it came—the wind blowing and wailing like a banshee, beating the rain around the eaves of the van and battering it against the heavy plank door, making it rattle on its hinges.

By morning, the rain had stopped falling and the sun came out in full glory. The temperature rose to almost forty degrees, and the long icicles which hung from the eaves, dripped steadily onto the ground, creating puddles around the camp buildings. That night, the snow on the frozen river looked slushy and the men knew if the weather stayed "soft," it would be only a week or so before the Tittabawassee would be running free.

That week, while the river gradually cleared of snow and ice, the men worked like demons, coming in at night exhausted and wet to the skin—working from dawn until after dark in the heavy, wet snow. The sleighs were kept running over the thick ice that the sprinklers had built up in the sleigh tracks—but if the thaw continued, McTaggart knew the whole operation would break down just like it did last year.

Sunday finally came—and the men were glad. After dinner, most of the crew went back to the bunk-camp to sleep, while the foreman and a couple of the road monkeys went off on an inspection tour of the sleigh roads.

Cap and Jack Driscoll went on an excursion of their own—to look at some of the cookee's traps which were near the south end of a place called Devil's Island (so-called by the rivermen because so many logs got hung up there during the log drives).

The small island was about one and one-half miles up river from the camp. It was a swampy piece of land about six rods wide and three-quarters of a mile long—only separated from the far bank by a narrow stretch of river that had branched off from the main stream to make the island.

The place was wild—covered with almost impenetrable growths of small trees, thick clumps of cedars, pines, rotting windfalls and bushes which hid from view the west bank of the river. In the summertime the point at the south end of the island was over-grown with reeds and cattails, making it a haven for the muskrat and beaver.

When Cap and Jack Driscoll reached a place on the river which was across from the island, they scrambled down the sloping side of the bank to the water's edge. Most of the ice had broken up; and some of the big cork pine logs, that had been rolled into the river to await the spring log drives, were now floating free in the water. (A boom stretched across the stream to keep the valuable property from straying too far away). The men wore needle-like calks in their boots, and with the agility that came from years of practice, they hopped aboard one of the logs—Jack pushing out from the shore with a peavey, which he carried on river jaunts. The log lazily left the shore with its passengers balanced on its broad back—the ride across the river ending when the log bumped against a tangle of tree roots that were exposed above a skim of ice that still clung, in sheltered places, to the shore of the island. Without a word, Cap went off to inspect his traps. To keep out of his way, Driscoll turned in the opposite direction to poke around the island by himself.

The top-loader's path was aimless as he walked through thickets of trees, ducking under low branches and around bushes. He was enjoying the quiet and peace of the afternoon and decided to sit down and have a smoke. Jack looked around for a fallen tree trunk that would make a good seat. He spied one and started toward it, taking his tobacco pouch and pipe out of his pocket as he walked, not paying too much attention to where he was stepping. Suddenly he stumbled. The toe of his boot had tripped over a solid object that was half concealed by the low spreading branches of some cedars.

"What the deuce!" he exclaimed, catching hold of a branch of one of the trees. He regained his balance, only to discover that he had lost his pipe. Provoked, the top-loader hunched down to look around for the corncob, but saw instead something under the cedars that made him forget all about his pipe. It was a sawed-off end of a log.

Upon closer examination, Driscoll saw that the slice of pine was about four or five inches thick—a piece that had been sawed off a log that was almost four feet in diameter. Quickly, he brushed away the wet snow that still clung to the huge circle and saw John Shannon's log mark—a figure "3" within the outline of a keg—clearly stamped seven times on the surface of the piece.

"Log thieves! And I'll be damned if they ain't operating right under our noses."[3] The top-loader stood up to kick around in the underbrush. He found more sawed-off log ends, all of them cut from Shannon's logs and hidden under the grove of cedars.

Jack looked over his shoulder. Was he alone? Had someone heard him knocking around in the bush? Then, for the first time, he was aware of footprints in the melting snow. They were trampled into the wet earth in such a way he couldn't tell his own prints from those of the thieves. Driscoll had seen enough. He went to find Cap.

The two friends could see that whoever was operating on the island didn't work on Sunday—or at least, on this par-

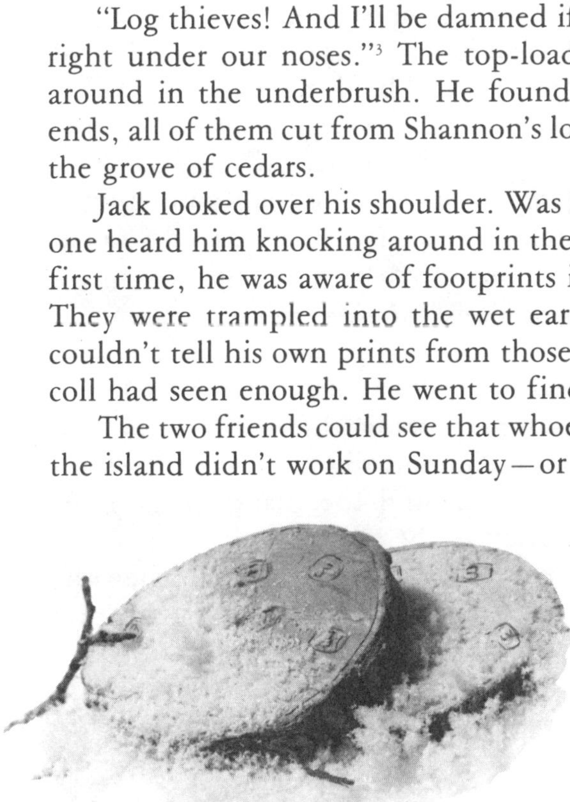

Driscoll saw John Shannon's log mark clearly stamped on the surface of the slice of pine.

ticular day—for there wasn't a sign of anyone around. It was kind of spooky, thought Jack. Out loud he said, "Let's look for the logs—they must be somewheres close."

Cap—his traps forgotten—helped Driscoll cover up the log ends, to leave them as they had found them, and then began their search around the hidden inlets along the swampy edge of the island. They inspected, as well, the mainland shore across the narrow stretch of river, following footprints where they could see them in the confusion of exposed tree roots, underbrush and fallen trees. Cedars and tamaracks grew to the river's edge on both sides of the stream, their branches touching overhead to form a natural arch. It was an ideal hiding place; an army detachment could have hidden there and no one would have been wiser. And that is where they found the logs—hidden under the over-hanging trees in the channel behind the island. Half of them had been made into a raft.

"I count forty," Driscoll said. "What do you get—?"

Cap rolled one giant over with his foot, watching it slowly turn in the water. "The same," he answered. "Guess whoever stole them is going to sell the whole bunch for mavericks. Most mills don't ask questions about where logs come from."

Driscoll didn't make any comment on Cap's rejoinder. He was trying to determine how the thieves worked—how they got the pine from Shannon's rollway, across the river and into the hiding place behind the island. They must have started their operation in the fall, several weeks before the big freezeup and continued with their work when the thaws opened up the river. They couldn't have worked after the stream froze over—hauling logs on the ice would have been a dead give away. He looked at Cap. "These must be the ones you were tellin' me about—the logs that was skidded direct to the bank from the prize clump. Look at them! Not one under twenty foot long and all about three to four foot acrost! The pirates sure got nerve—."

Cap grunted in agreement. He said, "Well—whoever is doin' the stealing has to be living somewheres near by—but there's not a sign of a camp." He was silent for almost a minute before adding, "The only place I know of that rents rooms is—." Another silence. The cookee then said what he and his companion were both thinking, "—They must be stayin' up at the saloon—."

The top-loader stood up, and while he was brushing an accumulation of dirt and snow off his knees, he asked, "How are we going to know if we don't go and find out?"

Cap waited—and then he heard Driscoll say, "You go on back to camp and I'll mosey on up there. Its less than a mile from here—."

Driscoll paused when he saw the look in Cap's eyes. "Don't worry," he added gruffly. "And don't say anything to McTaggart until I get back. No use in stirrin' up a hornet's nest."

As soon as the two friends reached the east shore again, they parted company—Cap going back to Camp 3—the top-loader walking in the opposite direction, toward the Red Keg Saloon.

The minute Jack pushed his way through the door of the barroom, a hush went over the room when the men recognized the top-loader (the grapevine had told them Silver Jack was working at Camp 3). But seeing Driscoll make no overtures, they went back to their glasses and playing cards.

Driscoll asked the bartender for a drink, and when it was put in front of him he picked it up and turned to lounge with his back to the bar—one elbow on the smooth pine top and a calked heel hooked over the brass rail. Casually, he looked over the occupants of the room. A half dozen Sunday customers were there—some standing at the bar, others sitting at tables drinking and socializing. Behind him Griswold, the bartender, slowly ran a damp rag over the polished surface of the counter.

Jack said, "Got tired of the same old grind, so I thought I'd come up for a drink." He took a sip of his beer. "So this is the famous Red Keg Saloon. Heard a lot about the place up in Canady."

Griswold replied (not without some curiosity), "I thought Robb's place was off limits to you boys down at Camp 3."

"That's what I heard too," answered the top-loader with a pleasant smile.

The woodsmen who overheard the remark laughed at the top-loader flaunting a superior's orders, and Jack knew then he had laid to rest any suspicions that thieves might have had (if indeed, the log stealers were in the barroom) on seeing a member of Camp 3 come into the saloon.

Now, more at ease, Driscoll observed that most of the woodsmen in the room looked like they were cut from the same cloth—the usual lumber crew on leave from weekly duties. But before he had time to speculate further on the whereabouts of the log thieves, he saw two men come down the steep stairway from rooms above. With their appearance, the top-loader turned back to face the bar where he could watch, unobserved, the men whose images were reflected in the large plate glass mirror. Over the rim of his glass he saw them walk over to join another man, who was obviously an acquaintance.

Although the three men were booted and dressed in warm clothing like all the others in the room, there was something about their looks and manner that made Jack watch them closely. Unlike the other woodsmen, their faces were pale—as if their activities were nocturnal or restricted to indoors. He noted too, that the men were huskily built. One was balding and wore thin moustaches down each side of his mouth. The other two had longish dark hair—one man being closely shaven and the other looking like he hadn't seen a razor in a month of Sundays.

The three companions stuck together, taking their drinks to one of the tables at the far end of the barroom where they hunched over their glasses and talked quietly among themselves.

In a low voice the top-loader asked Griswold, "Who are the gents sitting at the table next to the window?"

The bartender looked over to where the three men were sitting and then answered quietly, as if taking a cue from Driscoll, "They say they're land lookers workin' for some lumberman out East. I never heard of cruisers living in a tavern while on a job, tho—."

Griswold stopped talking to wait on another customer—afterwards coming back to where Jack was standing. He continued, "They come up in September, not leaving until about the middle of November. About two weeks ago they showed up again—."

"What handles do they go by?"

"The bald one's name is Dan'l Chriscaden; the one that needs a shave is Pete Norman and the dude with his back to us is Christy Dodge."

"Robb know them?" asked Jack.

"Not that I know of. They come out of nowheres—. Pay on time and mind their own business. That's all Ross cares about." Griswold continued to polish the bar.

"They got a rig or any horses?"

The reply was affirmative.

It was getting late (the choreboy had lighted the lamps in the barroom), so after making a few more inquiries of the genial bartender, Driscoll said "good night" and ambled out the door, satisfied with his visit to the saloon. He was convinced that the three men of mystery were the ones stealing logs from Shannon's banking grounds.

While he hurried through the woods, Jack tried to think of ways to apprehend the log rustlers. "McTaggart will have some ideas on how to handle the problem," he thought as he saw the welcome lights of the camp buildings through the trees.

The top-loader broke into a run when he heard Cap's horn calling the crew to supper, hoping to catch the foreman before he started out for the cook-camp. But before he reached the van, he saw the big form of McTaggart appear out of the shadows. He heard him say, "Driscoll, come into the office, I want to talk to you." The foreman had been waiting for him.

The sound of McTaggart's voice told the top-loader that he had been informed about the visit to the forbidden place up river.

The two men stepped into the van. Uriah Bell was sitting slouched behind his work table.

McTaggart spoke. "Bell tells me he saw you going down the trail to Ross Robb's place. Were you at the saloon?"

Driscoll looked over at the scaler, but the man avoided his eyes.

"Yes I was—," Jack said, looking at Bell. He turned back to the foreman and continued speaking, "—but I can explain—."

Behind him, McTaggart heard the scaler make a noise in his throat. He didn't like the sound of it—but on the other hand, he couldn't allow his authority to be questioned. He said, "As far as I can see, there can't be any good reason for you or anyone else to go up there. In the morning pick up your time." Then he added, disappointment sounding in his voice, "I thought I made myself clear about Robb's place. You're the best man I've got loading, but if I let you get away with this, I'll have more trouble on my hands than I need right now."

Jack, who had his pride to consider, stiffly nodded his head and turned to go without saying another word. But, before closing the door he paused to say in a voice so low that Bell couldn't hear, "Boss, you are dead wrong about me, and I'll be provin' it to you."

After the door closed behind the top-loader, the scaler got up from his chair and McTaggart heard him say, "Well, you can't say I didn't warn you about Driscoll—."

Irritated, McTaggart turned around and snapped, "Shut up, Bell, just keep your trap shut. I don't want to hear anymore about this, understand?" and slammed out the door.

Red Keg's log mark.

11

K is for the KNAVES who rustled the logs

Upon rising the next morning, the men noticed that Jack's bunk had not been slept in. And later on, as they tramped to the woods, there was speculation as to what had happened to him.

The only one who seemed unconcerned about the top-loader's disappearance (besides Uriah Bell), was Cap Wilkes, who all of a sudden seemed to be displaying an unusual amount of industry in the cook-camp. The night before, he had kept Kelly awake by doing chores he usually did in the morning—like filling the water barrel and banging in and out of the door to bring in arm-loads of stove wood. Kelly finally growled at him to douse the lamp so he could get some shut-eye.

The next morning, the cookee's energy had not lessened. He flew around taking smoking pans out of the cook's hands almost before Kelly had time to get them off the stove—and then snatched the dishes off the tables from under the men's noses while they were still chewing their food.

"What's got into you—you takin' a new tonic or did you find Kelly's jug?" complained one of the crew who reached for a donut only to see his fingers close over air as Cap whipped the bowl off the table.

If Kelly noticed anything unusual about the cookee's behavior, he didn't make any comment about it, one way or the other.

After breakfast, when everything in the cook-camp was cleaned up, Cap took off his apron and hung it on its customary nail behind the door. He put on his cap and when Kelly's back was turned, he stuffed something bulky inside his jacket before buttoning it up.

"Goin' over to see how Angus is coming with my new pung," he said. "I'll be back in time to take out flaggins." And before Kelly had time to run a critical eye around the room to inspect his helper's work, Cap had already ducked out the door and was hurrying, not in the direction of the blacksmith shop, but in the opposite way, through the woods, taking a path he generally took when he visited Devil's Island on his Sunday expeditions.

Forty minutes later, out of breath, having run most of the way, he arrived at his destination—the high banks opposite the island. He crouched there between two great tiers of logs until he could get his wind back. When his breathing came easier, he cautiously looked around. Making certain he was alone, he cupped his hands around his mouth and made sounds imitating the harsh calls of a jay bird. He waited—listening intently as the call died away over the water. On the island he could hear measured strokes of a crosscut saw biting into the wood of a pine log. Behind him, he heard the sounds the choppers made as their axes thudded into pine a half mile away.

Suddenly, the sawing on the island stopped. Why? Cap held his breath—and waited. Then, he was aware of the sound of tinkling bells. Again, the cookee looked out from his hiding place and saw, coming down the road, a sleigh-load of logs slowly approaching the spot where he was hidden. The teamster LaRiviere was perched on top of the load. His team was stepping carefully along the road—their movements making the bells that hung from their collars ring and the harness trappings jingle. Cap could hear Frenchy talking gently to the horses, "Gee Peppaire! Jacques! Allons grandes bêtes!"and making chirruping sounds with his lips as he guided the ponderous sleigh over the melting tracks of ice.

The sleigh was almost opposite Cap's hiding place when he heard a return call of a jay over the noise made by the Arcadian and his team. His signal had been answered. The cook's helper waited until the big sleigh lumbered by, then scrambled down the steep embankment. Reaching the bottom, he stepped out on a log and poled it across the river to the island.

When Cap jumped ashore, he was immediately lost to view from the mainland—the jungle-like tangle of cedars seemed to swallow him whole. He stood still for a moment-listening. The sawing had resumed. He was about to take another step when suddenly Jack Driscoll appeared before him. The top-loader's hiding place was so well concealed that even Cap, the old woodsman, hadn't seen it.

Jack grinned, then put his finger to his lips, indicating

they weren't alone. Softly he said, "How do you like my digs?"

Cap answered by tossing the package he had been carrying inside his jacket to his friend (food, purloined from under the eagle eye of Duff Kelly). He whispered, "All the boys are wonderin' where you went to. If it makes you feel any better, McTaggart looked mighty blue when he went out this mornin'."

Driscoll frowned, "Not much satisfaction in that. I just want to find out how those dudes work the logs over yonder so we can spill it to him. That's when I'll be feelin' better."

Cap nodded in agreement. He looked around the place where the top-loader had set up temporary housekeeping. "Seen 'em yet?" he asked—his voice low.

"Yup, this morning. The same crew I saw at the saloon. Just the three of 'em seem to be workin' together. I sure admire their nerve!"

Cap said, "They must be pretty experienced in handling pine to get those big logs behind the island. How do they manage it?"

"Well, for one thing—I found out they do their stealin' at night and their sawin' by day. Since our own operation is over the ridge, back about a half mile in the woods, the only thing they have to worry about in the daytime is bein' heard by our teams when they drive by on the road followin' the shore. They know a sleigh load of logs comes down from the cutting areas only every so often, heading for the banking grounds south of here, so they hold up their operation until they see the sleigh go by—and then they work like they was possessed when the coast is clear. They don't have to worry about the empty sleigh returning since the teamsters take the go-back-road on their trips back."

Cap had seen some log stealing in his day, but these

men had the nerve to take the biggest—the best logs that Shannon had banked. He said, "How the Sam Hill do they manage to get logs that size over behind the island in the dark and all?"

Driscoll brushed aside some snow on the damp ground, smoothing away sticks and pine needles. And while he explained the operation to Cap, he drew a crude diagram in the dirt.

"Their rig is simple—but it fits their job to a 'T.' It includes a capstan, or windlass that they made out of a log—and a length of rope what's operated by a horse that's hitched to the capstan. Here's where the rig is, in this clearing on the mainland acrost the river. Somehow, they brought in the horse—which they stable in a log hovel over there. I suppose they brought in their tools, rope, chain and raftin' pins in the boat they're using to steal the logs with.

"They wait until dark when they know the coast is clear, then two of 'em rows the scow over to Shannon's rollway, pulling the rope after them. The rope's attached to the capstan. When they reach the banking ground, they work a chain under and around the end of a prize log what's floatin' free—passing the rope through a ring in the chain. Then, they signal the man stationed in the woods—and the horse begins turning the capstan to wind up the rope—pulling the log acrost the river, so its in line with the river channel behind the island. Then, while the one man in the boat operates the oars, the other one guides the log with a pike pole—and with the help of the current, the log floats the rest of the way in behind the island. When the log is here—right where they want it, the two men maneuver it to this spot where the water is shallow and almost level with the shore. They saw off Shannon's marks then roll the log back into deeper water where they keep all the logs until they make them into a raft." Jack looked up, saying incredulously, "They even got the nerve to use torches at night!"

Cap whistled softly. Driscoll stood up, brushing off pine needles, wet leaves and dirt from his knees. He said, "Now that we know about the operation they won't get the raft of logs out, but until they're stopped, they'll be cuttin' off four or more inches from the ends of every log they steal."

"Listen!" Jack held up his hand, cocking his head to one side. "They quit sawin' again. We've been so busy jawin' we didn't notice another sleigh going by. But *they* did." He looked in the direction of the thieves' operation and shook his head.

Cap said, "What you going to do now?"

"Not sure yet. Even though we've got the goods on 'em, McTaggart's so blame mad about me goin' up to Robb's saloon, he won't listen to me—."

Driscoll shook his head again. "It's up to you, Cap. I can

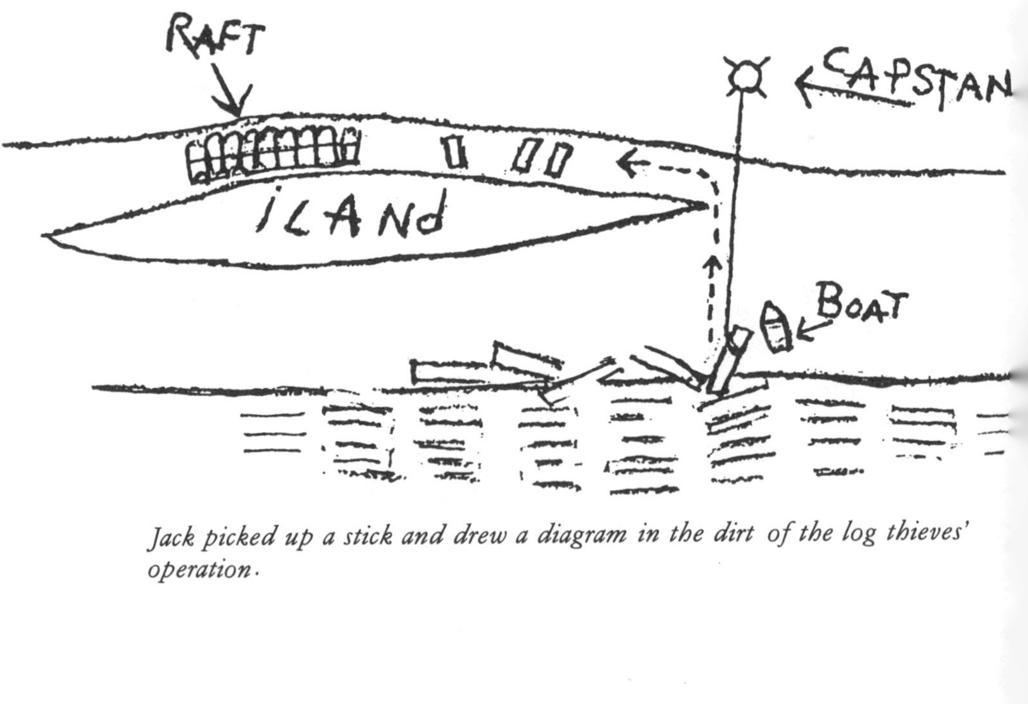

Jack picked up a stick and drew a diagram in the dirt of the log thieves' operation.

stay here and keep watch while you go and let McTaggart in on the show." Warming up to this idea, Jack continued, "Bring him here tomorrow night—alone. That'll give me time to make certain of movements and timing."

Cap liked the plan. He was interested not only in catching the log rustlers in the act, but in clearing his friend of any wrong doing. Besides, he knew that McTaggart needed the services of his best top-loader—and the sooner, the better.

After the cookee left the island, Jack spent the rest of the day watching and waiting. There seemed to be no change in the thieves' routine. The three men sawed off log ends and worked on the raft until about three o'clock in the afternoon when they stopped to eat a lunch they had brought with them. Then they laid around resting—only commencing their night's labors after the Red Keg crew left the woods for the bunk-camp.

The rustlers worked on until midnight when at last Driscoll heard them get into their boat to pick up Dodge on the west shore. Only when he heard oars creaking in oarlocks, did Jack go to inspect the thieves' work. They had added two more logs to the raft.

The next night, Driscoll listened for Cap's signal—this time, the "hoot, hoot" of the owl. And after the top-loader returned the signal, it was only minutes before he was standing face to face with Alex McTaggart.

In a low voice, the foreman said, "Well, Driscoll, what's going on here? Cap tells me there's thieves working on our logs."

Jack's reply was stiff—his pride still smarting. "Cap and me come acrost their work on Sunday." Then, unnecessarily, perhaps, he added, "That's why I went up to Robb's place—to see if any strangers were around there."

In the shadows, Driscoll saw the foreman move impatiently, and he interpreted the gesture as being an expression of censor at his going to the saloon without the foreman's consent. Determined to set things right, he hurried on before McTaggart had a chance to voice his disapproval a second time.

"For two nights now, I've been watchin' the same three I saw at Robb's, stealin' logs from Shannon's rollway. In the daytime, I watched them sawin' off the Keg-3 mark—."

The top-loader abruptly broke off speaking when, without warning, a flare of ruddy light turned his and his companions' attention to a lurid scene that suddenly came to light between the trees—a sight that made further explaining, on Jack's part, unnecessary. Dumfounded, the foreman watched the dark form of a man secure a smoking pine-knot torch to a pole which had been thrust upright into the soggy ground. The torch revealed the inky surface of the river shining in the erie light—and in the shadows beyond, a fringe of tamaracks growing almost to the water's edge. A rowboat was in the narrow channel, and Driscoll and his comrades watched in silence while another man jumped from the boat and secured it to an overhanging branch of a tree.

The two thieves promptly directed their attention to a twenty foot-long stick of pine that they had towed across the river from the rollway. They detached the chain and rope from the log, noisily dropping the chain into the boat; then grabbed cant hooks and began to wrestle with the pine, slowly working it with great exertion (and many oaths) to the edge of the bank.

The three watchers moved with caution to a position closer to the thieves' operation. In a whisper, Jack informed the foreman, "The two men sweatin' over the log are Dan

Chriscadin and Pete Norman. There's a third man workin' with 'em called Christy Dodge. He's at the windlass over on the west shore."

The top-loader did not go into more detail, knowing that Cap Wilkes had already described to McTaggart the covert activity on the island.

Peering from their hiding place, the trio watched the two thieves cant the massive log close to swampy low-lying ground—their chests heaving with the struggle. After a half dozen minutes, they heard Chriscaden say, "I think we can get one more log over before we call it a night."

His partner groaned. "We got time to build us a fire and dry out a little," he said. "We'll freeze up stiffer 'n a board crossin' the river again if we don't." He added, "Shannon's logs won't do us much good if we go to our glory with pneumonia before we get a chance to sell 'em."

McTaggart, Driscoll and Cap watched in envy while the rustlers built a fire and proceeded to dry their clothing and warm themselves in the heat the flames provided.

Thirty minutes—then forty minutes went by. At last Chriscaden and Norman put out the fire and made ready for the night's final river crossing. One man stepped into the boat while the other doused the torch—and in the dark, the three men watching heard the thieves push out from the shore to row down the narrow river channel.

McTaggart, Driscoll and Cap quickly made their way to the east shore and in the silvery light of the moon, now drifting from behind a curtain of clouds, they watched the scow cross the river, trailing the long rope behind in the icy water.

When the boat and its passengers were out of hearing, Driscoll said, "I've thought of a way we can catch the three of 'em—." He paused, thinking perhaps the foreman would have an idea of his own on how to handle the situation. But McTaggart, with a wave of his hand, encouraged the loader to continue speaking—all the time listening to the sound of oars dipping into the water as the log stealers rowed toward Shannon's banking ground for another log. The foreman was getting madder by the minute.

Jack said, "I think it would be a good idea to get the sheriff up here from Midland to make a legal arrest. These birds are nervy, and slippery to boot. Without Pritchert, they might get off this whole thing with a $5 fine." Driscoll paused, waiting for some comment from McTaggart. Encouraged by the foreman's silence, the top-loader continued.

"In the mornin', you could post some of our boys at each end of the island—and on the mainland too—with instructions to keep out of sight. At noon have somebody blow Cap's horn over at the cutting area so that Chriscaden and his boys'll think everythings goin' on as usual. While the pirates are busy sawin', you an' Pritchart, Cap and me can move in and catch 'em red-handed with all the evidence right there. The boys, posted on the island, will make sure the thieves don't get away in case the three of 'em try skinning out."

McTaggart liked the idea. The only flaw in the plan was, they wouldn't have time to go to Midland for the sheriff and get back again by tomorrow noon. So, the day after was decided upon as the time to make the arrest.

After agreeing upon signals, the foreman and Cap departed, leaving Driscoll in his hiding place to keep an eye on the log rustlers. He was to report on any changes in their movements.

Before going to sleep that night, for the first time in his life, Alex McTaggart prayed for a warm day in January. "At least, please keep the weather soft until we catch those thievin' log stealers."

In spite of the foreman's entreaty to the Lord, the temperature went down well below freezing that night. And as if the thieves had a premonition that something was in the wind, they worked all night long sawing off log ends and adding the unmarked logs to their raft. It was only an hour before dawn when the torch was put out.

Driscoll moved in as close to the log pirates as he dared, watching them make preparations to leave the island—and he was startled to hear one of them say, "Well, that's it. We're ready to move her out tomorrow night." The other man spoke up, "About time too—it's gettin' colder 'n Siberia. Just hope we can get the raft well past McTaggart's camp before the river freezes over again."

Driscoll could barely make out the rest of the conversation because of the noise the thieves made pushing off in their boat—but before they disappeared into the shadows, Jack heard them rejoicing over the fact that they would be able to sleep all day tomorrow.

"What they mean is—today," muttered the top-loader to himself. It was already morning.

"So the plan is changed," he thought. "I better warn McTaggart not to come up here at noon. We can pick those birds up on the raft as they start out tonight."

So, while the unsuspecting log rustlers were heading for their beds at the saloon, Jack Driscoll was beating it through the woods in the opposite direction, arriving at Camp 3 in time to waylay Cap, who was coming out of the cook-camp with two water buckets hanging from a yoke over his shoulders. The cookee was walking in the direction of the water hole when Jack caught up with him.

The top-loader hastily told Cap about the change in plans; then he asked him to wait until Uriah Bell was out of the van before relaying the message to the foreman. In the meantime, he said, he would be at the cook-camp getting some hot food in him before the horn sounded. He would try to keep out of sight while awaiting word from McTaggart. Kelly could be trusted to keep his mouth shut, he hoped.

After breakfast, there was a brief meeting with Cap and the foreman—and then the exhausted Driscoll went back to his hiding place on Devil's Island to keep watch on the stolen logs and try to get forty winks. There wasn't a sign of the three log pirates all day.

That night after supper, several members of the Red Keg crew were pulled aside by the foreman who told them about the stolen logs and the planned interception of the raft. The men were posted at strategic places along the river. It wouldn't be easy for the rustlers to escape into the woods or down the shore now. To insure the safety of the raft of logs, the men stretched another boom across the river, farther down stream.

The night was cold and sable-black. The wind was rising in the north—ruffling the surface of the river. The woodsmen, in their silent vigil, could feel the sting of snowflakes on their faces as they waited. McTaggart thought to himself, "We better get them tonight. If the temperature keeps dropping, the river will freeze over and we'll never see Chriscaden and Co. again."

An hour went by. Still, there was no sign of the thieves. The foreman shivered in the dark, wishing he could see his watch. Suddenly, a disturbing thought crossed his mind. "Cap," he whispered.

"Right here, Boss." The cookee was at McTaggart's elbow.

"Did you see anything of Bell after supper?"

Cap at once caught the foreman's meaning. But before he had time to answer, he heard Jack Driscoll's voice. "You don't suppose he's in on the deal with those buzzards? I'll bet he got word of our plan—."

McTaggart interrupted him. "If Bell made a deal with that bunch, he could make a nice profit on the logs. He knows their worth better than they do."

"Cap, go check out the island and vicinity—and Jack—you go on up to the saloon to see if Chriscaden and his chums are anywhere around there." He paused, and then added, "Watch out for Bell while you're about it."

Before the words were out of the foreman's mouth, the cookee was on his way—and Jack was off down the trail going toward the Red Keg Saloon.

The night being dark and stormy, the going was slow. Half way to the saloon, the top-loader thought he felt someone or something brush past him. Could it have been Uriah Bell on his way back to camp after warning those scamps? Well—he would soon know when he got to Robb's.

Before long, Driscoll saw the lights of the saloon winking through the trees.

There were only three or four customers in the barroom—it being a week night. After pausing inside the door and not seeing any sign of the log thieves, the top-loader walked over to the bartender. Griswold seemed to be busy straightening the array of bottles lined up along the shelf behind the bar. He didn't turn around, even though he saw Driscoll's reflection in the mirror.

Jack tapped lightly on the polished surface of the counter. When its attendant finally turned to face him, the top-loader noticed the man seemed jumpy and (as Jack would tell later on), "a little green around the gills."

"I'm looking for Chriscaden and his two pals. Have you seen 'em tonight?"

Griswold hesitated. The man seemed reluctant to speak. Then, he bobbed his head toward the front door and blurted out, "They went out. Don't know where—." But, having said that much, he seemed resolved to pass along something else. In a voice so low and shaky (Jack found himself leaning over the bar so he could catch the words) he heard the bartender say, "—Saw your scaler today—. Bell—. He was talkin' to the three of them—." The man didn't finish, but broke off abruptly to swing around and resume his show of lining up the whiskey bottles.

Dumfounded, Driscoll stared at the bartender's back. He waited, and when he saw that Griswold was not going to speak to him again, he gave up and walked out of the saloon.

Jack stood for a moment on the narrow porch. He was sorely disappointed at the way his plan had turned out. "After Bell's tip-off, those birds are half way to Bay City by now," he thought. Then he cheered up a little, thinking, "At least, the logs ain't going nowheres—that's for sure. And McTaggart knows now I wasn't drinking at the saloon for the pleasure of it." The top-loader shrugged his shoulders, reflecting on the evening's outcome. There was nothing to do now but go back and tell the crew the log pirates had gotten clean away.

He was dead tired—but his boss was waiting for him. Driscoll stepped off the porch and into the woods—setting out for Camp 3 at a fast pace. But he hadn't gone a hundred yards when he heard a noise behind him. He stopped in his tracks, turning to see the feeble light of a lantern swaying crazily from side to side coming toward him. He heard a voice shouting, "—Wait up! Jack! Wait for me!"

It was Griswold. When the man caught up to him (and could get his breath), he said, "Those men—the ones you was askin' me about—they're in the saloon—upstairs in room number 4. All the time you was inquirin' about their whereabouts, the three of them was behind the office door holdin' a horse pistol pointed at my head. I couldn't say nothing without getting it blowed off."

Without a word, Jack turned on his heel—and within minutes he was back at the Red Keg Saloon, bounding three steps at a time up the steep wooden stairs to the second floor. Without hesitating, he put his boot to the door marked "4" and burst into the shadowy room. In the dim light that shown from the wall lamp in the hallway, he saw the three log stealers standing one behind the other. The man in front was holding a pistol with both hands, pointing it at arm's length at Driscoll's head. Without thinking, Jack swung his foot up to knock the gun from the man's hands just as it went off. The pistol hadn't hit the floor before the top-loader's other boot found its way to the chest of the gun wielder, knocking him against his two companions. All three men went down like a column of dominos under the two hundred pound-weight of Jack Driscoll—the last man falling against the wooden dresser, shoving it against the wall and shattering the mirror to smithereens.

The woodsmen in the barroom came on the run when they heard the gun go off—and when they crowded into the room and saw only mayhem—not murder—they were a disappointed lot. The fracus seemed to be over. The only thing they saw was Jack bumping two of the thieves' heads together—one balding and the other covered with black hair, which looked like a scattered hay stack as it collided with the head of his partner. The third man was lying stiff as a corpse with broken glass sprinkled over him.

"These birds have been stealin' our logs," said the top-loader, in the way of an explanation to the onlookers. "That right, boys?"

"All right, so we lifted a few logs. Old Shannon wouldn't never of missed 'em," mumbled Dodge as Jack bunched the three together and tumbled them down the flight of stairs.

Ross Robb, who had come up from Bay City that day, was standing at the foot of the stairway. He looked glum when he saw Driscoll bouncing his steady-paying boarders down the steps—but seeing which way the wind was blowing, he said, "The cellar's dry—you can lock up the three of them down there until the law gets here."

And that's where Jack Driscoll left the log rustlers that night. In a way he felt kind of sorry for them. "They worked so dang hard for nuthin'," he thought to himself as he hiked back to the camp.

McTaggart puts into operation a plan to move even more pine to the banking grounds — which means additional hard work for his crew.

Woodsman with cant hook.

12

L is for the LOADERS
Red Keg's "top dogs"

The night Jack Driscoll put the log rustlers out of business, the mercury plunged toward the zero mark on the big thermometer that hung outside the van door—and the north wind, which shifted easterly, brought with it a foot of new snow.

The road crew wasted no time plowing out and repairing the sleigh roads.

During the weeks of soft weather, all operations had slowed down. The foreman kept the sleighs running, but the teamsters made fewer trips to the banking grounds, hauling reduced loads to keep the roads from breaking down completely. By the time the sprinkler could be gotten back on the job, the skidways were full of logs waiting for the trip to the river.

Every day McTaggart was on the job before daylight working out problems as they arose and directing operations to get more logs on the banks. He stayed awake at night trying to think of ways to improve operations at Red Keg. One of the first improvements he made after the freeze-up was to send two of the boys, who were more partial to "fat pork and sundown" than work, over the tote road with their pay slips in hand. Two new men, handy with the cant hook—and a new scaler, to replace Uriah Bell (who had not been heard from since the capture of the log thieves), were sent up from Saginaw City by Shannon's mill superintendent, Charlie Burleson. McTaggart stepped up the pace to make up for time lost.

Morale was high after the log stealing episode and the foreman wanted to keep it that way, so when some of the men started to kick about the steady diet of salt meat that was being dished out three times a day on a non-stop basis, McTaggart decided to have a few words in private with the cook.

"Kelly, you keeping those pigs penned up yonder for any special reason? Seems like you don't need any more than a couple for eating the slops."

The cook knew what was coming, but stubbornly kept on stirring his batter without replying.

McTaggart persisted. "Adam told me last night that he's been noticing the boys looking kind of funny at his ox team, and he says if some morning he finds Bright or Buck keeled over with a knot on either of their heads, there's going to be the very devil to pay. Plain speakin', Kelly, the boys have a hankering for a taste of fresh meat, so before we lose another of our bulls 'by accident,' you better do something quick." The foreman ended his speech by nodding in the general direction of the yard where the cook's pet porkers were kept.

Kelly got the message. For Christmas dinner, the cook had served the hind quarters of ox-driver Able Fry's "old Babe," who had mysteriously met his fate in the barn one night just before the holiday. All the boys, who were questioned about the untimely death of the off-ox, professed an elaborate innocence—but there was evidence to show that Babe hadn't just run amuck into the side of the stable in a fit—but had met with a blunt object expertly placed by human hand, smack between the eyes. The reproachful looks that Kelly got from Able Fry when he served up Christmas dinner was enough to un-nerve the most satanic "stomach-robber."

When McTaggart stepped out the door—the interview being over—Duff Kelly heaved a deep sigh of resignation. Later in the day, a little butchering was done out back, and for a few days, the crew enjoyed roast pork loin and fresh ham.

It also relieved the monotony around camp when Cap Wilkes and Driscoll stepped up their Sunday trapping and hunting expeditions—the results of which found their way into the iron stew pot on the back of the stove.

By this time, everyone in camp, from Kelly on down to Patrick Toohey, the lowly chickadee, was pulling together as a team to win the prize bonus that the boss lumberman promised to the camp that dumped the most logs into the Tittabawassee in the spring. No one worked harder than the loaders.

One of the most dangerous jobs in the lumber camp (where teamwork was most essential) was performed by these skilled men.

After the ox teams "trailed down" or dragged the logs from the cutting areas to the skidways, the loaders, with the help of horse teams, piled the logs in tiers at these way stations along the sleigh roads. This is where the pine awaited transportation to the river, and where more loaders and other horse teams worked to remove the logs from the sleighs and deck them in great piles along the river banks.

On Sunday, when there is a pause in the camp operation, one might stroll along the skidways and stop to look at the pyramids of pine resting so serenely at the sides of the sleigh roads. Try as one might it is difficult to imagine how lethel is each log in each great heap. But visit the skidway and banking grounds on Monday and watch the loaders sweating to work the logs into piles—or to move them onto sleighs for hauling to the river, and one will quickly understand why the woodsmen had such healthy respect for the unpredictable monster saw logs.

Most of the pines cut at Camp 3 were from two, three to five feet in diameter and were sawed into sixteen and twenty foot lengths—which meant that each log weighed from two to eight tons each.[1] Moving them was a feat that required nimble minds and limbs, as well as great strength and courage.

The tool used by the catty-footed loaders was called a cant hook. In the hands of a greenhorn, the great dangling steel hook that was attached to the stout, five-foot long, tapered ash pole made the device an awkward, frustrating thing to manipulate. But in the hands of men like Jack Driscoll or Whispering Bill, the big swinging claw-hook made the heavy stick a very effective tool.

At Red Keg, two teams of men worked at the skidway. One team piled the logs on skids after the oxen hauled them from the stump. The other team, headed by Jack Driscoll, maneuvered the logs from the piles to the sleighs. Loaders Matt Small and Tom Giddy were his "sendup" men. Ira MacDonald and his pair of sorrels (Belle and Jem) completed the team.

Whispering Bill was the top-loader who worked at the river. The two new men in camp were his sendup men. Tom Poole and his dappled grays, Doc and Bob, supplied the horse power.

Three sleighs were kept in motion hauling logs from the skidway to the banking grounds. Every morning before dawn (except on Sundays), the teamsters were up and on their way, driving their teams to the places where they had left the sleighs the night before. And all day long, while one sleigh was being loaded at the skidway, there was another one unloading at the river—and a third making its trip back on the go-back road for another load of logs. On good hauling days, each sleigh made the two and one-half mile round trip three times, carrying (each trip) about 4,800 feet of logs to the river banks.[2]

This was a pretty good showing for the size of the sleighs that were used to haul logs at Camp 3. But McTaggart wasn't satisfied with the results they were getting.

The foreman decided that some action was called for—some radical change in the operations that would put more logs on the banks. And after mulling over several ideas in that direction, a plan took shape in his head. He chose the next Saturday night to announce his new scheme to the crew.

The day had been a good one—the weather clear, cold and bright. There had been no accidents in the woods for almost a week. The men had eaten a hot supper and now they could relax, rejoicing in the fact that tomorrow was Sunday, the one day in the week they could take off from their labors in the woods. For most of the crew, Saturday night was the best time of the week—the night that promised "two nights in the straw and three meals ahead."

One by one the men came back to the fire-lighted bunk-camp. Some gathered around the camboose—others congregated on the deacon's seat, while a few men rested contentedly in their bunks. All were waiting for Jimmy MacDonald to come in from the barn and tune up his fiddle for some Saturday night fun when they saw the foreman enter the smoky room.

McTaggart walked to the end of the shanty and stepped up on the deacon's bench. He stood there with his hands on his hips and waited for the attention of all the men.

The rumble of voices gradually quieted. The whir of the grindstone stopped, and the room was silent except for the snapping of burning logs and the sizzle of a hundred or more pairs of wet socks, boots and leg rags that sagged in bunches from the lines strung up around the fire.

"Everybody here?" the foreman asked.

There was a general head-turning—men looking to see who in the crew was missing. Only Kelly and the cookee, the new scaler, the blacksmith, the barn-boss and his helper, and the teamster Jim MacDonald were not present—having duties elsewhere.

McTaggart said, "I know you're waiting for Jim to come in from the barn to tune up, so I'll take just a minute to tell you about some changes I'm going to make in our operation—starting Monday." He paused, to let his opening remark sink in.

"I got a message today about how camps 1 and 2 are doing. The report is, they're a little ahead of us."

Protests were heard here and there around the dusky room, but McTaggart waved them aside and went on with hardly a pause.

"Whether they're out in front at this point is not important. What we want to do is win out in the end."

Lively sounds of agreement.

"Now then, this past week I've been making a study of our whole operation, and I think I've come up with a plan that will put us on top of the heap and keep us there until the end of the season. It's not that you boys aren't working your tails off—you're doing all you can with the teams and sleighs we got to work with. But we've *got to move more logs*."

Now that the foreman had the full attention of his crew, he stepped down from the bench, propping one boot on the seat before continuing his speech.

"Since I've come to Michigan, I've seen some changes in methods used in the woods—and I'm sure you boys who come from the Penobscot, the Kennebec, St. John or other lumber districts out East have seen them too: bigger sleighs, road grading, icing—to name a few of the changes that have come about since the war."

Again, the sound of men's voices were heard here and there in the room.

"—I kin remember times when the whole dang crew slept on nothin' but spruce boughs laid on the dirt floor of the hovel. We all hunkered up close to each other, covered up under one big spread to keep from freezin' to death—."

Another voice broke in. "Aye, and we ate nothin' but fat pork, ship biscuit, codfish and beans the whole winter—and if you wanted tea you paid a dollar a month for it."

"That ain't nothin'—Billy and me kin remember the first time we seen a cant hook. It was right here at Red Keg. All we had to cant logs in them old days was the old swing dog."

McTaggart motioned for silence. The "recollecting" was getting out of hand.

When the room was quiet again he continued speaking.

"Since Christmas, all the camps in the region have quit cutting and are concentrating on getting what logs they've cut to the river. Now I know it's taking a long chance on leaving our logs in the woods for fire or timber borers, come spring—but I've decided to keep our best cutting crew busy until the last minute to get every log on skids we can. Thanks to our road crew, our sleigh roads are better than the ones at the other camps. The ice in the tracks is back to what it was before the thaw—and even with some more thawing, I think we can take a chance on getting the logs from the skidways to the river in time for the drives, even if there isn't a lick of snow on the ground."[3]

McTaggart paused, and then said, "Now, I'm coming to the part in my plan that will mean extra muscle on the part of the loading teams."

Whispering Bill looked over at Jack Driscoll as if to say, "Here it comes."

Jack only blew out a long thin stream of smoke from between teeth that clutched a corncob pipe and waited for the foreman to continue.

"Up until now we've been using five and a half foot wide bunks on our sleighs, and no matter how we load them we can't get much more than 5,000 feet of logs on at a time. So, I've decided to increase the size of our sleigh bunks to twelve feet."[4]

There was a general sound of astonishment in the room—but the foreman ignored it, carrying on as if he had not heard the stir.

"Using the wider bunks, we can send more than double the amount of logs to the river at one time. McCloud is already squaring up the timbers for one of the sleighs from some straight-grained oak logs he's been saving for sleigh runners. With some help from you boys, the sleigh should be fitted out by Monday night. We'll give her a trial run on Tuesday."

The chorus of men's voices rose once again in the dusky room. In later years, wider bunks on sleighs became common—but in 1868, a twelve-foot wide load was unheard of. Here was, indeed, a challenge, and the men were already talking about how to meet it head on. There were questions too. Would the sleighs hold up the bigger loads without breaking down? What about the wear and tear on the roads? And what about going up and down hill with the additional weight on the sleighs?

Each question was answered by the foreman. No, the sleigh runners wouldn't break down. McCloud had made them of solid oak and five inches wide. They would hold up any size load.[5] The road crews would have to bend their backs a little more than usual to keep the roads in shape. There would be a snatch team of horses used ahead of the regular teams to help pull the sleighs up rises in the ground (or any other place where needed)—and yes, the old method of warping down or using snubbing ropes and chains to control the loads going down hill, would work the same way with twelve foot sleigh bunks as it did with sleighs carrying smaller loads.

Jimmy came in from the barn just as McTaggart answered the last question, and while the teamster was pulling his fiddle out of the meal sack (where he stored it) and began making noises with bow over cat gut to tune up—the foreman added a brusque warning to every man in the room, to keep an even sharper lookout for trouble from now on.

As McTaggart stepped out the door to go back to the van, the teamster MacDonald struck up a lively quadrille on his squawk box, and a couple of the boys were already kicking up their heels—the cares of the day—and the thought of more hard work to come—forgotten.

With the help of swampers Duncan McPhee and Bill MacDonald, the blacksmith worked all day Sunday on the new sleigh and by nightfall on Monday, it was completed and fitted with its over-size bunks.

The evening star was already bright in the sky when the crew came in from the woods that night—but tired as they were, they stopped off at the blacksmith shop to see if the three Scots had finished the sleigh. By the light of lanterns, they saw the curiosity resting in the yard outside the forge—its great hand-hewn crossbeams reaching out on each side—three feet over the sleigh runners.

McCloud watched the men as they inspected the sleigh, and he couldn't help but think about the stout two-runner bob sleds he used to make for camps up in Canada and in Maine. The end of a sixty-foot stick of pine would be chained on the single beam of the bobs; and with the rossed end of the log trailing behind in the snow, it was hauled by six pair of straining oxen up hill and down, over the winding road to the stream, prodded on by a dogged ox-driver whose patience and picturesque profanity has never been equaled before or since.[6]

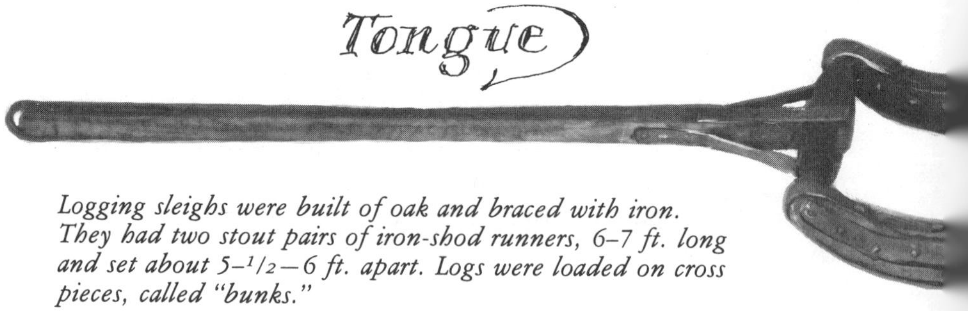

Logging sleighs were built of oak and braced with iron. They had two stout pairs of iron-shod runners, 6–7 ft. long and set about 5–1/2–6 ft. apart. Logs were loaded on cross pieces, called "bunks."

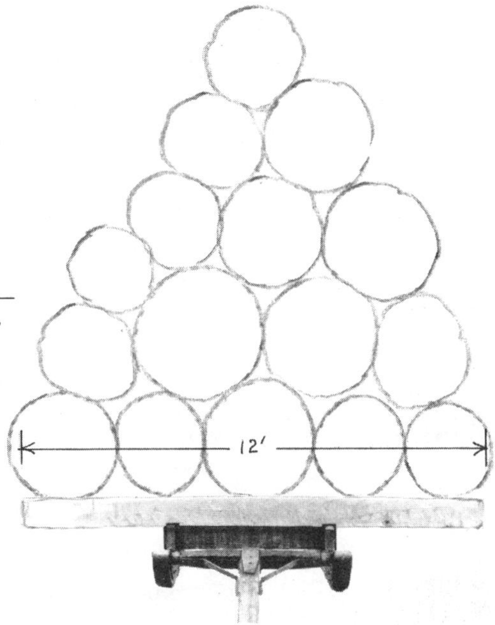

Two views of a logging sleigh to compare the number of logs (2–2½–4 ft. dia.) that could be loaded on 5 ft. and 12 ft.-wide bunks.

"Well," thought Angus, "the boys will try out the new sleigh, and if it works out to McTaggart's satisfaction, I'll fit the other two sleds with wider bunks—but not before." Those were his orders.

The next morning, Cap Wilkes called out the teamsters at the usual hour and while Ed Flatt and Elias January were driving their teams to waiting sleighs in the woods, Frenchy LaRiviere, who had the honor of driving the trial load, waited in the barn, fussing over his horses—checking their harness and chains, until the loaders were ready to move out to the designated place in the woods.

The sun was already flushing the sky before the loading teams (both man and beast) were in their places at the skidway ready to begin loading the big sleigh. When it was light enough to see, Jack Driscoll signed that he was ready to direct the movement of the first logs to the new sleigh bunks.[7]

McTaggart stood on the sidelines with the new scaler. The men watched intently as Frenchy commanded his horse team to pull the sleigh parallel to one of the log piles. When it was in position, blocks were shoved under the ends of the bunks (as a precaution against tipping when the first logs were loaded), and two long poles were propped at an angle, about ten feet apart, against the side of the low-slung vehicle.

Without further ceremony, the two send-up men, Matt Small and Tom Giddy, began to pull with their cant hooks the first log from the face of the pile and position it at the foot of the poles. The log would be hauled up the pole runners to the sleigh by means of a long, slim decking chain which was attached to the evener of a cross-haul team of horses who waited with the teamster Ira MacDonald on a path or clearing across the sleigh road.

When the log was in position Tom Giddy said, "All set. Pass the chain over."

The teamster handed the end of the decking chain to Jack Driscoll, who in turn, tossed it across the sleigh to Giddy. Tom drew the cable over and then under the center of the log (which rested a little off the ground on skids), before passing it back to Jack. Driscoll anchored the chain by driving the swamp-hook into the heavy cross-piece of the sleigh. The pine log now rested in a loop of chain. It was ready to be rolled up onto the new twelve-foot sleigh bunks.

"You can start her up, Ira," the top-loader said to the teamster.

MacDonald, who stood at the side of his team, chirruped to the horses. At his signal, they leaned into their collars, slowly moving off across the road, taking up the slack in the decking chain. When it was taut, the log started moving up the slanting poles. Slowly, steadily, it inched up, guided by the cant hooks of the two men standing at each end of the ascending pine log. Several times the log started to turn sidewise—but one or the other of the send-up men would dig the metal shoe of his cant hook under the troublesome log end to straighten it—then continued to guide the log until it reached the top of the poles.

Another signal was given.

"Whoa," said the driver. The well-trained horses stopped in their tracks. They stood motionless, their feet set firm, leaning forward to hold the chain rigid. Jack Driscoll, with cant hook in hand was ready, poised at the opposite side of the sleigh to meet the log as it moved from the slanting poles and onto the sleigh bunks. He gave another signal to the teamster, who again chirruped to his team. The horses gave a final tug on the chain, and when the log dropped onto the sleigh, the teamster released the trip-hook (a levered device that fastened the end of the decking chain to the horses' evener)—thus releasing the chain without stopping the animals.

The sleigh creaked as the great pine log settled on the bunks. Slowly, it began to roll toward Jack—but he thrust the end of his cant hook hard against its side and stopped it in the center of the sleigh.

"Come back!" Jack shouted. That was the signal to the cross-haul teamster to return for the next trip.

The top-loader pulled the decking chain from around the stick of pine and while the horses got into position again, the send-up men made ready another log to be loaded.

Thus, four more enormous logs—each one weighing from five to six thousand pounds—were hauled, inch by inch, up the skid poles and onto the sleigh. When they were chained into position, Driscoll climbed on the tier and waited there to direct the movement of more logs that would make up the second layer. He would use his judgment where the pine would best balance the load.

And so, each Goliath in its turn was rolled up the slanting poles. The decking chain—its end hook now anchored in one of the logs chained on the sleigh—snaked over the load to the cross-haul team, who upon command continued to pull the logs up one by one, until there were four tiers loaded and chained in place. There was room for one more log—the peaker. It would rest on top of the load.

The peak log was selected by the ground men and positioned at the foot of the poles (which now rested against the fourth tier of logs). Once again, the top-loader threw down the decking chain to the send-up men. They looped it around the log and passed the chain back up to Jack. He drove the swamp-hook into one of the logs in the fourth tier. Once more, a signal was given to the teamster. Ira chirruped to his horses and the peaker began to move.

McTaggart, the scaler and several cant hook men, who were watching the proceedings from the side lines, held their breath. The sleigh was groaning in protest at the increased load. Would it take another four ton weight? Would the chains part, releasing the logs, to squash out and endanger life and limb of man and beast?

Jack stood high on the load waiting for the top log to reach him. If the horses should suddenly lunge or jerk too hard on the chain, the log would go over the top and down the other side of the load, knocking the top-loader off the pile before the teamster could release the trip-hook.

But, this wasn't the day for accidents at the skidway. Belle and Jem took their time, pulling steadily on the decking chain as they walked off down the beaten path with Ira at their rumps, encouraging them with an even flow of words to keep their pace steady. The great pine log moved up the skid poles until at last it reached the toes of Jack Driscoll's corked boots.

The loader stepped back as far as he could without falling off the load, and then the watchers saw the steel on his cant hook flash as he caught the log in its jaws, maneuvering the stick of pine into the space on top of the load. The log dropped into place with a thud.

The top-loader straightened. Then he hopped on top of the peaker—and waving his cant hook over his head, he shouted, "We did it, boys—we did it!"

There was a sound of cheering from the side lines—but it was cut short by the foreman.

McTaggart, who had been standing with his breath

sucked in while the last log was loaded, was not ready to celebrate. He said curtly, "Driscoll, get the chains over that top log—there'll be time enough for hand shaking and jumping up and down after LaRiviere delivers the load to the river."

The foreman was right, of course, and without another word, Jack secured the peaker log with chains and then jumped down to help Matt and Tom twist springy poles into the binding chains at the sides of the load to take up the slack.

While the sleigh was being loaded, the scaler measured the logs as they were chained into position. Using a blue lumber crayon, he marked on the log ends the number of board feet of lumber that each stick of pine would produce at the mill, while McTaggart added up the figures in his head. When the last log was put on the sleigh, the total came to over 17,000 feet.

"A little over three times the amount we've been able to move at one time on the old sleighs," thought the foreman. He was well pleased.

Angus McCloud had come to the skidway that morning, to watch the proceedings. He stood in the background observing the great logs as they were rolled up the slanting poles and onto the new sleigh bunks. There was no doubt in his mind that the mountain of pine would reach the river safely. McCloud recognized the milestone. He knew another step had been taken in the advancement of the pine lumber industry. Without waiting for orders from the foreman, he turned to go back to the blacksmith shop to begin dressing oak timbers for bunks for the other two sleighs. As he turned to go he caught sight of Frenchy LaRiviere (in a bearskin coat) standing on top of the big sleigh load, straddling the top log like a Roman charioteer. Angus had to admit Frenchy was a sight to behold. In one big fist he held the long lines to his team and with the other, he flicked a long team whip just over the heads of the big Belgians—all the while shouting encouragement to the lunging beasts to overcome the inertia of the sleigh and get it moving without the help of block and tackle.

"Hue! Hue! Grandes bêtes! Move! Sacre bleu! Allons, allons!" cried LaRiviere.

The voice became softer, more coaxing, "Ah-h-h, c'est bon—c'est bon! Voila! C'est ca, c'est ca, Peppaire—Jacques—."

The blacksmith did not linger to watch the sleigh disappear down the iced road, but he could hear the screech and rumble of the wide, iron-bound sleigh runners weighed down by tons of pine, scraping over the track as the sleigh began its journey down the logging road to the river. McCloud's step quickened when he thought about Shannon's other camps. They wouldn't be slow in rebuilding their sleighs to handle bigger loads. By burning a little more midnight oil, he would have the other sleighs fitted out with wider bunks by the end of the week.

• • •

It was the foreman's plan to follow the big sleigh (McTaggart believed in personal followup), but before he climbed into the cutter, he stood in the roadway and watched Frenchy LaRiviere start off down the iced road, his legs astraddle the top log on the big sleigh load. Like the blacksmith, the foreman was certain the idea of using the wider bunks on the sleigh was going to work. And why shouldn't it. The sleigh was stout and he had working for him the best road monkeys in Michigan. They had used everything from ax and adz, shovel and grub hoe to make the roads as level and as straight as was possible.[8] They had cut through tall timber, scrubby trees and tangles of brush and vines. They had grubbed out stumps and removed stones and corduroyed low swampy areas when it was impossible to go around them. And on each side of the eight-foot wide road bed, the crew had cut back projecting or overhanging tree limbs and brush so the ponderous loads on the sleighs could pass without snagging them.

The exact year when lumbermen began to ice their sleigh roads is a matter of speculation—but, in Michigan, it was after the Civil War. As road icing became accepted, camps developed their own methods of preparing roads for log hauling. Some lumber camps iced the full width of their roads—preparing them by grading down the road bed, leaving dirt ridges at the sides to prevent the sleighs from turning sideways or slewing off the road. Other camps used horse-drawn groovers that cut tracks in the ice for the sleigh runners.

McTaggart cut deep ruts into the road bed with implements that looked like large carpenter planes—devices that were attached to the front runners of a sled. Only the gouged out tracks were iced—not the whole width of the road. At the first signs of freezing weather, the foreman ordered out the big wooden water tank (built on sleigh runners) and as long as the weather permitted, the teamster George Pepper made several tours each night, driving the sprinkler over the logging roads to build up the ice in the tracks.

The foreman watched Frenchy and the big sleigh load until it disappeared around a curve in the road. Then, after speaking a word or two with the loaders, McTaggart climbed into the cutter beside the scaler who would accompany him to the river.

Frenchy LaRiviere and his team plodded on ahead, successfully rounding the first bend in the road–carefully taking the gentle curve at a slower pace then usual. A group of road monkeys came into sight. They were patrolling the roadway with brooms and shovels—removing twigs, sticks, bark and horse droppings—and building up with snow, places in the sleigh tracks where the ice was wearing thin. When the men saw the big sleigh, they stopped their work to wave to the driver.

The teamster shouted, "Bonjour! You boys goin' to help Frenchee down the heel yondaire?"

One of the road crew shouted back with good humor, "First you got to get to the top of the ridge, Frenchy! We'll be there by the time you get there. George's team is waitin' to help pull you up."

The teamster shook his head and laughed, then chirruped to the horses. They moved on, the great sleigh and its load of pine logs creaking and lumbering behind.

The journey to the river, except for a few slight down grades, was almost level. At times the road was winding and then straight for a distance, going through groves of mixed hardwoods and through stands of Norways. Everything in sight—the ground, as well as thickets of small trees and bushes—was covered with a glittering mantle of snow. But the good natured Arcadian, perched high on the groaning sleigh, was oblivious to the beauty of his surroundings. The wind gently blew through the tree tops, and every so often it sent down small showers of snow from over-laden branches. Experience had taught the teamster that at any moment these lethal puffs of drifting snow or a falling tree limb might startle the horses. The heavy chains binding the logs on the sleigh could part, releasing their hold on the tons of pine logs, causing them to squash out, the sleigh to tip over and perhaps kill both horse team and driver. With experienced ear, the teamster LaRiviere listened to every sound the sleigh made and kept an eye on the binding chains to see that the fid hooks didn't work out. At the same time, he urged his team forward with a constant stream of words, "Gee, Peppaire! Haw, Jacques! Allons, grandes bêtes. Allons—whoap—whoap!"

And, so, the great sleigh made its way through the woods until at last it approached a hill. The incline presented no problems for the lighter sleigh loads. But, anticipating the difficulty that the big sleigh might have there, McTaggart had ordered another team to be ready and waiting. When Frenchy reached the troublesome spot, he saw George Pepper standing patiently at the foot of the hill holding the lines to two big greys.

But the Frenchman had other ideas. He shouted, "Get out of the way, Jorge—these two great fellows can pull thees sleigh up that leetle hill—that ees nothing! You should 'ave seen the mountains we clim een Quebec!"

George stayed where he was, and when Frenchy hauled up alongside him, the barn boss's helper said bluffly, "I got my orders, Frenchy. What McTaggart says—goes. He give me orders to pull you up that hill, and I'm a-goin' to pull you up that dang hill whether you like it or not!"

LaRiviere laughed as he watched George Pepper, his nose pinched with the cold and little icicles hanging from his moustache, doggedly hitch his team to the sleigh ahead of the Belgians. When ready, the two teams leaned their shoulders into their collars and with the urgent commands of the men in their ears, the horses slowly made their way, inch by inch, up the hill—the chains over the logs straining at each step. At last, the sleigh reached the top and when LaRiviere was satisfied that the load was still secure, he directed his horse team to move on, shouting over his shoulder, "Au revoir, Jorge! Merci! See you next treep!"

The sleigh was now on the top of a natural sand ridge that was a half mile to three quarters of a mile wide and followed the course of the river for some distance. For three centuries, pine trees had stood, first as seedlings and then like green sentinals, on the broad elevation overlooking the Tittabawassee. But now the section that was owned by Shannon was almost denuded of the great trees, and only ugly salamander stumps and discarded tree tops poked up through the snow to show where the giant pines had once stood.

The sleigh road did not follow a straight course over the ridge and down to the river. The road crew had run the track along the top and parallel to the stream for a quarter of a mile before they found a place where sleighs could be eased down the hill with some safety. Even at this place, the crew had lowered the elevation by digging out with shovels a ten percent grade[9] for the sleighs to travel over. The smaller sleigh loads had no trouble doing down the hill; but if the big load were allowed to go unchecked down the incline, the sleigh and its burden of pine might plunge down like an avalanche, forcing the horses to race ahead of it as the sleigh gathered momentum, until finally the logs and sleigh would run over the team—killing the animals and hurtling the logs and driver down the slope. It would be a terrible scene and one to be avoided at all cost. This was the reason why the big sleigh loads were to be eased down the grade manually. The procedure was called "warping down."

As Frenchy approached the place where the sleigh would leave the sand ridge, he could see Alex ("Scotty") Scott and Jack Ireland (two members of the road crew) waiting for him. They had heard him coming and were in position, ready to help him down the incline. The men were standing on opposite sides of the road, each one holding the end of a stout rope in his hand. There were evidences on the ground to show that they had sanded the ice in the sleigh tracks leading down to the bottom of the hill. Sprinklings of sand and footprints, leading to and from a big barrel that was resting near one of the pine stumps, indicated as much.

"Hey Frenchy! What took you so long?" hollered one of the men. "We're dang near froze waiting for you."

The Canadian only smiled and waved at Scotty as he commanded his team to halt.

There would be no more small talk. The serious business of easing the enormous sleigh load down the hill would occupy all the energy, skill and good judgment the men and horses could muster. It was a difficult job even without a critical audience—but now, McTaggart and the new scaler, having caught up to the big sleigh, were standing off to one side ready to observe the business at first hand. The two observers watched intently as the road men looped and secured the ends of their long coils of rope through the iron bracing on the back runners of the sleigh. Pulling their ropes after them, the men walked to opposite sides of the road where each one took several bites (loops) with their lines around a stump. When this was accomplished, the sleigh was held secure between two lines of hemp, braked, and ready for its descent down the perilous grade.

While Ireland and Scotty were fastening their ropes, LaRiviere took the opportunity to walk around the sleigh and check all the hooks and binding chains once more. He pulled on this or that link and inspected each hook to make certain that none had worked out.

When the teamster was satisfied that all was secure, he climbed back on the load. He gathered up the lines, wrapping them around each calloused fist. Then he indicated with a nod of his head that he was ready. To his team he said, "Allons! Giddap!"

The horses moved, and the on-lookers heard the Frenchman say to the road monkeys, "Let her go!" To the horses he called out, "Easy—easy—e-a-s-y beeg fellows—," as they started down the incline.

Scotty and Ireland braced their feet against the tree stumps and slowly let out the ropes—all the time keeping a tension on the lines while the sleigh started easing down the hill. Both ropes were taut with the pulling force of the load, and they began smoking as the hemp dug into the tree stumps. The great sleigh now leaned forward at a dangerous angle, and McTaggart and the scaler watched intently as it moved—the chains over the logs straining at every link. They watched Frenchy—knees bent, body leaning backwards with powerful arms holding the lines to the big Belgians. Clouds of vapor billowed from the horses' dilated nostrils. They squatted almost on their hind quarters, digging into the road with calked shoes that made small showers of ice and snow fly out from under them as they worked to hold back the tremendous load. Haltingly—amid scrambling hoofs, rattling harness, the shouting of the men and the sound of iron-shod runners scraping over sand-covered tracks—the mountain of logs moved down the grade. Slowly—slowly the tons of pine moved toward the bottom of the hill, until at last the big sleigh was brought to a stand-still on level ground.

There was no attempt on McTaggart's part to smother the sigh of relief that exploded from his chest when the load came to a halt. The scaler said nothing, but his pale face showed his concern.

Without a word, Frenchy got down from the load and

untied the restraining ropes from the back of the sleigh—and then went directly to calm his team.

In silence, the foreman and the scaler climbed back into the cutter and drove down the hill. As they came abreast of the big sleigh, McTaggart pulled on the lines to slow the horse. He leaned out of the two-seater, and with voice warm with praise, he said, "A good job, Frenchy, a damn good job!" Then he chirruped to the little mare. And before his riding companion had time to do more than nod his head in recognition to the teamster, the cutter passed the sleigh and drove off down the road.

McTaggart and the scaler Hoops were there to watch when the logs were pulled off the sleigh. The drama of the morning ended with the releasing of the chains over the big load. From now on a twelve-foot bunk of logs would be treated as nothing out of the ordinary at Camp 3.

The foreman, satisfied that the bigger sleigh loads would be a success, took off down the go-back road on other camp business, leaving the scaler behind. Frenchy LaRiviere followed after McTaggart at a slower pace, letting his team take it easy after their unaccustomed exertions.

13

M is for MARKING HAMMER for branding the pine

Isaac Lee Hoops was the name of the man who had been called up from Saginaw City to replace Uriah Bell. Before the foreman drove off, he handed the new man two tools peculiar to a scaler's trade. One device looked like a sledge hammer—the other, like a common yard stick.

After leaning the hammer against a stump, Hoops walked over to one of the big log piles and stood there for a moment seeming to size up the logs that towered over him. The scaler was not a small man, but he looked frail standing near the great pyramid of logs. He was dressed much like the other members of the crew, except his clothing didn't look faded, patched or shabby. Unlike the other woodsmen, he wore a neat navy blue pea jacket over the usual layers of wool shirts. His boots were calked. They were ankle high and very stout—and over their tops bulged brown and black striped German socks which were pulled up over the legs of his pants and fastened below his knees with leather straps.

The ear flaps of the scaler's felt cap were pulled down against the bitter cold. Only a sandy-colored mustache under a sunburned nose gave a clue to the fairness of the

Hoops walked over to one of the big piles and stood there for a moment seeming to size up the logs towering over him.

man's skin. In the rude setting of the banking grounds, Isaac Hoops looked like quality.

The stick the scaler held in his hand deserves some comment—in that so much depended upon the accuracy of its use and the honesty of the man using it. It was a measuring device called a log rule.[1] Rather than figure from tables or by algebraic equations the number of board feet of lum-

ber in each log, the scaler simplified his labors by using the measuring stick.

He measured the length of a log by eye (all pine logs at Red Keg were cut to standard lengths) and its diameter (inside the bark) by hooking the L-shaped brass tip of the scale over the smallest end. He looked at the table of figures printed on the wooden stick (which coincided with the given length and given diameter) and presto! There, at a glance, the scaler read the volume direct from the scale.[2]

Simple?

"Hardly," Isaac Hoops would say, and then go on to explain that his job might have been a simple one if the measurement of logs was an exact science—if a single rule had been adopted by the lumber industry—if all logs were completely sound—if logs didn't taper—if all sawmills used the same gauge saws—if all mill hands had the same efficiency—and so on, *ad infinitum*.

The fact was, there was no standard device for measuring logs and there was no logger's scale that allowed for flaws or taper in pine logs.[3] These and other factors were what scalers had to consider whenever they faced a mountain of pine, as Hoops was about to examine on this cold day in January, 1868. (You can see why scaling logs was a very complicated business—one that made lumbermen like John Shannon

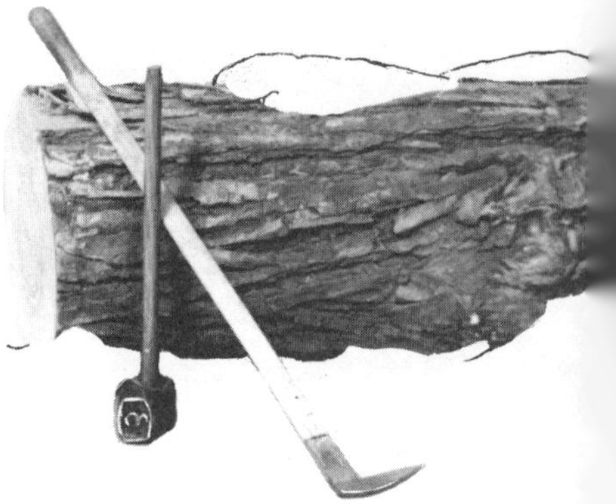

The marking hammer, and logger's scale

very dependent upon the experience, good judgment and honesty of their scalers—and incidently, made them highly suspicious of them as well).

Upon Shannon's orders, only "uppers" (logs that would yield only the clearest and highest grades of lumber) were to be scaled. All others found in the piles were to be rejected as culls—logs not fit to be measured. These instructions should have made the scaler's job easier. But the lumberman's orders only served to complicate his work. Anxious to make a good showing on the foreman's tally sheets, some of the woodsmen were including logs in the piles that were less than perfect, and they were using every trick they knew to disguise them[4]—hoping that Hoops would overlook the culls and include them in his count. It had been easy to fool Uriah Bell, who was not overly ambitious and didn't look for imperfections. But it wasn't so simple to put something over on Isaac Hoops, whose past experience included watching hundreds of logs go through saws at the mill—where defects soon became apparent. For this reason a hostility had grown between the crew and the new scaler.

With a sigh, Hoops noticed, as he walked around the deck of pine, that there were some logs placed far into one pile (an old trick). He shook his head as he scrambled to the top of the deck, wondering why the men insisted on hauling out of the woods logs that they knew would be rejected in the end count.

After reaching the top of the heap, the scaler forgot about the crew and went to work. Now standing—now kneeling—now stretching far over the side of the log pile, he swiftly examined and measured each log in its turn, making deductions from full scale to allow for taper or slight defects. With a pencil, he wrote down the results of his calculations in neat columns on smooth beech-wood tablets he carried in his pocket. After recording each measurement, Hoops marked a big blue figure "8" with a lumberman's crayon on the end of the logs that passed his critical inspection. By noontime, when the welcome sound of the cookee's horn sounded through the woods, many of the logs in the great pile were marked with the crayon.

Isaac tramped to the dinner hole for the nooning with the other woodsmen and, as he expected, the boys glared at him over their beans and tea, guessing that the new man had rejected the cull logs they had included in the pile. But he paid no attention. Gulping down the last dregs of his tea, he handed his empty plate and tea dish to Cap and returned to the rollway, alone.

By mid-afternoon the scaler paused in his work to survey the deck of pine to make certain all the logs had been accounted for. Satisfied that all were measured, he laid aside the log rule and picked up the company marking hammer (the sledge that he had propped against the stump earlier in the day)—and until dusk, the cold air of late afternoon was filled with whacking sounds.

Hoops brought the raised pattern of the marking iron down five or more times on the smooth, soft pine surface of every log end that he had marked with the blue crayon. When he finished, there was no doubt about who owned the deck of logs. The Keg-3 mark was John Shannon's mark of ownership and like all log marks in Michigan, it was registered, by law, in counties through which the logs floated on their way to the mill. Shannon's log mark was protected just like a copyright or trademark. Like the cattle brand in the west, it wouldn't come off in the wash. If the marks were altered or cut off by rustlers, there would be the very deuce to pay—as log thieves Chriscaden, Norman and Dodge found out when they were discovered operating on Shannon's property.[5]

Stealing logs was one of the oldest occupations in American and Canadian forests. Like some western cattlemen who got their start in business by rustling cows, there were men who got started in the lumber game by cutting off or altering another log owner's marks and substituting their own on the stolen property. There was hardly a settler who owned land through which a logging stream passed, who didn't have a pike pole handy to steer logs from the river onto his property. Every season, while the drives were on, he usually had logs to sell to the nearest sawmill for much needed cash or to provide the lumber to build an addition onto his house.

The log mark, that little symbol of ownership, kept the thievery down to about wholesale and a kind of order in a rough and tumble industry that operated on a no-holds-barred basis.

Being something of a scholar, the scaler's interest in marking logs had led him to make inquiry into the subject. In his research he learned that the stamping hammer had been in use for centuries by European foresters (who called the device a *marteau*) long before the tool was thought of in Michigan—or on the North American continent.[6]

Early New England and Canadian woodsmen followed the old English tradition of using the ax to cut their log marks into the sides of logs[7]—and when eastern lumbermen migrated to the Lakes states, in the early half of the nineteenth century, they continued to follow the old tradition until they were forced to find another way to mark their logs.

With an ever increasing number of lumbermen dumping pine into Michigan rivers and streams, it became more and more difficult to design a bark mark that could be readily distinguished from the marks of other individuals who were operating on the same stream.[8] Duplications of log marks led to angry disputes, law suits and fighting among log owners until 1842 when a Michigan State law was passed that required owners to register their marks in the county where their logs were manufactured into lumber.[9]

The law only partially solved the lumbermen's dilemma. Log marks had become so complicated and so difficult to identify, jamming at the sorting grounds became almost a daily occurance. This difficulty, combined with the old method of hacking marks on the sides of logs (which required sorters to roll over almost every log to see the marks on them), further slowed down the sorting process—until exasperated lumbermen operating on the Muskegon River did something to relieve the situation in their area.

In 1859, through local legislation, they brought about pressure to pass a law that required log owners, who floated logs on the Muskegon or any of its tributaries, to *stamp* the *ends* of their logs and to register their log marks in Newaygo and Muskegon counties.[10] The stamping iron or marking hammer, with its raised pattern cast on the striking surface, was adopted by the Muskegon lumbermen for marking their logs.[11] Thereafter, no matter how a log floated in a stream, a log mark would be visible to the men whose task it was to sort them according to their brands.

Log owners in other lumbering areas were not slow to adopt the stamping iron, but the lumber industry in Michigan would become a little giant before a State law was passed which spelled out log marking regulations in detail. This finally came about in 1867 when an Act "to protect the title of the owners of floating logs and lumber" was passed by the Michigan Legislature.[12] The law included instructions for registering log marks "in the office of the clerk of each county through which such logs or timber shall be floated

The famous
"Circle M"
CHARLES
MERRILL & CO.

RUST
BROTHERS

WM. PETERS

WRIGHT
&
KETCHAM

E. H. HALL

JOHN
McDONALD

EDDY BROS.

MARTHA
HAY

A. T. BLISS

MEECHER
&
NERRITER

*Like the cattle brand in the West, the log mark wouldn't come off in the wash! The marks shown here are typical of the thousands that were recorded during the log-driving days—copied from a Tittabawassee Boom Company account book (1890–1893)**

**Eddy Historical Collection, Hoyt Library, Saginaw.*

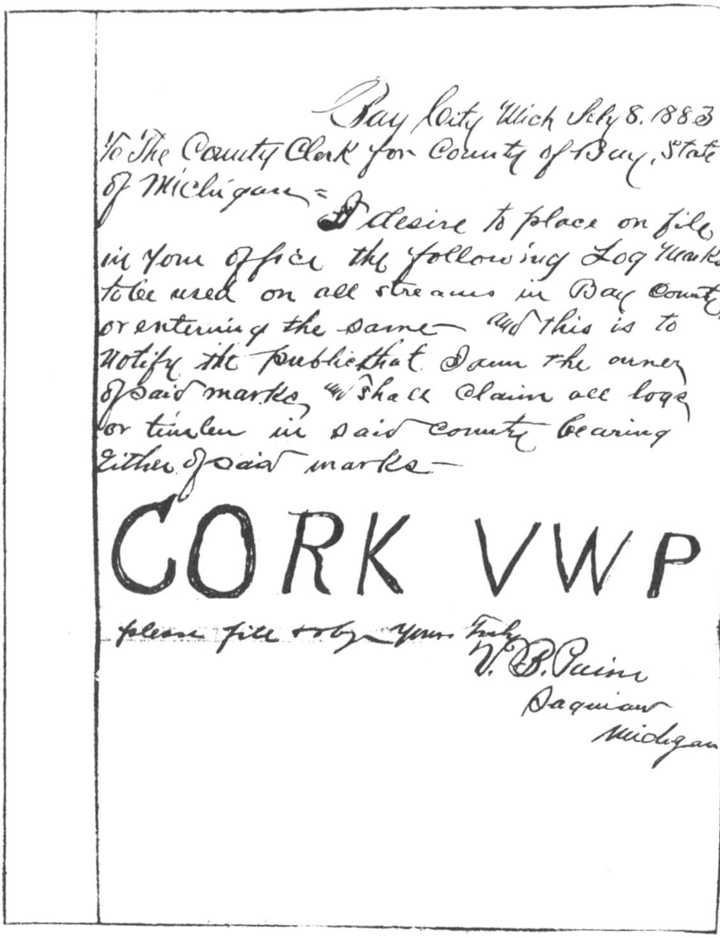

Bay City Mich July 8. 1883
To The County Clerk for County of Bay, State of Michigan =
I desire to place on file in Your office the following Log Marks to be used on all streams in Bay County or entering the same — and this is to notify the public that I am the owner of said marks and shall claim all logs or timber in said County bearing either of said marks —

CORK VWP

please file & oblige Yours Truly
V. B. Paine
Saginaw
Michigan

Log owners, in response to a law passed by the Michigan Legislature in 1867, protected the title of their floating logs and lumber by registering their marks at the Clerk's office in the County through which the logs floated. The above record was "placed on file" by V. B. Paine of Saginaw in Bay City in 1883.

for manufacture or sale." In 1869, the Act was amended to include penalties for forging marks.[13]

The 1867 Act required Michigan log owners to mark their logs "in a conspicuous place upon the ends or surface of each log." Many lumbermen had been doing this for years—but it was good to have a law on the books to back you up in court when an altercation arose over the ownership of logs.

The marking hammer that Isaac Hoops was swinging that afternoon weighed about six pounds and, as the day wore on, it began to feel very heavy.[14] But the scaler worked until the light failed—until he could no longer distinguish the marks on the logs before him.

The loaders, who had been laboring near by, were getting ready to call it a day, too, but the scaler did not wait for them. He picked up his tools and tramped back to camp as dusk turned rapidly to dark.

The sky was black when Hoops stamped the snow from his boots and let himself into the warmth of the van. He was surprised to see McTaggart there ahead of him sitting at his desk studying the camp books in a yellow circle of lamp light. The foreman looked up when the scaler came into the room and before Isaac had time to peel off his jacket, he said, "Isaac, I'd like to go over the tally books with you after supper; there seems to be a big difference between what you're scaling and what the boys are reporting on their cutting and hauling. Either they're trying to cut every dang tree in the woods and calling them uppers—or you're way off in your scaling."

The scaler smiled wryly, "I think Shannon's prize bonus offer is making the men a little careless in their cutting and hauling. I rejected six logs as culls today. If the logs had been top grade, they would have scaled over two thousand feet. That'll give you an idea what they're up to. If you like, I'll show you the logs that I didn't bother to measure last week—then you can judge for yourself where the discrepancy—."

McTaggart interrupted the scaler. "We'll talk some more about the situation after supper. I've been suspecting a skunk in the hen house ever since Bell left. I have a feeling a lot of the logs he marked will turn up as rejects at the mill. If that happens, we'll hear John Shannon holler all the way from Albany!" The foreman hesitated, and then he said, "Isaac, it'll mean more work for you—but I think we better take another look at all the logs Bell scaled."

Hoops didn't answer, but continued to go about the business of removing his boots and outer clothing.

At supper, some of the men made it a point to elaborately avoid meeting the scaler's eyes when he looked at them across the table; and through the meal all the serving dishes seemed to be just out of his reach. In spite of the no talking rule, Hoops finally said to the man sitting at his left, "Push the beans and bread over will you, Bill?"

The man answered without looking up. "If ye want beans and bread, git 'em yerself."

Isaac thought to himself, "So that's the way the wind blows. If the boys are sore now, I wonder how they'll feel tomorrow when they see McTaggart and me looking over all the decks of logs that Bell scaled."

A quiet Sunday afternoon
in camp ends in tragedy.
The foreman steps up the pace.
Tempers flare until the
situation at Red Keg is eased
by thoughts of spring —
and an episode at the bee tree.

14

N is for the darning NEEDLE with which the boys spend time (Sunday in Camp)

All week long the scaler Hoops divided his time between his regular duties and re-measuring the logs that Uriah Bell scaled. He found an average of two to three logs in each of the old piles that would have to be rejected. When he found a cull, Hoops drew with his blue crayon a large "X" across both ends of the log.

The crew was kept posted (unofficially) on the scaler's activity. The teamsters, the only men who could observe Hoops without letting on what they were doing, drove their sleighs past the log piles and tried to count how many of the logs were X-marked. But their attention couldn't be taken away from their teams long enough to make an actual count without coming to a full stop. So, the crew, not knowing how many logs were being discarded, and believing the worst, manifested their resentment by giving the scaler the silent treatment at the dinner hole and ignored him at breakfast and suppertime. They knew Hoops was reporting the losses to McTaggart, who in turn was writing down the figures in the camp books.

When the rejects began to reach 70,000 feet, the foreman became alarmed. The labor lost on the cull logs was not something John Shannon would take lightly.

By Saturday, all of the log piles that had been scaled by Bell were re-examined, and that night McTaggart made a brief appearance in the bunk-camp to tell a sullen crew what their trickery might cost the camp.

Finishing his speech, the foreman said, "The scaler isn't responsible for the losses. Put the blame where it belongs!"

McTaggart saw some frowns and head-shaking here and there in the dusky room. Heating up, he continued, "I say, put the blame on your own shoulders. If it wasn't for the road crews and their good work this season—and for the cold weather and bigger sleigh loads, we'd be out of the running for Shannon's prize altogether. Considering the big loads, we might still have a chance—but it'll be a mighty slim one." To emphasize his point, McTaggart banged a fist into his palm. "If I get one more report of culls at the rollways, there's going to be the very hell to pay!"

The men listened in silence to the foreman's talk—and it was sometime after he closed the door behind him before the usual Saturday night fun began. As soon as McTaggart pulled the door shut, noisy accusations, denials, name calling and finger pointing went the rounds of the room. But finally, the sound of angry voices died away. Most of the crew were tired of the whole business and wanted to get on with the evening's festivities.

But there were two among the crew who stubbornly maintained their innocence in the matter of the cull logs, blaming the scaler to the last for lessening their chances of winning the lumberman's prize. And when Jim struck up a few opening chords on his fiddle, the two sawyers, Sam Bean and Daniel Blackburn, went off in a corner to nurse their grievances. They took no part in the fun that evening.

On that note, Saturday night melted into Sunday—and for once, McTaggart was as glad as his men to see the day arrive.

There is an old lumbermen's prayer that sums up in a few words a woodsman's feelings about the days he spent laboring in the pine—and about Sunday, his one day of rest.

Blue Monday,
Bitter Tuesday,
Long Wednesday,
Everlasting Thursday,
Friday, will you never go?
Sweet, silver Saturday
in the afternoon,
Sunday, may you last forever.
Amen.
Two nights in the straw and
three meals ahead![1]

For most of the men, the brief respite from early morning rising and hard labor began on Saturday night after supper. And after the usual tomfoolery and hijinks in the bunk-camp, they fell into their bunks and dreamed their dreams of home and family, or of the girls on the White Row in East Saginaw, or Mattie's girls at the Domino Saloon down in Bay City—confident that Cap's horn wouldn't break into their dreams until an hour after the usual waking time.

Sunday came along just in time to break the six-day monotony of exhausting manual labor. For some, it was a day in the hay. For others it meant exchanging cant hook for darning needle—the ax for soap and wash tub. As one lumberman put it, Sunday in camp was dedicated to "cleanliness rather than to godliness."[2] Only occasionally did a "flying parson" appear on snowshoes to sermonize on matters of salvation and temperance.

While the foreman did not strictly forbid religious discussion in his camp (as did some lumbermen), he was secretly happy that visits from itinerant preachers were few and far between, feeling that religious discussions (as well as those of a political nature), only led to trouble among the men.[3]

There were only three "hard and fast" rules in camp that McTaggart enforced, number one being the off-limits rule regarding the saloon up river. Number two was his rule about drinking whiskey or any other alcoholic beverage while in camp. The third law laid down by the foreman was the one that was most often resisted by the men—it being, all crew members were required to wash their underwear at least once every fortnight.

The reason for the order should have been obvious to the forty-odd human beings who were nightly packed into one room—humans who had a mutual aversion to soap and water—who either smoked malodorous pipes until the room was a blue fog, or chewed tobacco (the juices flew from the deacon's bench towards the camboose with a regularity that was astounding). The "fragrance" of the room (well seasoned by the array of wet boots, unwashed socks and foot rags drying on lines around the fireplace) was enough to stagger any faintheart who might have wandered into the ill-ventilated bunk-camp from the outside and inhaled a lungful before stumbling out into the fresh air again. Then, too, there was the matter of the louse—that pestiferous pest called the "grayback," who flourished in the seams of the woodsmen's clothing and in their blankets.

There was a time, before McTaggart's rule about washing was put into effect, when the boys, before getting into their bunks at night, simply turned their underwear inside out and tried to fall asleep before the graybacks had time to crawl back inside again. And there were those in the crew who shamelessly wore their flannels a month, then turned them inside out to wear them "clean" to the end of the next month when they turned them back to begin the routine all over again. But the old tricks were a thing of the past since the foreman made boil-up day a regular Sunday occupation at Red Keg.

Sunday was fix-up day as well as clean-up day—when socks were darned and clothes were patched or mended—when whiskers were trimmed and hair was cut. There was always plenty of time left in the day for an afternoon's snooze, for letter writing, or the reading of much-thumbed Testaments, aged almanacs or dime thrillers that were pulled out of turkeys when chores were out of the way.

The routine on Sunday never changed, and this morning, like all the others, Cap grudgingly filled kettles, dipping water out of the water barrel with a tin wash basin, and set them to heat over hot embers in the fireplace. Scrub boards, wooden tubs and buckets were brought into the room from the dingle, and yellow soap, that was more lye than tallow, was produced from the cook-camp; and as soon as breakfast was out of the way, the bunk-camp turned into a lively scene.

The place measured about thirty by fifty feet and was lighted only by the fire burning in the camboose. Against the log walls, bunks (two tiers high) stretched along three sides of the room. A number of the bunks were occupied by men who had performed the rite of washing the week before. Some were sleeping—others were plumping up or replenishing the straw under their blankets. The voices of men who were occupied with domestic duties blended with the snores of those who were unconscious in their bunks.

Heaps of dirty clothing rested here and there on the floor and on the deacon's bench. Men stuffed clothes, stiff with sweat and dirt, into wooden tubs of hot water and rubbed flannel and home-spun over the corregated tin of washboards with calloused hands that were more used to handling cant hooks than soap. The soft yellow stuff was laboriously worked into cloth while the water in the tubs gradually turned from gray to black. When satisfied that enough grime was rubbed out, the garments were wrung into long damp ropes and given a shake (an action that set off a string of protests from the loungers) and then draped over sagging lines that were strung up around the fire.

The cookee was kept humping. While one man called to him to "Git a move on and put more water on the fire," another was yelling, "Where in thunder did you hide the soap." Their orders made Cap Wilkes hustle from the water barrel to the fire, to the dingle or out to the cook-camp and back again. Every time the cookee went out of the room and came back in again—or when somebody opened the door to toss out a bucketful of dirty water, a shivering woodsman standing in his bare feet and drawers invariably cried, "Shutthedamndoorthinkitsummerinhere?"

Outside, alone in the cheerless cold of the winter morn, the swamper, Orville Tigg, stood in the trampled yard near a big iron kettle that was steaming over a smoky fire. Every once in a while he plunged a long-handled contraption (made out of a large tin can and a long stick), up and down in the kettle and then peered into the rising steam as if to see the results of his work. Orvey was boiling "crumbs" out of his blankets and underwear, and with each plunge of the squeegee, he uttered an oath between clinched teeth, swearing vengence on the "boomer" who had infested his clothing.

The bunk-camp had been almost free of graybacks that winter, but earlier in the week an itinerant woodsman had put in an appearance looking for work. McTaggart, knowing

A shivering woodsman, standing in his bare feet and drawers, cried out, "Shutthedamdoorthinkitsummerinhere?"

the man would stay only long enough to get a whiskey stake and move on, gave him the usual camp hospitality of board and bed with instructions to leave after breakfast. The next morning the man left camp—leaving behind a colony of lice, who immediately attached themselves to Orville, the unfortunate member of the crew with whom the boomer shared sleeping quarters.

Orvey knew only one way to get rid of his uninvited guests, and he went about the business with all the expertise of one who had lived the better part of his life in lumber camps. When the water in the kettle started to boil, he dropped into it a fist-full of short-cut chewing tobacco (after an old recipe for sheep dip) and added a liberal quantity of snuff (his own idea), and chunks of yellow soap which he cut up with his jacknife. He stirred the mixture well and when the soap melted, he dumped his infested blankets and clothing into the kettle—all the time working the squeegee and keeping an eye on the boiling surface of the water to see if any "life" ventured to the top. Satisfied at last that the last grayback was either drowned or boiled, the swamper wrung out his blankets and underclothing, shook them vigorously, and draped each smoking piece over the wash line where they froze almost instantly. (Orvey had not bothered to rinse the mixture from his belongings, hoping the highly scented odor would discourage further visitations of the blood-sucking vermin).

It was about mid-morning when McTaggart made a brief appearance in the bunk-camp. After he closed the door, he stood there for a moment glancing around the room. He seemed to be looking for someone. He said, "Anybody seen Toohey this morning?"

When one of the men jerked a wet thumb toward one of the bunks, the foreman followed the direction, stalking across the room, neatly side-stepping wash tubs and buckets until he found the man he was looking for. McTaggart reached into the shadowy recesses of a bunk and locating the bare foot of a sleeping woodsman; he gave it a tug.

"Pat!" he said.

There was no answer. The snoring went on—even and unbroken.

"Pat Toohey!" The foreman's voice was louder this time.

The snoring stopped and the sleeping man struggled out of his dream. He sat up, blinking in the firelight, surprised to see McTaggart looking at him over the rough-board edge of the bunk.

"Toohey, how many shirts you got on by now?"

The chickadee screwed up his face, raising his eyes heavenward, as if in deep thought. He was about to speak but the foreman stepped in to say, "I'll bet you haven't seen your drawers by the light of day since last September." He sniffed the air and grimaced. "You smell like its been two years since you had a wash."

Toohey started to protest, but stopped when he saw the foreman point stiffly to the wash tubs and heard him say with a finality that the chickadee couldn't ignore, "You know the rules around here. Now, get moving!"

McTaggart closed the door after him amid the boisterous laughter of the men—but not before he saw Pat Toohey begin to peel off his shirts and under-drawers.

The road monkey had lost count of how many shirts he had on. When one got too dirty, he put on a clean one over it—and if McTaggart hadn't caught on to his game, he would have added several more layers before Spring, when he would have reversed the procedure—peeling off shirts as the days became warmer, until he was down to his flannels again.

"You might as well take a bath while yer at it, Toohey," said Scotty, who was cutting his partner's hair.

The Irishman shivered, "Not on yer life—I'll be warshed often enough in the Tittabawass' whin it gits to be spring—!"

The woodsman-turned-barber laughed and watched Toohey out of the corner of his eye while he snipped away at the black curley hair of his "customer."

Overhead, the growing assortment of laundry dripped steadily into the fire and onto the men who were sitting on the bench placed along one side of the camboose.

"Wonder what it'd be like in here on a Sunday without a wet wash hittin' you in the face 'er drippin' down the back of yer neck," commented Cully Wills.

Bill Bragg was sitting near by. He didn't look up from his work. Bill was cutting patches for his pants. He needed one for "fore" and one for "aft"—and he squinted near-sightedly as he tried unsuccessfully to cut, with a jackknife, squares of cloth from a sleezy grain sack.

"Damnation!" he exploded. "Wish you'd let me borry them scissors a minute, Scotty."

The Sunday barber paused only for an instant to say, "I got three heads to cut. If you want to wait, you can use them." He looked over at the loader. "How come you always has to borrow everything, Bill? Bet you ain't got the thread to sew the patches on yer pants."

"I have, too—pulled some threads outen a old sack—and Ben is goin' to loan me his needle, ain't you Ben?"

The man addressed was Ben Sweeney, who was leaning toward the fire to better see the big hole he was laboriously mending in the palm of a hand-knitted mitten. As he wove his needle in and out of the yarn, he answered Bragg, "You can use my needle as soon as I get done mendin' my mitts and a couple pair of socks." Looking up from his work, the usually good-natured Sweeney added a little testily, "If you wasn't so dang cheap you'd buy a pack of needles and some thread. You been moochin' snoose fer months. I swear Bill, I think you'd do anything to keep yer van bill down, just so's you can have a big stake to blow in town come spring."

Another voice was heard across the room. "Bill knows what he's doin'. He thinks all the gals in Saginaw is waitin' fer him—an' he's saving his money fer a yeller satin vest and dude suit he seen at Zivi's last August—hopin' they'll impress the ladies."

The teasing voice came from a woodsman who was scraping the last remnants of soapy lather from a clean-shaven jaw, carefully missing with his straight razor the six-inch mustache that curled up on each side of his nose.

Bill rose up in a wrath—but before he had time to do more than stick out his chin, Cap came in and made a timely announcement.

"There ain't no water left in the bar'l. If you got any more washin' to do, it'll have to wait till next week. I ain't got time to fill it again." With those words he banged out the door to go back to the cook-camp to help Kelly with the last minute preparations for Sunday dinner.

Sunday dinner was always special with some kind of treat on the table. Today, it was sourdough donuts—a repast that disappeared from the tables like snow in May—the boys pocketing a few sinkers for an afternoon snack before they ambled back to the damp bunk-camp to sleep, or to while away the afternoon at some other kind of relaxation.

Two or three of the men got Kelly's grudging permission to use the tables in the cook-camp to write letters—others, like Jimmy-the-Chopper, went off into the woods in search of wood for new ax helves. John Tree went along with Jim to look for spruce gum to take home in the spring as a treat for his children. Each man sought his own diversion, which for some did not differ much from week-day occupations.

McTaggart and the scaler walked back to the van, and without further ado, the foreman sat down at his desk, adjusted his lamp and was immediately engrossed with columns of figures in his account books interrupted, at first, by the brisk business of men coming in to purchase van goods. Isaac Hoops completed tallying his books and then took from a rude shelf over his bunk a thick volume of historical content. He was soon absorbed in battles of another day—the battle of the sawlog forgotten for the moment.

In the blacksmith shop, Angus McCloud was, as usual, busy at his forge—but today he had company. Several of the crew sat around in the shop in various relaxed attitudes "chewing the fat"—some helping the blacksmith to hold hot iron or taking turns pulling on the bellows rope.

The activity in the barn was as leisurely as it was in the blacksmith shop. The barn-boss, his helper and the teamsters worked at a slower tempo than usual. Some tossed down hay from the upper half-story of the barn; others cleaned manure out of the stables while George Pepper spread fresh straw on the floor. The barn-boss, Billy Flynn, while keeping an eye on his helpers, busied himself looking after the welfare of the animals, nursing the sores, cuts and bruises they had acquired during the week.

The shadowy interior of the low log building was agreeably warm. A not unpleasant blend of odors—manure, harness-leather, neet's foot oil, grains, horse linaments and fresh hay permeated the barn interior and spread to the outdoors. The sound of switching tails, the occasional clomp of hooves on the plank floor, and the gentle nickering sounds of horses blended with the voices of the teamsters, who talked "shop" while they mended harness or brushed and curried their animals.

Ox driver, Able Fry, bent to look at a cloven hoof of a member of his team. The foot felt hot to his touch. Straightening, he looked around the stable until his eyes found the barn-boss, who was working by the light of a lantern pulverizing resin in a mortar with a stone pestle. The lid of the stout pine medicine chest was his table.

"Billy, can you come over 'n take a look at ol' Bright? All week he's been favoring his near forefoot. Maybe that poultice you made up for him a while back would take the fever down."

The barn-boss looked up from his work. "I'll take a look, Able, just as soon as I git this pesky resin powdered."

Every Sunday, the horses were fed hot bran mash, into which the barn-boss mixed Epsom salts, pulverized resin, a little sweet spirits of nitre and anything else he could think of that would keep the animals in good condition. The tonic was only one of Billy Flynn's "receipts" (home remedies) for doctoring cuts, sores, bruises, sprains and most ailments to which horses and oxen in the lumber camp were prone. The big medicine box (called "Billy's pharmacy" by the crew), from which the remedies were dispensed, was kept under lock and key near the barn-boss's bunk in a cubical where harness, horse collars and sweat pads hung in a row on wooden pegs. The medicine box was well stocked with an odd assortment of jars, bottles, and boxes that contained "yarbs," salts, spices, spirits, compounds and substances such as witch hazel, peroxide, linseed oil, ammonia and glycerin (for treating sores, cuts and sprains); turpentine, mustard and pine tar (for colds); lard and hemlock water (for chafed necks and galled shoulders). There were spices in the medicine box like ginger and ginseng, as well as caraway seed, sulphate of iron, Epsom salts and resins (for formulating tonics).[4] The "patents"—Dr. Bennett's Golden Liniment and Hostetters Celebrated Stomach Bitters (good for man or beast)—were in the chest as well. Resting on the long shelf built above the "pharmacy" was a dusty collection of empty patent medicine bottles as well as mixing spoons, kettles, a barn lantern and an old coffee grinder (useful for grinding up spices). A grimy, much-thumbed Hostetter's Almanac (compliments of the bitters maker) hung by its familiar string from a hook at the end of the shelf and provided the barn-boss with the closest thing to a veterinary journal that he would ever see.

Billy Flynn was a combination of veterinarian and nursemaid to the camp animals. He loved every hair on their thick hides and wouldn't allow a teamster in camp who would abuse a horse or an ox—and he saw to it that the men took proper care of their teams. John Shannon purchased heavy draft horses for his camps from the Chicago horse market (fourteen head for Red Keg)—each horse weighing from 1600 to 1800 pounds each.[5] They were tough and strong but very susceptible to cuts and bruises. Most of the horses suffered at one time or another from harness chafing and galled shoulders (caused by heavy collars), and sometimes from bone spavin, a disease of the hind leg brought on by undue strain.[6] Billy slept in the barn at night and his ear was always tuned up, listening for a rattling cough or heavy breathing. If an epizootic epidemic started in the barn, it meant closing down the camp.[7] And his eyes were always on the lookout for a horse resting uneasily on one of its hind toes while standing—a sign of the spavins.

The barn-boss carried another key on a chain besides the

Billy Flynn, the barn boss, was a combination veterinarian and nursemaid to the camp animals. He loved every hair on their thick hides and wouldn't allow a teamster in camp who would abuse a horse or an ox.

one to the medicine chest. It was the key to the granary. Anyone who knows anything about horses knows that if a horse gets loose, he'll head for the oat bin and founder himself on grain (like a woodsman would "founder" himself on 25-proof horse liniment if he found the medicine chest unlocked).

To show how far Billy Flynn would go to care for the animals, there was a small box stove near his bunk. It was not there for the comfort of Billy, but for the purpose of heating water for the camp horses! The horses at Red Keg wouldn't drink water unless it had the chill taken off it.[8] Is it any wonder the woodsmen complained that the beasts in camp were better treated than they were? Like it or not, it was one of the hard facts of life in a lumber camp that men could be hired for twenty dollars (or less) a month while a good team of horses (without harness) cost five hundred to six hundred dollars.[9]

And so -- Sunday afternoon at Camp 3 wore away in a leisurely fashion. Even Kelly, after pushing kettles of stew (ready for supper) to the back of the stove, shooed the letter writers out of the cook-camp at four o'clock, banged the bar across the door, and without bothering to remove his boots or grimy apron, he flopped into his bunk, shoved his bowler over his face and was asleep almost instantly.

Outside, it was as still as night. And except for wood smoke that rose in straight columns into the frigid air from smoke stacks and camboose chimney, there was no movement abroad. Not a living thing was visible. The only sound was the faint ring of Angus's hammer on the anvil.

Inside the bunk-camp, the snores of unconscious woodsmen, lost to oblivion, the subdued voices of men swapping stories on the deacon's bench, the now and then "pop" of a pitchy pine knot bursting to send showers of sparks up the chimney—were the only sounds that broke the silence of that drowsy afternoon.

Suddenly, without warning, the occupants of the bunk-camp were startled out of their lethargy by the sounds of pounding on the door and the muffled, but urgent cries of a man in distress.

The contrast between the quiet of the moment before and the alarming noise they heard outside electrified everyone in the room. Omer Fuller was the first one to reach the door. He wrenched it open and was almost knocked down by the sawyer Sam Bean, who fell headlong into the bunk-camp. He would have fallen to the floor if Omer hadn't caught hold of him.

Sam's usually weathered face was colorless and twisted with pain. He was hatless—and the men closest to him saw that the sawyer's left arm hung lose in his reefer sleeve—his hand looking like it was smashed to a bloody pulp. An ugly red welt ran across his forehead and down his cheek. Sam's chest was heaving, and he was having difficulty getting his breath.

Fuller helped the sawyer to the deacon's bench—only turning his head to say sharply, "Somebody—anybody—go get McTaggart—quick!" He heard the door open and slam shut as one of the men responded to his bidding. The others crowded around the sawyer—out of sympathy and curiosity. Omer held his arm around Sam's shoulders (to keep him from falling off the bench) and while he waited for the man to catch his breath, he issued another order to one of the men standing in the semi-circle, "Go ask Kelly for some red-eye. This man's going to pass out afore we can find out what happened."

Before Fuller's words were out of his mouth, McTaggart burst into the room. He had Kelly's flask in his hand. He pushed his way through the men and as soon as he reached Sam, he knelt down and held the liquor to the man's lips—at the same time, took stock of the situation. Bean's arm was broken, and his hand was almost unrecognizable. But when there was a move to find Scotty's scissors to cut off the sawyer's sleeve so they could remove his coat, Sam, speaking with an effort, managed to say, "Don't—bother with me—. Git to—the banking grounds—quick. Dan'l—is under a—pile of logs. I been nearly two hours gettin' back—. If the logs ain't kilt him—." The sawyer barely stifled a sob before continuing, "If the logs ain't kilt him—he'll be froze stiff by the time you get there."

Immediately, McTaggart began issuing orders to those standing by.

"Go to the barn and tell Frenchy to hitch up the box sled. Put in some hay and some blankets—and tell Hoops—." He stopped speaking when he saw one of the men point over his shoulder. The scaler was standing at his elbow with the camp medicine chest in his hands. He laid it on the bench. "This is what you want," he said.

The foreman nodded and then turned his attention back to the injured man.

"Tell me quick as you can what happened and where you were when it happened."

Suddenly, Bean lost his pallor. The color in the man's face turned to pink when he saw the scaler standing near by. Almost in a whisper he gave his account.

"Dan'l Blackburn and me was sore about—'bout Mr. Hoops markin' the logs—and we was just tryin' to see—." The sawyer paused, seeming to choke on his words.

"Go on!" the foreman said impatiently. Time was wasting.

"We just wanted to see how many was marked with the X.

We was at the last pile that Bell scaled. Dan'l had a peavey and was pryin' up on one of the logs high on the face of the pile—when—when the whole damn thing broke loose. I remember reachin' fer him just before I blacked out. When I come to, I was layin' on the ground with my arm under the end of one of the logs what broke out. I tried to git loose—." The man shuddered and covered his face with his good hand. "I got out—but—. I seen him then. There was Dan'l with one of them monsters acrost his chest. I couldn't move it. He was still alive an' in terrible pain. He was moanin' pitiful-like. I stuffed my cap under his head—." With tears in his eyes, Sam Bean looked beseechingly at McTaggart. "That's all I could do for him, Boss. I had to leave Dan'l like that—poor old chap."

The foreman's face looked stern. He was impatient to leave—but he managed to say, kindly, "You'll have to wait till I get back to have that arm set. The bleeding seems to have stopped on your hand. One of the men can wash it for you."

He turned to face the scaler. "Billy will have some laudanum out in the barn—it'll ease the pain some—."

With those last words of instruction, McTaggart rushed out the door with the medicine box under his arm. He was thinking to himself as he ran across the frozen yard to the waiting sleigh, "What in Sam Hill can happen next?"

The camp medicine box.

15

Oh-h-h-h-h moans the woodsman, hurt in the bush

Frenchy LaRiviere and his team were waiting when the foreman came on the run out the door of the bunk-camp. The light of the mid-winter afternoon was already waning, and McTaggart, after putting the medicine chest in the back of the bob sled, looked to see if the teamster had thought about bringing a lantern. One was hanging from a hook at the side of the sled. The teamster had also brought a couple of cant hooks and a length of decking chain. They were lying on top of a buffalo robe and some horse blankets that Billy Flynn had thrown on the straw that covered the bottom of the sleigh box.

Big George Pepper had insisted on coming along. He was sitting grimly in the back of the sled, and McTaggart acknowledged the man's presence with a nod of his head as he climbed up on the wagon seat next to the teamster. Frenchy clucked to his horse team, signaling them to start up. In an instant the men were on their way to look for the unlucky Daniel Blackburn—only eight minutes after the sawyer's partner had staggered into the bunk-camp with his tragic news.

Both the foreman and the barn boss's helper were silent as the Belgians stepped off, going west down the snowy go-back road. The big horses had not been purchased for speed—but because of the teamster's urgent pleading, "Come Peppair, Jacques—ALLEZ! Il faut aller plus vite! ALLEZ!" they pressed forward at a pace that was unusual for the big animals.

The early darkening of the sky foretold of a coming storm—and, on occasion, the wind whirled powdery snow around the sleigh and its occupants, almost blotting from view the horses which looked, to the men in the sled, like great bobbing ghosts hurrying before them.

McTaggart, in spite of his concern for the hurt woodsmen, couldn't help wondering, as they hurried down the road, what the loss of two good workers would mean at this critical point in the camp's operation. Even if they found Blackburn alive, his injuries were almost certain to be of such a nature he would be of no further use to the camp for the rest of the season. And Sam Bean—what to do about him? His arm would not heal for at least a month, and from the looks of his hand—. The foreman shook his head. "He's a good riverman—one of the best there is. But how the devil can I justify keeping him on until spring breakup?"

These were the thoughts that were running through McTaggart's mind. He knew well the lumberman-owner's philosophy in regard to woodsmen who, while in his employ, became sick or disabled. And he knew, too, that Shannon's feelings were no different from any other businessman of the time. He had heard him say on several occasions, "I cannot see, because a man works for me for pay, I am forever bound to take care of him."[1] John Shannon, like most lumbermen, firmly believed his responsibility toward his employees ended with the payment of wages.

The light had almost faded by the time the teamster directed the horses over the sand ridge and then north on the sleigh road that followed the river. Before it failed completely, Frenchy hauled up on the lines to stop his team. Without a word, he got out of the sleigh to light the lantern. Snow enveloped his figure as he hunched to shield the matches from the wind. He struck the sulphur tips once, twice, three times against the rough side of the sleigh box. Each time, maddening gusts of wind blew out the flames before he could light the lantern wick. Finally, a flame caught hold—wavering fitfully, tantalizingly, until the teamster released the wire lever, letting the lantern globe fall with its familiar grating sound into the groove on the metal base. Frenchy adjusted the wick; and the light, now protected behind glass, glowed steadily as the teamster hung the lantern by its wire bale at the side of the sleigh. If, by some miracle, Daniel Blackburn was still conscious, he would see the approaching light through the rising storm. It would give him hope.

While the bob sled made its way down the go-back road, the men in camp waited restlessly for its return. Some were lying in their bunks. Others sat in bunches on the deacon's bench, while two or three men paced around the room. Each one was worrying about how the foreman and teamsters would find Blackburn at the rollway—dead or alive.

"Wonder if they found him yet?"

It was Ben Sweeney who asked the question that was on everybody's mind. He was standing at the side of Crosscut Bill holding a smoking oil lamp so his partner, who was washing the wounded hand of Sam Bean, could better perform his delicate task. Bill sat astraddle the deacon's bench while Sam sat facing him in the same attitude. A tin wash basin, filled with warm water, rested on the seat between the two men.

Bill looked up at his partner and said impatiently, "Just hold the lamp still will you Ben, and pay attention to what yer doin'. We'll know soon enough if Dan'l is dead or alive."

The injured man, winced—whether from pain or the untimely remark made by Bill.

Bean's heavy jacket had been removed and his left arm, now bent at a right angle against his chest, was neatly encased in a supporting sling made out of a clean grain sack. The ugly red welts and deep scratches on his forehead and cheek were highlighted in the firelight by applications of a greasey medication.

Behind Bill, Isaac Hoops bent over the barn boss's medicine chest studying its contents. He peered through a pair of steel-rimmed spectacles at the various colored, printed and hand-written labels pasted on the outside of an assortment of tins, boxes and bottles. The scaler lifted them, one by one, out of the chest as he made his selection, putting back those he did not want. He closed the lid and opened one of the tin boxes, taking out a small lump of alum to drop into the mortar. With several deft motions of the pestle, he crushed the hard substance to powder, then added portions of brimstone and gun powder to it. He briskly stirred the dry ingredients into a measure of vinegar and water.[2] When satisfied it was well mixed, he poured a small amount of the combustible concoction into a glass vessel and held it out to the injured man.

"Drink this, Sam. It will stop the mortification of flesh on your hand."

The sawyer made a face and turned his head away from the potion.

Hoops remonstrated, "It has always done the trick before. Drink it—then I'll give you some laudanum to deaden the pain."

The scaler's words had a quiet note of authority in them that was not lost on the woodsmen who, in grudging admiration, had watched him set the broken bones in Sam Bean's arm and then go about caring for his other wounds with all the confidence of a "Philadelphia doctor."[3] He had given them orders for basins, clean rags and hot water—and when the scaler asked for help to hold Sam to better cut off his coat sleeve and set his arm, the men quickly responded. There wasn't one in the bunk-camp who was not wondering where Hoops had learned to "doctor" so skillfully.

Isaac Lee Hoops was the mystery man in camp. Even McTaggart had no idea about the scaler's past. One day he showed up in camp and went to work. And that was that. If anyone had asked him about his former life, the scaler would have obliged the inquirer. He would have said that he was a New Brunswick man who had grown up in the Saint John River Valley. For three generations, members of his family had built ships at the great port of Saint John—ships in which the firm exported furs to the British Isles, and later on, lumber to the West Indies and Great Britain. Isaac was to go into the family business. He would have, too—except for the fact that the family fortunes took a sudden, disastrous turn the second year he was at the University of New Brunswick. Three vessels owned by Hoops Export, Ltd.—vessels loaded with cargoes of ship's timbers, lumber, staves and shingles, sank with all hands in gales—one ship off the coast of Cornwall and the other two off Cape Hatteras. The calamity brought ruination to the old firm. It collapsed—and with it, the end of Isaac's formal education. He immediately took to the pine woods to earn his living.

The year was 1847. He rolled up a few books and personal belongings in a blanket, said goodby to his family and went north with his bindle under his arm to begin a new life as clerk and assistant to the scaler for several lumber camps operating on a branch of the Tobique River in Victoria County. He was stationed at the depot or headquarters camp, which, like its satelite camps, was situated in an isolated wilderness.

In lumber camps, especially those far removed from a medical practitioner, the bush superintendent (or foreman) not only acted as the manager of the lumber camp but performed the duties of physician and surgeon as well. Who would doctor the boss of the outfit if he were injured in the line of duty was a matter of conjecture—but that is how Isaac Hoops discovered he had talents other than those necessary to keep the camp books in order.

One day he was roused from his ledgers by one of the teamsters who rushed into the shanty where the clerk was working to tell him the superintendent had been struck down by a falling tree. The bull-whacker had brought the hurt man all the way back to camp on a make-shift travois. Now he looked to the young scaler-clerk to perform a miracle.

While the teamster and others looked on, the clerk examined the superintendent. Then—after hastily reviewing pages in a copy of *The Family Receipt Book & Useful Medical Advisor* (among the books Isaac had rolled up in his bindle when he left home), he asked for the camp medicine chest and without further fuss, went to work patching up his employer. After that, the scaler-clerk found himself with an additional job—and every winter thereafter, he dispensed medicines, treated frozen toes, set broken limbs and sewed up wounds—all the way from New Brunswick to Michigan—the copy of the *Useful Medical Advisor* falling apart long before he reached McTaggart's Red Keg.

Hoops wasn't always successful in his doctoring, but he did the best he knew how under the circumstances. The rest was left up to the Lord, he said. If the result was fatal, the scaler saw to it that his patient got a decent burial.

The lumber industry was the most dangerous of all occupations. Hundreds of woodsmen were crushed, mangled, or knocked into eternity by dislodged tree limbs or the irresistible force of tumbling, rolling logs. Men were blinded by chips flying from a chopper's ax, cut by axes glancing off frozen tree trunks, or by broadaxes and adzes when squaring ship's timbers and other "long stuff." It's no wonder that the medicine chest was an indispensable article in the lumber camp.

Brief notices appeared now and then in mill town newspapers: "A man named Aaron Aten, camp of John Murray, Higgins Lake, injured while loading logs."[4] But most of the injuries (and deaths) that took place in the woods were not brought to the public's attention unless the townspeople happened to see a woodsman hobbling down the street, his wounds bound up in dirty rags. He swallowed whiskey (ten cents worth of Epsom salts in a quart of whiskey was supposed to cure almost anything) and patent pain-killers in backrooms of cheap boarding houses while "convalescing," hoping to make it back to camp before becoming a ward of the county.

Up to the early 1870's, the woodsman who survived an accident in the woods received only the sympathy of his comrades, and what they dropped into the "you-know"—the hat or tin basin that was passed around the bunk-camp. With his collection of pennies and nickels knotted in a bandana and his time slip tucked away in his turkey, the woodsman was sent to town in the back of the tote wagon to fend for himself.

The hospital ticket—a kind of hospital insurance for woodsmen—changed all that. The plan was introduced in East Saginaw in 1872 by a newspaper man.[5] With the purchase of a $5 "ticket," a woodsman was assured of one year's treatment (and drugs) in a participating hospital. A widow-maker might put him out of commission for a while, but he could rest assured he wouldn't be heading for the county poor-farm while he was recuperating from his injuries. A hurt woodsman, resting between clean sheets in a hospital, might be glad he had listened to the hospital agent who stayed over in camp one winter's night to give his spiel to the boys.[6]

While they sat around the fire, impatiently waiting for the return of the foreman, Hoops and the rest of the crew tried not to think of the bleak future that was in store for Sam Bean and Daniel Blackburn—for the year was 1868. The hospital ticket was not yet heard of.

The time seemed to drag after the scaler closed the lid on the medicine chest. He took his time stoppering bottles and rinsing glasses and basins before putting them away.

Sam had insisted on staying where he was, on the deacon's bench, and not resting in his bunk like Hoops wanted him to do. While he sat staring at the door, his companions pulled out pocket watches—snapping them open to see the time and clicking them shut again, unsatisfied at what they saw on watch dials. And those who didn't have time pieces kept asking, "What time is it now?" Other than that, the room was silent except for the everlasting cracking sounds of the fire in the camboose.

Six o'clock—the time when Cap usually blew the horn for supper on Sunday night—came and went.

Hoops thought to himself, "If the cookee would only blow the supper horn, at least eating would give the men something to do. Wonder what's holding him up?"

There was a movement at the door. The men looked toward it, expecting to see McTaggart. Instead they saw Cap standing before the open doorway. The darkness of the winter's night was like a velvet backdrop behind the apron-clad figure. A gust of wind blew in from the outside, making papers and playing cards scoot over the floor and a shower of sparks blow up the chimney.

The cook's helper slowly closed the door behind him. In the firelight, Cap's face looked wretched as he made an effort to speak. At last, he said, "I got something to tell you. You might as well know it now as after supper—." He stopped, seeming to choke on his words. Regaining his composure, he resumed speaking. "Jack and me was coming back from the island a while ago, walkin' along the sleigh road following the river. We seen one of the log piles in a heap all over the ground—and—went to see what happened—." Again, Cap's voice faltered.

The men leaned forward—holding their breath. Sam Bean stood up and would have gone to the cookee, if the scaler hadn't pulled him back to the bench.

Cap looked at Sam and said, simply, "He's dead, Sam."

The sawyer's eyes filled with tears, but he savagely

brushed them away with his good arm. A husky voice from one of the bunks said, "Go on Cap—."

The cookee took a deep breath and continued his speech. "Jack and me didn't know anybody had been with Dan'l, so Jack stayed there and I come on back to tell McTaggart what happened. But when I went by the barn on the way to the van, Flynn saw me and told me you already knew about the accident and that McTaggart and Frenchy was on their way to the river."

Tragedy was commonplace in the pine woods—but the death of a fellow crew member never failed to surprise or sadden the ones left to carry on. The distressing news silenced the men. There was nothing else to say—no more questions to be asked. Cap waited for almost a minute, then sadly turned to go out the door. He hesitated, came back—and said, almost in a whisper, "I near forgot—supper's ready." He closed the door noiselessly behind him.

At that moment, the sleigh carrying the would-be rescuers was drawing near the place at the banking grounds where the logs had broken out. The snow squall had waned, and as the men peered into the darkness looking for the pile of logs, they saw instead, a figure of a man running towards them with arms raised. He was signaling them to halt.

Frenchy hauled up on the lines. With rattling harness and clouds of vapor blowing from distended nostrils, the horses stopped in their tracks.

The foreman had already jumped from the moving sleigh. He grabbed the lantern—holding it high. The figure belonged to Jack Driscoll.

"How in thunder did you get here so quick!" hollered the top-loader.

Without answering, McTaggart pushed the man aside to run toward the heap of logs that he could see but dimly. But he was checked by Jack's voice.

"He's dead, Boss. Froze stiff. Good thing too—. Chest caved in—back broke. What the hell happened?"

The nagging doubt in the foreman's mind that they would arrive in time to save the sawyer's life had persisted all the way out to the river bank—and now, seeing in the circle of lantern light the ghastly sight of the man lying so still under the massive weight of a great pine log, momentarily drained the strength from his legs. He breathed a silent prayer of thanks for the freezing cold which had mercifully saved the sawyer from further agony.

Bending over the body, he noticed that Daniel's head still lay on Sam Bean's fur cap. Jack Driscoll saw the cap at the same moment.

"Who was with him," he said.

It was George Pepper who answered him. "Sam Bean. Sam's arm got broke and his hand is smashed up—but he got back to tell us what happened. 'Cordin' to his story, him 'n Dan'l was trying to see how many logs Hoops was throwin' out as culls, when the whole tier went haywire."

Driscoll slowly shook his head. Death seemed to him a high price to pay for not minding your own business. He heard McTaggart say, "We might as well pull the log off and get him back to camp." The foreman's voice was flat—matter of fact.

With cant hooks and decking chain, the men, with the help of the horses, rolled the log from Daniel's body. With gentle hands, they wrapped the broken remains of their dead comrade in a rough shroud of blankets and placed him on the buffalo robe in the sleigh box.

George picked up the tools, and he and Jack climbed into the back of the sleigh to sit with their legs hanging over the tail gate. The foreman and teamster were already aboard.

Frenchy chirruped to his team, who turned the sleigh around. They were on their way back to camp with their sad burden.

It was after eight o'clock when the four men sighted the glow of McCloud's forge fire, and they knew they had arrived back at camp. They were home. But there was still a job to do.

McTaggart said, "We'll bury him in the grove of Norways yonder beyond the stable."

The men knew, without having to ask, the foreman's reason for selecting that particular site for Daniel's final resting place. The ground beneath the ancient trees was insulated with over a hundred years' accumulation of pine needles. The grave diggers would have a hard enough time this stormy night performing their gruesome task without having to go through eighteen or more inches of frozen soil in another place before they dug below the frost level.

The horses were brought to a halt in front of the barn door where Billy Flynn stood waiting with a lantern. The men got out of the sleigh; and Frenchy, the barn-boss and his helper went about the business of unhitching the team and leading the horses to their stalls. The team would eat even while the men were removing their harness.

Before George Pepper disappeared into the barn, the foreman collared him. "As soon as you eat some supper," he said, "round up a couple of the boys and start digging. I'll mark the spot. We'll bury Blackburn tonight, just as soon as you get the hole dug." Having passed along these instructions, McTaggart abruptly turned on his heel and started off for the bunk-camp to tell the crew that Daniel had been brought back to camp.

It was almost midnight before the grave under the Norways was made ready for Daniel. In spite of the early hour when the crew would rise the next day, no one went to bed. Some helped the grave diggers. Others sat around the fire in the bunk-camp and talked in low tones about the sawyer, reminiscing about former times and other places where they had worked or caroused with Blackburn. Sam Bean stood alone beside the dead sawyer's bunk sadly putting in order his friend's personal effects. He had orders from McTaggart to search through Daniel's meager belongings for a name or an address where the foreman might write to inform relatives or other interested parties about the man's death.

While Sam and the others were thus occupied, McCloud was in the blacksmith shop hastily fashioning a make-shift coffin for the unfortunate sawyer. The blacksmith went about his job by knocking the heads out of two flour barrels and fastening them together, end to end—leaving one end open to receive the body. McCloud was finishing the task when the foreman stuck his head in the door.

"How you coming—about done?"

"It's all set," the smith answered. "You can help me carry it out to the sleigh."

"Good. I'll call the boys."

It was a curious looking group that assembled around Daniel's grave that black night. The yellow glow of two or three lanterns lighted the downcast faces of the crew and brought to view the excavation and the mound of dirt so hastily thrown up from the rude opening. The Norways loomed over the men—their black trunks looking like towering pillars in a dimly-lighted cathedral. The wind sighed through the tree tops—a mournful sound that heightened the melancholy of the scene—while snow sifted noiselessly over Daniel's coffin and down the necks of the silent congregation.

Because of the lateness of the hour, the ceremony was

brief. McTaggart repeated from memory the 23rd Psalm of David—his voice at times fading in the fitful wind. The men stood shivering, with heads bowed, listening to the timeless song. When the words were said, the foreman paused (as if to lend more meaning to the verse), then nodded to the four woodsmen standing at the sides of the excavation. At the signal, they slowly paid out the ropes that held up the wooden curiosity. The semi-circle of men watched in silence as the coffin disappeared into its resting place.

The foreman waited until he heard several spadefulls of frozen earth rattle onto the wooden sides of the barrels. Then he said quietly, with no outward emotion to betray his feelings, "Get to bed boys—it'll be morning soon. We've still got logs to move."

Before McTaggart left the grave site, Sam Bean came to him and thrust a scrap of paper into his hand. The foreman sighed as he crumpled it into a ball. Without being told, he knew the paper would have an address written on it. The most difficult part of his job—the part he disliked more than anything—was writing to loved ones to tell about the death or injury of a sweetheart, husband or son. Thus, it was past midnight before McTaggart propped an envelope (with the return address, "John Shannon, Logs & Lumber, Saginaw City, Michigan," printed in bold type face in the upper left corner) against the oil lamp before he blew it out. The letter would go out with the tote teamster tomorrow. When it would reach its destination was anybody's guess.

The wind sighed through the tree tops—a mournful sound that heightened the melancholy of the scene, while swirling snow sifted over Daniel's coffin and down the necks of the silent congregation.

Zip's bee smoker.

16

P is for the fast PACE set by the "push"

It seemed to the foreman his head had only touched his pillow when he felt a familiar tap on his shoulder. Out of the gloom he heard Cap's voice say in an undertone, "It's four o'clock Boss."

McTaggart smothered a groan as he swung his feet to the cold floor—and while the cookee lighted a lamp, he wearily reached for his boots.

Cap adjusted the draft openings on the stove and jammed wood into the firebox. Oblivious to the racket, Hoops snored on. He would sleep until the five-minute horn sounded for breakfast.

"I expect Shannon up again sometime this week," the foreman informed the handy man. "Keep your eye peeled for him and give me a signal when you see his rig." Jerking a thumb in the direction of the door where the jack towel hung, he added, "Don't forget to change—."

Cap nodded his head to indicate he had gotten the message about the towel and disappeared into the polar darkness.

McTaggart splashed the icy contents of the water bucket into the tin wash basin, and while he performed the rites of shaving, washing face and hands and grooming hair and mustache, he tried to concentrate on the business before him.

Shannon wanted four million board feet of top grade pine from each of his three camps, ready for floating down the Tittabawassee at break-up time. The camp that floated out the most logs over the four million would win the boss lumberman's prize.

The foreman had told his men they had a slender chance of winning—but he hadn't told them how slim a chance it was. McTaggart shook his head. The next few weeks weren't going to be pleasant ones. He was already short-handed—and with the constant pushing, there was certain to be carelessness in the woods. And short-cuts. Tempers would flare up over nothing. The men would bear watching. Getting over four million feet of logs on the banks looked almost hopeless. But he'd be damned if he would let the other camps win out over Red Keg—not after devising the wider sleigh bunks and all the risks the men were taking loading and driving the sleighs to the banks.

McTaggart dried his hands on the towel. They'd get their four million-plus out, or he'd know the reasons why not!

That week, knowing the weather would be the biggest aid or obstacle to his lumbering operations, the foreman took to looking for signs that might give him a hint as to what he could expect in the way of continuing cold. He even visited the barn to consult the pages in Billy Flynn's almanac for the forecasts printed there.

In spite of his waking to hear thunder in the early hours of Candlemas Day, a snow storm had blown in leaving behind another two feet of snow. But McTaggart wasn't fooled by the show of wintery weather and pushed the crew just that much harder to take advantage of the snow and cold.

John Shannon showed up in camp as expected. The news about the cull logs and Daniel Blackburn's death had been reported to him in Saginaw City, and when he alighted from his cutter, McTaggart could see the lumberman wasn't too pleased with the way things had been going at Red Keg. His face was stern—his black eyes cold. And even before he removed his coat and cap upon entering the van, he said, "Is the sawyer—," the lumberman paused, "what's his name—the man who was hurt on the logs—."

"Sam Bean," interjected the foreman.

"Yes—Bean." Shannon made an impatient gesture. "Is Bean still in camp? You know my feelings about paying for dead weight."

"I know your feelings on the subject." McTaggart tried to sound respectful, yet firm. "The man is still in camp. Sam is a good riverman and I'll be needing him come spring. He's carrying his own weight around here—been helping all over the place—doing everything he can—."

"That can't be much—with the use of only one arm," the lumberman said testily.

Before his employer could continue, the foreman hurried on. "Besides being a good riverman, Sam is one of the best "powder-monkeys" on the Tittabawassee. He can handle No. 1 dynamite and can untangle a jam quicker than most rivermen I know around here."

McTaggart saw a faint look of resignation in the lumberman's eyes. He added, "Sam's arm'll be healed by the time the ice breaks up—it was a clean break and set good by Hoops."

John Shannon said, "All right, all right—keep him on." His voice became hard. "But no pay—not a cent from the time he and Blackburn broke out the log pile until the drive begins. And make certain his van bill comes out of what's coming to him at the end of the season. The work he does around camp will just about pay for his bunk and found."

The lumberman, now relieved of his heavy outer garments, sat down heavily in the swing rocker and lighted a cigar. "Now then," he said, "I'd like to hear about the logs that Bell scaled. Frankly, I don't know or care how you're going to make up the difference, I just want my four million from Camp 3 to roll off the banks when the river drive begins."

And so the talk went on. After supper, the books were inspected—the men working over them until almost eleven o'clock, when Cap came in with the tea pot.

Early the next morning, John Shannon climbed into his cutter and pulled the buffalo robe over his lap. McTaggart and Isaac Hoops stood beside the rig waiting for his last-minute instructions.

"I won't be up any more this winter," said the lumberman, "but I'll be watching for your reports. As to your final figures, send them down to the mill office before the drive begins." He made another adjustment to the bulky robe. "As to the date when the winners of the prize bonus will be announced—I will let you know more about that before the month is out."

Shannon pulled off a glove and reached out and shook hands with his foreman and scaler. With the gesture, the visit was brought to a close. He chirruped to the chestnut mare who took off at a light touch of the buggy whip on her flank.

As they watched the cutter disappear around a bend in the frozen river, McTaggart and Hoops both felt an easing of strain—a lessening of tension that was always present when the lumberman was in camp. Keeping Sam Bean in camp had been a small battle won, thought the foreman. Now, for the big battle.

It was a relief when Sunday came around again. All week long McTaggart had kept close on the men's heels. Each night, when the crew came to the van to report their day's work, he scrutinized the figures they handed him, comparing them with those in the scaler's reports. Every day the foreman posted each team's figures on the bucking-board—tacking it to the bunk-camp door—to keep the rivalry up to a driving pitch.

As the week-end neared, McTaggart instructed Isaac Hoops to give him an estimate on the number of logs in the woods that were still awaiting transportation to the river. On Saturday night the scaler reported that they had made up their losses and were closing in on the four million mark. Still afraid of an early thaw, the foreman decided to pull all but one team of choppers, sawyers and skidders from their usual work, and send them to help load sleighs at the skidways, and pile logs at the rollways.

By the time Sunday rolled around, there wasn't a man in camp who wasn't bone-tired—exhausted to the point of numbness. Tempers were worn thin by the hustle and competition, and it didn't take much to provoke a fight. Like, for instance, the incident that took place in the cook-camp during dinner time. Hilly James had reached out with his fork to spear a slice of bread from the plate as it went down the line and accidentally knocked over a pitcher of molasses into the lap of Bill Bragg. In an instant the air was blue with the loud vituperations of the enraged swamper.

"You don't think I git enough of molasses *in* and *on* every dang thing I eat—now you gotta *POUR IT ALL OVER ME*!" hollered Bill, who watched helplessly as the brown syrup ran down his right pants leg. "Fer two cents I'd knock yer teeth down yer throat, you—you—!"

Cap came running to the rescue with a wet rag, and McTaggart quieted the rumblings that were breaking out here and there around the room by banging his fist on the table. The noise settled down. But it was a close call. If the foreman hadn't been there, the whole place might have erupted into a free-for-all.

Tempers were worn thin by the hustle and competition, and it didn't take much to provoke a fight.

Cap cleaned up the molasses and went back to where Kelly was working. While he rinsed out the sticky cloth, he said, "I think I'll go out this afternoon and look for that bee tree I seen last summer. If there's any honey left in the hive, it'd be a real treat for the boys. They're mighty tired of molasses. Might relieve a little of the edginess around here."

So after the remains of dinner were cleared away and the crew was snoring in the bunk-camp, Cap Wilkes sought out his friend Jack Driscoll.

"I'm goin' out to try to find a bee tree I seen some time ago. Want to come along?"

"Where's it at?" asked Jack.

"She's down river about two miles—in a grove of hardwoods—'tween Camp 2 and here."

"Be a nice walk," said the top-loader. And off the two companions went on their snowshoes, pulling after them the cookee's pung. On it was an ax, a length of rope, a couple of wooden wash tubs and several lard buckets stacked one inside the other.

The woods was silent except for occasional sounds of chattering squirrels and scolding jays. The men must have walked about a mile before Cap broke the quiet to say, "Ever see anybody in the woods lookin' for a bee tree?"[1]

Driscoll shook his head. "Can't say that I have."

The cookee laughed. "It really can shake you up, seein' fer the first time some fool chasin' through the woods with his head throwed back and his nose stuck up in the air, bumpin' into trees and fallin' over logs and such—running to see where some bee goes with her load of honey."

It was Jack's turn to laugh. His imagination let him see the bee hunter chasing through the woods—pausing, out of breath, to scan the heavens and then taking off again in the same helter-skelter pattern.

"When did you find the hive you're lookin' for?"

"Last summer. Some of the boys come up to camp early on account of buildin' on the cook-camp and bringing in the new cook stove. The Sunday after they got in, I thought I'd go out and have a look in the woods to see if I could locate a holler log or a tree where a swarm of honey bees might have located. Kelly likes honey to bake with," he explained, "and it sure relieves the monotony of eatin' molasses all winter long to have a taste of honey on your biscuits or in your tea."

"It was a pertickler hot day I picked for traipsing in the woods, and after searchin' for about a hour without any success, I picked a spot in a place where I thought honey bees might be workin' and set down with my back to the trunk of a tree to wait for some activity.

"I sat, quiet-like, almost hidden by some wild blackberry bushes. I tried to keep my eyes open—but a patch of sun wandered down through a opening in the trees acrost my shoulders, makin' me fall asleep."

Cap paused in his story to neatly skirt what looked like a big snow drift—but what 'might have been a clump of cedars completely concealed by snow. The cookee had once walked into such a trap—falling into a well of snow and a tangle of small cedar trees—a predicament in which he nearly perished, not being able to burrow out of the deep snow or climb the small cedars.

The cookee continued. "All of a sudden, I heard somethin' what woke me up. Something or somebody was crashin' through the woods—sounding for all the world like the U.S. Cavalry going after a tribe of Modocs. I stood up, careful to stay behind the trunk of the tree, but instead of a bear—what I thought was makin' all the noise—there was old Zip Somers, the cookee from down at Camp 2, a-runnin' with two empty buckets swingin' from a neck yoke—goin' through the woods hell-bent-fer-leather. His face, what was pointed straight up, was all red and wet with sweat—and as he run, the pails kept bangin' up against tree trunks and catchin' in vines and on the bushes. In spite of myself I laughed right out loud at the sight—but old Zip didn't even hear me, he was so busy. He turned off all of a sudden and then stopped, not fifty feet off. I seen him walk slow-like around a big old beech tree. From where I was standin' I could see the tree was pretty good sized—must of been at least two hundred fifty year old with a lot of spreadin' branches almost all the way down to the butt. The next thing that happens is, I hear Zip say out loud, "So—that's where you was a-goin' you little so-and-so!" I knowed for sure then, that he had found a bee tree.

"I still couldn't see no hole or a entrance for the bees to get in—so I moved up a bit—careful to stay out of sight. From my hiding place, I could see the trunk of the beech was holler all right, and the bee hole was about eighteen feet off the ground, where one of the branches fell off the punky tree a long time ago. I could also see bee guards flyin' and crawlin' around the hole, protecting the entrance.

"I thought to myself, 'Zip must not know much about honey bees, 'because he was wearin' a old felt hat."

Cap looked back at his friend and said as an aside, "Bees hate felt, and they attack it by stingin' it. The other bees smell the bee pizen, and smellin' it makes them excited—and when that happens—," Cap paused to find the right phrase, "when that happens, I jest as soon be as far from a bee hive as I can git!

"Anyways," he continued, "besides wearin' a felt hat, I seen he had on a wool shirt—which the bees like just about as much as they like felt. And he didn't have no ax to get at the honey. I made up my mind right there that I would wait around to see the fun."

The cookee stopped talking long enough to pause in his tracks and slowly look around. He saw a blaze mark on a tree and took off again with the top-loader behind him. Jack noticed they were now in a part of the woods he wasn't familiar with. The trees were mostly hardwoods—maples, yellow birch, elms, white ash and a few oaks. "The beech tree must be near by," thought Driscoll. Out loud he said, "What happened then?"

"Well," said Cap, "what happened next, happened so fast I don't know yet whether I seen it or dreamed it. The only clear thing I remember was wonderin' why Zip didn't light the bee smoker he had with him, instead of tuckin' it in his pocket afore he started climbing up to the bee hole—when I saw the bear."

Cap chose that particular moment to stop and explain, "Smoke quiets bees—my Pa told me once that bee keepers has used smoke for hundreds and hundreds of years to pacify the little buggers, so they can handle 'em without getting stung—."

Exasperated, Jack interrupted the cookee, "Never mind what yer Pa told you—what happened to Zip?"

"Well," Cap resumed, unconcerned at his friend's impatience, "the bees what was at the entrance, come right down to investigate, just like I thought they would, and got a sniff of Zip's hat. They started stingin' it, and that fool Zip, not knowin' a bee from a canary, took off his hat and started slappin 'em with it. Of course, that only made 'em madder. They really went on a rampage then. Well, he finally give up and started down the tree—a-clawin' and a-tearin' from branch to branch with what seemed like the whole hive a-roarin' after him. Fact is, he come down so fast he didn't see the bear till he fell on top of him!

"I don't know who was more surprised, the bear or Zip. I seen their eyes meet and the next thing I knowed, old Zip picked up the bee smoker, what had fell to the ground, and threw it at the bear—and then took off fer Camp 2 with bruin and the bees right behind him. They was goin' so fast, all's I saw was a glimpse of Zip's red suspenders and a brown streak when they went by me."

Jack was still laughing at the thought of the cookee racing through the woods with both bees and bear in hot pursuit when he saw Cap stop and point to a great beech tree that stood in an opening in the woods. Its branches stretched far from its short, thick trunk.

"There she is," the cookee said. He started walking toward the tree.

"Wait a minute Cap—what happened then? Why didn't you collect the honey yourself? You must of had your hatchet with you—and Zip—he left behind his buckets—."

Cap walked back to where the top-loader was standing and said patiently, "Jack—once bees is stirred up, they've been known to stay on the alert fer as long as twenty-four hours, and if anybody even goes near the entrance to their hive while they're mad—watch out!

"So when Zip and the bear was out of sight, I just picked up his buckets, neck yoke and bee smoker and stuffed 'em under some bushes, then took off for camp, blazing a trail all the way—in case I wanted to come back. I never did though—what with the boys comin' in and all."

"Did the bear catch the cookee or what?" persisted Jack.

"Nope—ol' Zip got back to camp all right—but pretty bad stung up by the bees. He outrun the bear, though—and later I heard that he was braggin' to the cook down there if he only had a gun, they would of had bear steaks to eat."

Cap smiled at the thought of Zip Somers confronting an angry bear, with or without a gun. He laughed outright.

While the old woodsman was telling the end of his story, he was digging down in the snow under some bushes. After a half a minute or so held aloft a wooden bucket. "They're still here—right where I left 'em! So, Zip didn't get back neither," he said with satisfaction. "Probably couldn't find the place again—."

Satisfied that Cap's story had come to an end, Driscoll went to the bee tree to look it over while his friend was examining the articles left behind by Zip Somers. Jack noticed that the smooth surface of the grayish colored bark on the trees was scarred by what looked like claw marks.

"Bears," said Cap, who had come up behind the top-loader, "lookin' for honey same as us."

Jack looked around the tree. "Squirrels and some pheasants and deer has been around, too, by the looks of the tracks," he said. "Probably looking for split nuts."

Both men looked up into the old tree and saw the entrance to the hive. Here and there, leathery, copper-colored leaves of the ancient beech tree still clung stubbornly to its branches.

"Don't see no sign of bees at all," said Driscoll.

"Bein' winter, they'll be clustering together in there to keep warm," explained the cookee. "Hope they ain't et all the honey," he added.

"If you'll give me a boost, "I'll go up and take a closer look at the entrance. I don't think we'll need smoke. Honey bees don't like the cold."

Cap kicked off his snowshoes and the top-loader gave his friend an assist to the first branch of the tree and then watched him go up from branch to branch, like a man half his age, until he reached the small hole that was the entrance to the bee hive.

"What do you see?" called Jack from below.

"Nothin'," answered his friend. "It looks like nobody's to home. Could be they was scared off by somethin'—or their queen might of died—hope the mice hasn't got in there."

Cap looked down at Jack. "Hand me the ax. I'll cut the trunk open right about here." The cookee pointed to a place about two feet below the small opening. "We'll see how much honey is left in there."

The ax was handed up to the honey-robber. He fixed his feet firmly on the branch he was standing on, and then carefully chopped a small hole in the side of the hollow tree trunk. Jack noticed that Cap didn't go about his task "wholesale," but worked carefully at the opening, enlarging it bit by bit so as not to destroy the honey comb—if there were any in the tree.

"What do you see now?" called Jack. "Any honey in there?"

"We really hit it rich," shouted Cap. He sounded excited. "Ain't a bee in here and the whole summer store is just like they left it. Can't chop down the tree to get it out—she's an awful size. I think if we throw a rope over a branch up here, we can tie a bucket to it and pull it up and down to fill it and empty it into the tubs."

It was a tiresome, messy, sticky job—but within an hour the two men had the wash tub and lard buckets filled almost to the top with the comb and honey.

"Cap, there ain't much more room on the pung for any more," Jack said.

The cookee answered, "There ain't but half a—."

But Jack didn't hear the end of Cap's sentence, for behind him he heard a voice say, "I think they got enough honey for us right there on the sled—don't you boys?" The voice was mocking.

Driscoll turned around, and saw standing only a few yards behind him a tall, skinny individual with stubbled chin stained with tobacco juice. His mouth was screwed up into a tantalizing smile. He wasn't alone—but was accompanied by two other men. One was holding a pull rope that was attached to a big hand sled on which rested several wooden wash tubs and buckets.

Even though most woodsmen operating on the Tittabawassee were out of touch during the long winter months (unless they chanced to meet at the forbidden saloon upriver), most of the men were acquainted with one another, having worked together at one time or another in the past. But Jack Driscoll, the Canadian, having only recently come to the Valley, did not recognize the men standing before him.

"This must be ol' Zip who Cap was talkin' about," he thought. "An' it looks like he's come for his honey."

As Driscoll eyed the newcomers, he saw they were big men. For a fleeting moment he felt at a disadvantage—feeling like a fool standing there with a sticky dipper in his hand. But he quickly recovered his composure, noting that outside of their size, the three men who were confronting him didn't look too formidable—although they seemed to be spoiling for a fight.

"Well now," said Jack with as much deliberation as he could muster, "it looks like we got company." Louder, he said, not taking his eyes off the intruders, "Cap! Your friend Zip is here—he's brought along some chums. Why don't you come on down so we can all have a talk."

Cap, who had heard the voices, was already making ready to climb down the tree. Jack would be needing his help, he thought.

17

Q is for the QUARRELS on which the foreman doth frown

Cap had scooped out almost the last of the honey bees' winter stores from the hollow in the big silver beech when he heard the voices below. Cautiously, he looked down, and when he saw who the voices belonged to, the cookee knew in an instant that he and his friend would not get back to Red Keg without a fight. His heart leapt at the thought. It had been a long time since he had felt his pulse quicken like that. From his elevated position, he could see the top of Zip Somers's knitted cap (his felt hat, the one that had so antagonized the bees, was somewhere under the snow). It was Zip all right. Sour-tempered, string-bean Zip Somers, with pinched nostrils and eyes sunk deep in his head, was standing there in front of Jack Driscoll with his chin stuck out a mile. He was backed up by two husky woodsmen. Cap recognized the men. Bill George—sawyer—hair cut short with a curly bang hanging down in front. The other man was the head chopper at Camp 2—Jack Black, with sandy sideburns and stubby beard—handsome, in spite of his rough appearance. All three woodsmen looked impatient for the fight that was certain to ensue. Only the victors would carry away the spoils of the battle, thought Cap—in this case, the honey from the honey tree.

The cookee carefully worked his way down from his perch until he stood at the base of the tree. He held a half-filled bucket in one hand and his ax in the other.

Somers was bellowing, "You ain't no right to that honey—I found that old beech last summer—and I mean to have what was in it!"

Cap couldn't resist getting in his two cents. He said, "How come you waited 'till now to clean out the hive? You had dang near all winter to do it."

" 'Cause Zip couldn't find the tree—."

Somers turned on the Camp 2 chopper who had offered the explanation and said angrily, "Shut up, you booby—I knowed where it was—I was just takin' my time findin' it. Besides, we didn't need the honey 'till now." His voice became accusing. He pointed to the wooden vessel Cap was still holding. "Is that one of my buckets you got there, Wilkes?"

Jack Driscoll was only half listening to the exchange between Cap and the men from Camp 2—wondering about the fuss that was being made over a little honey. He would gladly have divided it with the woodsmen from down river and was about to say so when he saw Zip reach for a hatchet that was in one of the empty tubs on top of the hand sleigh.

Jack said, "Hold on, Somers." He made a move toward the cookee. "Don't pick it up." Driscoll's voice was calm—but it had an edge to it.

The words made Zip hesitate—but only for a blink of an eye. Heating up, he shouted, "By thunder, anybody'd think you and that ol' buzzard Wilkes owned the woods!" At the word "woods," he picked up the hatchet. But before he had time to clasp it firmly, Cap swung the pail of honey over his head and let go.

"There's yer old bucket!" he said.

Jack saw the wooden missile leave Cap's fist feeling it zip by his ear. He expected the bucket to land near Somers's shinbones—but when it landed with deadly accuracy upside down on his head, and he saw the cookee of Camp 2 sprawled in the snow with honey running in rivulets into his collar and spreading over his greasy jacket, he began to laugh—his laughter rolling upwards and out of his throat, arresting the attention of the astonished woodsmen looking on. Even Cap looked at his friend as if he had suddenly lost his senses.

The bucket landed with deadly accuracy upside down on the cookee's head.

It would seem that the top-loader could have picked a better time to have his funny bone tickled. His untimely mirth certainly didn't improve the tense situation that he and Cap were in. There they were—one big, hard-fisted man (at the moment helpless with laughter) and his friend, although tough as a hickory nut—a man who had counted many birthdays.

Jack's education was sketchy at best, so it would be fair to say he had never read or heard the sound advice penned so long ago, "Laugh then, at any but at fools or foes." Nevertheless, if he had learned the words at his mother's knee and observed the sage's words every day of his life, the sight of the lard bucket sailing through the air and landing so perfectly on target would have put them out of his mind.

Jack's laughter subsided at last when he saw the murderous look on Zip's face. The cookee struggled to his feet, having at last ridded himself of the bucket. Shaking with wrath and indignation, he shouted, "Now you *done* it! You *really* done it! And with war whoops that would have done credit to the Sauks of the Saginaw Valley, the fight was on.

All five men squared off. Suddenly, the opening in the hardwoods was alive with the grunts, groans, thuds and cries of battle. And the old beech tree, that had felt the furies of lightning and fire and gales driven inland from stormy lakes, now witnessed the rage of men pitted against each other—releasing the tensions and passions pent up after months of unrelenting hard labor.

Jack managed to holler at Cap, "Throw away your ax! We don't need it. We can take 'em on one at a time!"

Cap reluctantly flung the tool onto the pung while Jack, who had finally gotten into the spirit of the battle, bore into the fray with results that astonished even him.

It was a tiring business—fighting in the deep snow. All five woodsmen endeavored to get the upper hand. Fists swung in the air, missing their mark. Blows aimed at noses connected with eyes or an ear—and the force behind the exchanges sent bodies flying into snowbanks or bushes.

Jack Black, at the first opportunity, threw Zip's hatchet at Driscoll, who dodged the weapon to see the blade bury itself in the tough bark of the beech. Cap Wilkes tried to keep the pung at his back, defending the prize at all cost. But at one point in the battle, he was backed up to the sled until his heels caught in the runners. A push by his antagonist sent him backwards into one of the tubs—but before he settled to the bottom, a long arm of Driscoll hauled him out.

The old beech and her neighbors looked on in silence, the only witnesses to the melee. As the shadows lengthened and the men's shouts became fewer, Jack Driscoll took time to catch his breath. He was surprised to see Zip Somers, with a bloody lip leaking into the brown stain on his chin, lying in the trampled snow with arms out-stretched like a child making a "snow-angel." Bill George, taking advantage of the breather, was trying to help his chum Black to his feet. Cap stood near by. He still looked game, but Driscoll could see the starch had gone out of his knees. It seemed like it was a good time to claim the prize. Towering over the two loggers from Camp 2, he said, "When you catch yer breath, you'll find Zip's stuff over there." He pointed to a clump of birches where the neck yoke and bee smoker lay in the snow. Jack was tired of the fighting—and it was getting late. What McTaggart was going to say about him and Cap coming into camp in the state they were in, was as certain as snow in December.

Bill George had given up trying to help his partner to his

feet. But he managed to say to the top-loader, "You ain't seen the last of us, Driscoll." The sawyer didn't sound convincing.

Cap, who had resisted the interlopers like the youthful "warrier" who had fought the "enemy" on the parade grounds of old Fort Saginaw, leaned against the trunk of the ancient beech. With heart pounding and chest still heaving, he whispered happily, "I guess we won after all." A smile creased his weathered face.

Jack bent over to scoop up some snow, and rub it over his hands and face. Cap followed suit, and while he was thus occupied, the top-loader took stock of the pung and its contents.

"Outside of a little brush falling in the tubs, and you sittin' in the middle of one," he laughed, "I guess the honey ain't hurt none."

Cap made a face. He gingerly felt the back of his pants and jacket and then tried to pick off the accumulation of twigs, snow and dirt and other debris that was matted on his clothing. "Do you think we ought to leave them any honey-comb?" he asked.

"Might as well leave some of it in Zip's bucket," answered Driscoll as he adjusted his snowshoes to his pacs. "Don't matter much though—they'll be just as mad with it as without it." He looked over at the three men from Camp 2. "Sure look peaceful, don't they?"

It was almost dark by the time Driscoll and Cap Wilkes came within sight of Red Keg. Teamsters Ed Flatt and Bill Cowing were leaving the barn for the bunk-camp when they spied the cookee and top-loader limping up the road pulling behind them the sled loaded with the battle prize. They couldn't believe their eyes. Both men looked dirty and disheveled—their faces bruised and scratched. Flecks of blood and something dark, shiny and sticky-looking stained their jackets and pants.

In no time the word about the top-loader and cookee spread through the camp, and Cap and Jack Driscoll were at once surrounded by an excited crew, who heard in snatches the barest outline of the afternoon's adventure. (The top-loader and his friend didn't have to tell much more than that—the men would fill in the story with their own embellishments). As Jack and Cap took turns telling their tale, the boys escorted them (and the prize) to the cook-camp. Duff Kelly, hearing the noise of the approaching procession, stood in the open doorway to survey the scene. With hands on his hips and a frown on his face, he shouted, "What's all the ruckus about?" He spied Cap. "And where in hell have you been? Supper's nearly ready to go to the tables and I ain't seen hide nor hair of you all afternoon." Then he saw the pung and its contents as the men drew it to the door. The expression on the cook's face changed. He looked almost benevolent. "Honeycomb!" he hollered. "Why you old son-of-a—! You found the bee tree!"

Kelly didn't see the cuts or bruises on his helper's face—only the honey in the wash tubs on the pung. "Set them here in the dingle boys, while I stir up somethin' to pour it on!"

An amazed top-loader whistled in disbelief as he watched the cook disappear into the smoky cook-camp. "I'll be danged! That's the first time I ever seen that old dough-pounder smile!"

Cap shrugged his shoulders (dismissing Kelly) and said happily, "It was a great fight, wasn't it Jack?"

"Yep," answered Driscoll, "and I have a feeling we ain't heard the last of yer friend Zip."

• • •

On the Sunday morning in February when McTaggart wakened to the rumble of thunder, he had gotten up to stand in the open doorway of the van—tense, fearful—half expecting to feel raindrops on his outstretched hand. A streak of lightning stabbed the black sky, and the foreman heard again a rumbling far off to the south. He stood in the dark—waiting. But only the sound of the wind—which seemed to be rising—met his ear. McTaggart softly closed the door so as not to waken the scaler and went back to his bunk to lie staring into the dark—waiting for the rain to fall. And so he fell asleep.

When the foreman woke again, he could feel frost hanging in the room like a pall. While he lay there trying to decide what time it was, he heard the wire bell of the shelf clock slowly strike five times. It was morning. Then he remembered the thunder and hastily crawled out from under the buffalo robe to draw on pants, boots and shirt. But before McTaggart had time to go outside and check on the weather, he heard Cap fumbling at the door. The instant the cookee lifted the latch, the door slammed back against the wall with hurricane force—the wind rushing in to hurl a shower of snow into the cold room.

Cap set the water pail and lantern down before leaning against the heavy planks to close it. "Ain't seen a Groundhog Day like this 'n in quite a spell!"

Hoops, who had also risen at the sound of the striking clock, said, "I used to hear my father say, 'If Candlemas be fair and clear there'll be two winters in the year.' "

McTaggart replied as he snapped galluses over his shoulders, "It looks like we got our second winter in spite of that old saying," and went outside into the storm to inspect the thermometer. The foreman braced himself against the wind, and while it caught at his shoulders and tore through his hair, he held the lantern so he could see how far the mercury had dropped in the night. The thin red line had almost disappeared from sight. It registered twenty-two degrees below the zero mark.

At eight o'clock, McTaggart checked the temperature again. It had risen ten degrees—and it stayed there for two days until the northeaster blew itself out.

All day Sunday the crew stayed indoors by the fire, only going out to help Cap shovel paths to the wood pile, the barn and privy—while snow drifts piled up around the camp buildings and obliterated the sleigh roads.

By Monday, the storm had gone off to some other place and the camp struggled to life again. Standing shoulder to shoulder, the men and oxen pushed aside the deeper drifts in the roads so the horses could pull the plow through without floundering.

Spring looked very far away indeed. But McTaggart was too familiar with the vagaries of Michigan's weather to be fooled by her capricious nature. While the cold spell held on, he would take advantage of every hour.

The altercation at the bee tree took place on the Sunday following the big storm, and although McTaggart frowned on fighting amongst the men, he grudgingly admitted to himself that the incident under the old beech couldn't have happened at a better time. It gave them something to talk about besides the untimely death of Daniel Blackburn and the monotony of hard labor in the woods. And the fight had given the crew an added impetus to beat out John Shannon's other camps for the lumberman's prize bonus. So he kept silent on the subject.

But Jack Driscoll, who had expected to get chewed out by the foreman, couldn't understand why he and Cap had escaped McTaggart's wrath.

"If he's going to give us hell, I wish he'd get it over with," the top-loader confided to Cap one night after supper. "It ain't like him to let a fight go by without saying something. I spoke to him today, starin' at him with both my eyes almost swoll shut an' black and blue—an' he just looked through me like I wasn't there."

Cap waved aside his friend's puzzlement. He said, "He's thinkin' about the drive that's coming up—."

"The river-drive? What are you talkin' about! The Tittabawass' is solid with ice from bank to bank—and in places, covered with three feet of snow!"

Cap stuck in his oar before Jack could continue. "Don't forget what happened in January, and I've seen that old river clear itself of ice in February before." He added, "The boss don't intend to get caught without plans if a early spring thaw hits camp."

The cookee was right about the foreman making early plans for the drive. In fact, the afternoon when Cap and Jack Driscoll limped into camp with their booty, McTaggart was hard at it, working out details for Red Keg's part in the up-coming log drive; and he didn't hear the to-do over the arrival of the camp heroes.

But in spite of his making plans for the drive, the foreman didn't let it interfere with his work in the woods. He never let up. Throughout the month of February he took every advantage of the snow and cold, driving the men and animals as hard as he dared.

March came in more lion-like than lamb—with the temperatures still well below the freezing mark on the camp thermometer. On the first day of the month, which happened to be on a Sunday, McTaggart and Hoops went on an inspection tour along the river. They saw almost a mile-long stretch of thirty-foot-high piles of logs stacked on the banks.

Still, the foreman was not satisfied. George Pepper had been down to the base camp and brought back the word that Camp 2 had 300,000 feet of logs over the quota on their rollways.

McTaggart confided to Isaac, "If Pepper has his information straight, we can't leave a log in the woods. I'll have to get out every stick of pine the choppers have cut to beat the camps down river. And I need at least twelve to fourteen more days to move them."

That night the foreman heard again the ominous sounds of thunder, and during the days that followed he waited on edge for the worst to happen. And it did.

On the seventh of March the mercury in the thermometer stretched all the way up to the forty degree mark. Everything in the woods started to run. But that night, when McTaggart checked the thermometer, it had fallen to freezing again. Then, on the ninth, without warning, it started to sprinkle. And before nightfall the shower turned into a driving rain. All night long it lashed the huddle of camp buildings, coming down so hard it put out the fire in the bunk-camp when it ran down the big chimney like a waterfall. The foreman's fears had become a reality.

The next morning, in spite of the downpour, the men went off to work—and when it was daylight, they could see how fast the snow was receding from openings and along the roads and trails. The thick ice in the sleigh tracks was holding up but it was melting fast. It would have disappeared altogether if the rain hadn't ceased. At noon, McTaggart stopped the sleighs in hopes that the temperature would drop later on. It did. By six o'clock in the evening, it went down to freezing again; and the foreman ordered the men and horses to their usual places. All night long they worked by the light of flaring torches. The road monkeys made trips into the denser parts of the woods (where the snow was more protected from the elements) and carried the snow on small hand sleighs to the roads where they mended the bare spots as best they could.

And so man and beast struggled. On the tenth, eleventh and twelfth of March the sleighs were able to move only at night when the temperature dropped. Most of the crew slept by day and worked by night. By Friday, which was the thirteenth, it took all the persuasion and tact the foreman could summon to get the men out to their places in the woods and on the riverbanks. They were exhausted to the point of mutiny. Being a superstitious lot, they worked slowly on Friday, taking more care than usual, afraid of any mishap that could send them to town in the back of the tote wagon. The end of their winter's labors was in sight, and they didn't intend to spend the summer laid up in some back room in a mill town.

That night the crew refused to work and McTaggart didn't insist. He knew that one more day's labor would clean out the skidways.

Saturday, the fourteenth, arrived cold and clear—and the men went at the cleanup with uplifted spirits. At three o'clock in the afternoon, Frenchy LaRiviere, with feet braced on the empty bunk of his sleigh, waved his hand at the loaders as he turned the horses into the go-back road for the last time. The other teamsters had finished at noon.

"See you tonight, mes amis! We'll 'ave a beeg partee, no?"

"You bet, Frenchie!" returned one of the loaders, and they all laughed when Whispering Bill called out to the Frenchman, "Can I have the first dance in honor of this bein' yer final trip for the season?"

The teamster called over a shoulder, "But of course—."

By seven o'clock that night the last saw log was peaked on top of the last pile at the banking grounds—and it was a jubilant bunch of woodsmen who tramped back to the camp. This was one Saturday night they would howl! Fatigue and aching muscles were already forgotten.

After supper McTaggart made a brief appearance in the bunk-camp. The men who hadn't signed up for the drive would be off down the tote road after breakfast the next day, so the foreman chose this time to say goodbye to the crew.

When McTaggart stepped into the blue haze of the room, he was greeted by an excited rumble of voices—voices of men suddenly released from hard labor. McTaggart could hear the sounds of Jim tuning up his fiddle over the noise Cully Wills was making blowing warm-up chords on his

The Saturday night whing-ding was about to begin.

mouth organ. The Saturday night whing-ding was about to begin.

The crew had been expecting the foreman to show up and when he braced one foot on the deacon's bench, the noise in the room quickly subsided. The men waited for him to speak his piece.

McTaggart got right to the point. He announced, "Just before I came over here, the scaler gave me the grand total of pine we cut and banked this season. Including the logs we retrieved from the thieves—counting them—we'll have over four million, six hundred thousand feet to run through the sorting gaps."

There was an instant sound of cheering and clapping heard around the room—but, characteristically, the foreman held up his hand to hold down the noise. He told them he didn't want to give them the idea that they had won the contest. They would have to wait until the figures from all the camps were released. If Camp 3 won the prize, there would be plenty of time to celebrate. He continued speaking.

"Mr Hoops is taking our books down to the mill on Monday. We got the word this week that the final figures and the winners of the bonus will be posted at the mill office on the 24th of April. All those who can meet me there on that day—which is on a Friday—be there about seven in the evening. That'll give the men who already got jobs in town time to get there. If any of you are sailing on the Lakes or can't be in Saginaw on the 24th, tell me where I can send you the results of the contest before you leave camp."

There was a long pause. The foreman looked down at the floor, seeming to study the toes of his boots. Then he looked up again to see the familiar faces he had come to know during the past months—some of whom he would never see again. He cleared his throat. "Well," he said, "I guess that's about all. If you want a bunk next winter, look me up at Shannon's mill in September. It's been a mighty fine year."

When the foreman turned to go, there was hardly a sound in the room. He was almost at the door when he heard a gruff voice somewhere in the shadows say, "Good luck Boss, bringin' down the rears." The voice seemed to speak for all the men in the room. McTaggart nodded his head—then ducked out before anyone had time to say more.

Later that night while the boys were sashaying in the bunk-camp and blowing off steam generally, the foreman stood on the river bank between two great piles of pine logs looking out over the surface of the Tittabawassee. The night was so still he could hear Jimmy MacDonald's fiddle bow scraping on cat gut, and the stomping of a dozen pairs of boots on pine boards keeping time to the teamster's jig. By the light of the moon, McTaggart could see dark spots staining the surface of the rubbery ice. He sighed. Another logging season had come and gone—and he couldn't help feeling a little sad at the passing.

BOOK TWO

LOG DRIVE

The Red Keg crew yields to drive fever when the river-boss O'Halloran shows up. The ice in the old Tittabawass' goes out; the camp breaks up and the drive begins.

Pike pole, peavey and "corked" riverman's boots.

18

R is for the RIVERMEN who send the logs down

Micah John O'Halloran was the one who would superintend the big drive for John Shannon's camps on the Tittabawassee. Micah John was the name his mother had given him—but on the log-driving river, where he was known by all, the good-natured riverman was simply called "The Mick." Muscular, nimble of foot and agile as a cat—the river-boss was alert in mind as well as body. Like the rest of him, his hair resisted confinement—obstinately poking up through holes in his old felt hat—an item that was as much a part of the riverman as was his dog Jupe.

The Mick bragged that he had been born with a pike pole in his hands somewhere up on the Penobscot in Maine. No one doubted it for a minute when they saw him riding the back of a saw-log—balanced superbly on the slippery, half-submerged conveyance—with three-eighth inch steel corks in the soles and heels of his driving boots set solidly in the thick bark of the pine, "himself" waving peavey or pike pole and roaring out instructions and encouragement to his riverhogs.

O'Halloran, like McTaggart, was familiar with every crook and bend in the old Tittabawassee. He knew her in drought and in flood, having tramped in all seasons over most of the eighty-mile long pine belt along the main stem of the log-driving river. Beginning at the river's headwaters at Lake George in Ogemaw County, he had followed her gently winding course through Gladwin, Midland and Saginaw counties. And he had been over the great pine country through which the upper river's tributaries flowed—namely, Roscommon, Clare and Osceola—knowing well how the Tittabawassee converged with two other well-known log-driving streams, the Cass and Shiawassee rivers, before easing into the Saginaw.

To an old hand like O'Halloran, who had run logs down the Penobscot and down the St. John to the Oldtown booms in Maine—the Tittabawassee might have seemed tame in comparison to the wild white water, the water falls and craggy channels that he knew in the East. But to the river-boss, a sixteen-foot saw-log was as ornery and unpredictable when floating four and one-half miles an hour down a placid stream as when rushing through rapid water or tumbling over water falls. When the Tittabawassee River was backed up fifteen miles or more, packed solid from bank to bank with a jam of pine grinding and groaning under the pressure of tons of water held back by thousands of logs—those were times when he and his crew "walked down the middle of the river" with caution, looking for key logs to relieve the jam and get the pine moving again. In the East, neither the foreman McTaggart or the riverman O'Halloran had ever seen such quantities of logs in motion as they had witnessed on the Tittabawassee in Michigan.

One evening, back in February, McTaggart was sitting at his desk working on the camp ledgers when the tote teamster, George Pepper, stuck his head in the doorway of the van and asked if he could come in. Across the room, the scaler, Isaac Hoops, sat at a small table totting up columns of figures.

The foreman watched the big man come into the room and shut the door behind him. He could feel it in his bones that George had come with word from John Shannon's river-boss—the teamster having returned that afternoon with supplies from the base camp where he heard all the news.

The Mick bragged that he had been born with a pike pole in his hand somewhere up on the Penobscot, and no one doubted it for a minute when they saw him riding the back of a saw-log.

Pepper was dressed in heavy outer garments, and he looked uncomfortable in the closeness of the little room. He loosened the top button of his great coat. "Got back kinda late—," he said.

Before the teamster could continue, McTaggart interrupted him with a question. "How does the river look down around Midland?" In spite of his spring fever itch, the foreman still hoped for more weeks of cold weather to clean out the skidways. The condition of the river down stream interested him almost as much as word from the river-boss.

Pepper shifted his weight and freed another button. He said, "Still froze solid, Boss. Come all the way up on the river with the supplies. Didn't see no holes."

The tote teamster saw McTaggart's face ease into a smile. Encouraged, he continued speaking. "I saw The Mick down to Midland when I stopped off to give the team a rest. He said to tell you he'd be up next Monday."

The boss of Camp 3 peered into the shadows to look at the big calendar hanging on the wall. Monday was on the 24th of the month. He turned back to the teamster. "That all?" he asked.

Pepper nodded. He was perspiring freely by this time, and when the foreman thanked him for bringing the message about the river-boss, he quickly made his escape to the outside.

As soon as the tote teamster went out the door, Hoops put down his pen and picked up some sheets of paper on which he had been jotting figures. He said, "I had a feeling O'Halloran would be up soon, so I checked out the skidways again—."

McTaggart looked up from his books.

"According to my figures," the scaler continued, "if we can haul to the river all the logs we have waiting in the woods, we'll have well over our four million banked by the time the drive begins." Hoops laid his papers down.

The foreman glanced again at the calendar, thinking to himself he was lately making a habit of looking at the monthly time piece. Aloud, he asked, "How many feet do you reckon we need to move out in order to dump in the river 600,000 feet *over* what our contract calls for?"

Hoops replied, "About a million feet."

The foreman drummed his fingers on the desk top and then got up to consult the calendar at close range. He studied the page for February—flipping it to March before saying, "According to that, we'll need at least twenty more days to get the logs out of the woods. That'll bring us to the 14th of March." He turned back to his desk, saying under his breath, "The weather's *got* to hold out 'till then!"

Hoops saw McTaggart's worried look. Thinking a change in subject might lighten the foreman's mood, he asked, "How many of the men plan to stay on for the drive?"

"Twenty-two have signed up so far," answered McTaggart. "That's not counting Kelly, Cap Wilkes and Pepper who'll do the cooking and the fetching for the sacking crew."

The mention of the log drive brought back the spring fever itch, and the foreman got up again—this time to exchange his wooden stool for a more comfortable seat in the swing rocker. As he settled into the chair, he said, "I haven't heard you mention what your plans are, Isaac, after we break camp. (Hoops had not signed up for the drive).

"We've been so busy around here I forgot to tell you—the last time Shannon was up he asked me if I would manage the new company store he's putting up near his mill. There'll be a house built behind the place where I can live." The scaler paused, and then said with a smile, "Finally, I think I'm going to settle down."

McTaggart was surprised to hear Isaac's news, but he was happy to hear that the scaler had found a place to put down some roots—and said as much.

"I'm glad about the store," he said, "but I'm sorry you won't be back next season." The foreman's regret was genuine. Hoops had proved to be a good scaler.

McTaggart went on to say, a little ruefully, "Things will be different next fall. The F & P M tracks are already laid to Midland[1] and as soon as the weather clears, they'll be running right through here and deeper into the pine country." He sighed. "I can see what's coming. Year 'round lumbering. Seems like they can't take it out fast enough with just horses and oxen. Next thing you know, logs will be delivered to the mills by rail, straight from the woods."[2]

Hoops interrupted his boss's gloomy predictions. "I don't think the river drives will ever go out—not around here anyway."[3] He saw McTaggart's face brighten a little.

"I hope not," said the foreman. "Seems like the big drives are a part of the whole picture—leastwise they are to me."

The Mick blew into camp on Monday evening just as the men were coming out of the woods. He rode in on horseback—his dog Jupe following close at the heels of his big liver-colored mare. And while the popular riverman was being greeted by the crew (all the men talking at once, clapping him on the back and shaking his fist), the dog stood off a ways—aloof. No one petted him, for everyone there understood that he was a one-man dog—O'Halloran's dog.

The story about how the two met on the river was one that had been told and retold in countless saloons and around camboose fires in lumber camps all over the region.

How The Mick, one summer day, was standing on the shore watching logs float past on their way down stream, when he heard the familiar cry, "Jam below!"

Instantly, O'Halloran yelled to the men operating the fly-boom (floating in the midst of the logs) to pull the long timbers across the river to halt the pine that was advancing down stream. He had only started off to investigate the seat of the trouble when he heard another shout, "Great Jupiter! Mick! Look at that crazy dog!"

The river-boss looked where the man was pointing. He couldn't believe his eyes. Riding on the back of a saw-log (not held back by the boom), was a lean, black dog with long legs, balancing there with all the casual indifference of a seasoned riverhog. The feathery tip of his tail waved in the light breeze like a banner, and the dog, seemingly untroubled by his predicament, paid no attention to the men who shouted at him from the shore.

They say that the river-boss hesitated for only an instant—before leaping from where he stood on the shore—to hop from one pine log to another to save the beast from being mangled in the pile-up below. O'Halloran reached the animal but as luck (or fate) would have it—when he made a grab for him, both man and dog were thrown into the river by the shock of logs colliding with others farther down stream. The Mick disappeared from view with the animal still clutched in his arms, and the rivermen who witnessed the scene remembered watching in horror when they saw the logs close over the spiky hair sticking out of the old felt hat. It seemed forever before they heard someone shout, "There they are! Yonder!"

With relief the men saw O'Halloran's head (minus the hat) bob to the surface of the water—the riverman having found a space between the logs some distance from where he went down. The dog was still in his arms. Without hesitation, The Mick heaved the dripping animal onto one of the sticks of timber where the dog crouched, sticking as tight to the bark as the burrs in his tail. And when the boss-riverman reached out again and again to grab the log so he could save himself, it was the dog who caught hold of the man's suspenders with his teeth—hanging on until at last O'Halloran struggled up and out of the water just as the river of logs ground together to become one solid jam.

The tellers of the story say the animal never once lost his composure throughout the episode, and when his rescuer started for the shore, the dog calmly shook himself, and then trotted over the logs after O'Halloran midst the shouts of the men on the river bank.

The Mick was adopted by the dog, who was given the handle "Great Jupitor" ("Jupe" for short) by the riverman who called out the oath when he first saw him on the logs. And from that day on, whether he liked it or not, wherever the boss-riverman went, the dog Jupe was not far behind.

After supper O'Halloran accompanied McTaggart and the scaler back to the van. Hoops promptly went to his table to work on his books, and The Mick settled into the rocker—the dog's muzzle resting on his master's knee.

Satisfied that the fire was roaring up the stove pipe, McTaggart pulled up his stool and sat down, anxious to hear about the plans for the up-coming river drive.

"We come most of the way up on the ice," said The Mick. "Started at the booms at Green Point.[4] Stopped off at Midland and at camps along the way. Jogged along, takin' our time lookin' for soggy spots and holes in the ice—." The riverman paused when he felt his dog's jaw slide off his knee and waited until he saw the animal comfortable, stretched out full length at the side of the rocking chair.

The Mick resumed his speech—the fingers of one horny hand gently pulling at the silky ears of his canine friend. "The river is looking a little soft down around Saginaw," he said, "but you sure are locked in up here! Shannon's so dead set on getting all his logs into the river before anybody else, I wouldn't be surprised to see him come up the first of March with a load of explosives tied on that cutter of his and blow out the ice himself!"

The river-boss laughed at the picture he had drawn, and Jupe thumped his tail on the floor as if to join in on the riverman's good humor.

McTaggart took the opportunity to say, "Is there any activity down at the booms yet?"

"Plenty," answered O'Halloran. "Even though the Company extended their works a mile and a half farther up river last year, I saw them getting ready to drive in spiling for more shore booms. They built another boarding house at the State Road boom for their men at that point—which makes four boom houses for the hands all told. What with ice houses, barns and pin houses situated at other points along the river,[5] the whole works will be lookin' pretty impressive by the time the Company is in full operation again."

The "Company" the river-boss was referring to was the Tittabawassee Boom Company, the organization that sorted and rafted out millions upon millions of board feet of pine that was dumped into the Tittabawassee and her tributaries every season. The Company's sorting and rafting operations began at a place on the river called Bryant's Trip.[6] This was where the sorting process began—where logs from hundreds of lumber companies were brought to a standstill before advancing to be separated and sorted according to their marks or brands. The "trip" was a great floating barrier made of connecting timbers that stretched across the stream. The boom was designed to hold back the pine—its operators only "tripping" (opening) the boom to allow logs to pass in such quantities as could be managed by the sorters and raft-makers below.

From Bryant's, the company operations extended ten miles down river to a place called Green Point, where the Tittabawssee (flowing from the northwest) and the Shiawassee (flowing from the southwest) met—the two rivers merging with the Saginaw at the Point.

At the height of the log driving season, it was not uncommon to see the great trip-boom holding back millions of feet of pine—logs that covered the surface of the river some fifteen miles back to Midland and beyond—a congestion that waited to pass down to the sorting gaps.

Up to the year 1856, it was the custom for log owners to sort their own logs—until the increasing quantity of pine coming down the Tittabawassee made their task too confusing and complicated. It took a canny lumberman from Maine—one Charles Merrill of "Circle M" log mark fame—to bring order out of chaos.

In 1856, Merrill formed a company under Michigan's first Act to provide rules for forming boom companies. But before a decade went by, even Charles Merrill and Co. was having trouble handling the ever increasing volume of logs floating down the Tittabawassee. In 1864, after sorting and rafting out nearly two billion feet of pine, a group of lumbermen, who had vast interests in the Saginaw pine region, took over the job from Merrill, forming a larger boom company[8] under the new general statute passed by the Michigan State Legislature in 1864. The new law provided more specific rules for the formation of capital stock booming companies.

The Tittabawassee Boom Company, with its veteran seven-man Board of Directors at the helm, gradually enlarged and improved upon the old Merrill booms. They acquired more shore property by lease and purchase and constructed and maintained miles of new booms and piers. They erected boarding houses and built dams where needed and hired each season a sizable crew of rivermen to work as sorters, raftsmen, pin-whackers, tallymen and pile-drivers. It wasn't long before the Tittabawassee Boom Company was vying for honors with the gigantic Beef Slough Boom Company located at the mouth of the Chippewa in Wisconsin—each company claiming they were the largest of their kind in the world.[9]

O'Halloran went on with his discourse. "The superintendent told me the boom company ain't hiring on their crews until the end of March or the first of April—so we'll drive Shannon's pine all the way down to Bryant's."

The usual practice was for the Tittabawassee Boom Company to take charge of all the logs at the boom limits, which was for many years located at the village of Midland. From Midland, boom company rivermen, working in teams, shuttled the pine from one "beat" to another until the logs reached Bryant's trip-boom.[10]

The scaler had been keeping an ear tuned in on the conversation between McTaggart and O'Halloran. Their talk about the booming operations interested him, he being a newcomer to the Saginaw region. Hoops asked, "What does the company charge for boomage?"

The river-boss shifted his weight in the rocker to better direct his attention to the scaler. He said, "Last season, the rates was 87-1/2 cents per thousand for logs thirty feet or less in length. But the log owners did so much kickin' about the rates being so high, it's rumored down at Green Point that they will be lowered to 80 cents this year."[11]

McTaggart, who was more concerned about dates than he was about boomage rates, asked, "What's the date we're to go in?"

"The 16th of March—ice or no ice. That's it. From the way it looks down to Saginaw, the river up here is bound to be open in places—at least it'll be soft by that time. If the ice is still solid between camps on the 16th, I want you to blast it open from here down to Camp 2. They got their orders to do the same. I'm dang certain the ice will be out from Midland down to the booms. If it ain't, we'll blast all the way down."

The foreman got up to poke at the fire again. Over his shoulder he asked, "You think the company will have enough men working at the gaps to handle our logs?"

"By the time we get there, they'll have a full crew hired on," O'Halloran assured him. He waited for McTaggart to get settled on his stool again before saying, "You got blastin' powder here, I hope—."

"Yes—and a good man to handle it," said the foreman, thinking about Sam Bean.

"Good." The river-boss pulled on his pipe for a second or two before saying, "I guess you already know I want you to take charge of the sacking, Alex. Shannon wants all of his logs in the booms this spring. He don't want to wait until next fall for the Company to bring down the rears."

McTaggart replied he had already laid out plans in this direction. But he wanted to choose his own crew. Sacking was the most laborious chore on the river. Salvaging the "draggers"—the laggard logs left behind by the main crew of rivermen—was strenuous work. He wanted the pick of the rivermen—men with strength to heave, pull, pry, push and roll hung-up logs off the river banks, the sandbars and other channel obstructions—to get them moving so they could join their numbers below.

"I'd like to take about eighteen of the boys from here—the rest can go with you." He added, "I'll get Angus started on the wangun this week. Kelly and Cap can do the cooking and cleaning up for our crew, and Pepper'll steer the scow and fetch provisions if we run out."

"Good! Well—I guess we're all set." The Mick settled back in his chair. He puffed on his pipe in silence and then said casually, with a sidelong look at McTaggart, "By the way, the boys down at Camp 2 was telling me they got Shannon's prize all sewed up. The scaler down there let it out of the bag they'll have 300,000 over their four million to go into the river."

Hoops looked up from his work. He smiled at the foreman and went back to his papers without saying a word.

McTaggart made no comment either, but he seemed to have taken a sudden interest in brushing pipe ashes from the front of his shirt.

O'Halloran, still hoping to get a rise out of his friend, continued to bait the hook. "They tell me your old scaler Uriah Bell pulled off some shenanigans up here and you're goin' to have to leave about a million feet of pine on the banks."

"Is that what they're saying?" smiled the foreman.

The Mick was fishing—but McTaggart wasn't going to bite. He changed the subject before the boss-riverman could pursue it further. The number of logs scaled at Red Keg would remain a secret until either he or Hoops handed the books to John Shannon or to his mill superintendent in Saginaw City. But the mention of his former scaler brought to mind the big logs left behind by the log thieves—still stuck in the ice behind Devil's Island. The foreman couldn't help but smile when he thought about their "ace in the hole" all cozy and snug, just waiting to float down the river with the other logs. He'd get Hoops over to the island tomorrow to scale them. "Must be at least 80,000 feet of prime cork pine back in there," he thought with some satisfaction.

The good-natured O'Halloran left camp the next morning—his shadow following close behind. The river-boss wouldn't be back until the drive was underway.

• • •

During those last, frantic days in camp, the foreman raced against time and the weather to get all of the logs his choppers had cut out of the woods and onto the river banks before the thaws came. But in spite of the long hours he put in running the camp, he managed to sandwich in making out lists of equipage and provisions for the drive. When the first days of March approached, he sent George Pepper and his team down the rutted tote road to the base camp (not now trusting the ice on the river for traveling) to pick up the supplies that had come up from Saginaw on the new railroad. And the foreman gave orders to the blacksmith to begin working on the wangun.

Able Fry and his ox team dragged twenty-foot pine logs to the river bank where Angus McCloud dressed down the logs for the rafts that would provide the sacking crew a place to bunk while working on the river. In his "spare" time, McCloud made necessary repairs to pike poles and peaveys—the tools the rivermen would use on the log drive.

During those last days in camp, the woodsmen, who would be working on the river, made their own preparations for the upcoming drive. Corked boots were pulled out from under bunks, and leather was rubbed with liberal applications of tallow. The steel corks (calks) in the soles of their boots were inspected, and if any were worn down, new ones were purchased in the van (at nine cents a dozen) and inserted with a calk set and punch.

The scaler, Hoops, was given the task of making out the crew's time. It was a very ticklish job. Not all the men worked for the same pay—or were they in camp the same amount of time. Too, there were their van bills that had to be totted up and subtracted from wages earned.

This season, John Shannon hired his unskilled laborers at $16 a month. The teamsters, choppers and loaders, who were considered to be skilled workers, were employed at $22. The cook and blacksmith were paid $45 a month (only $35 less than the amount McTaggart and the scaler received).

Although the men were hired by the month, they were paid by the day at the lumberman's scale rate[12]—which took into account the number of hours they worked, based on a six-day week. For example, the swamper, Cully Wills, was hired at the $16 rate—but he was paid according to Shannon's pay scale for unskilled laborers—61 cents a day.

The woodsmen would have no trouble cashing in their time when they got to town. Any bank, saloon, store or other business, where John Shannon was known, would perform the service—at a price—sometimes as much as ten percent of a man's pay.

By Sunday, the 15th of March (the day before the first logs would go into the river), all but twenty of McTaggart's logging crew had picked up their time and departed for Saginaw or to one of the "beat" camps on the river to which they had been assigned by the river-boss O'Halloran. (The Mick had set up four camps—about four miles apart—to accommodate twenty or more men). That afternoon, the

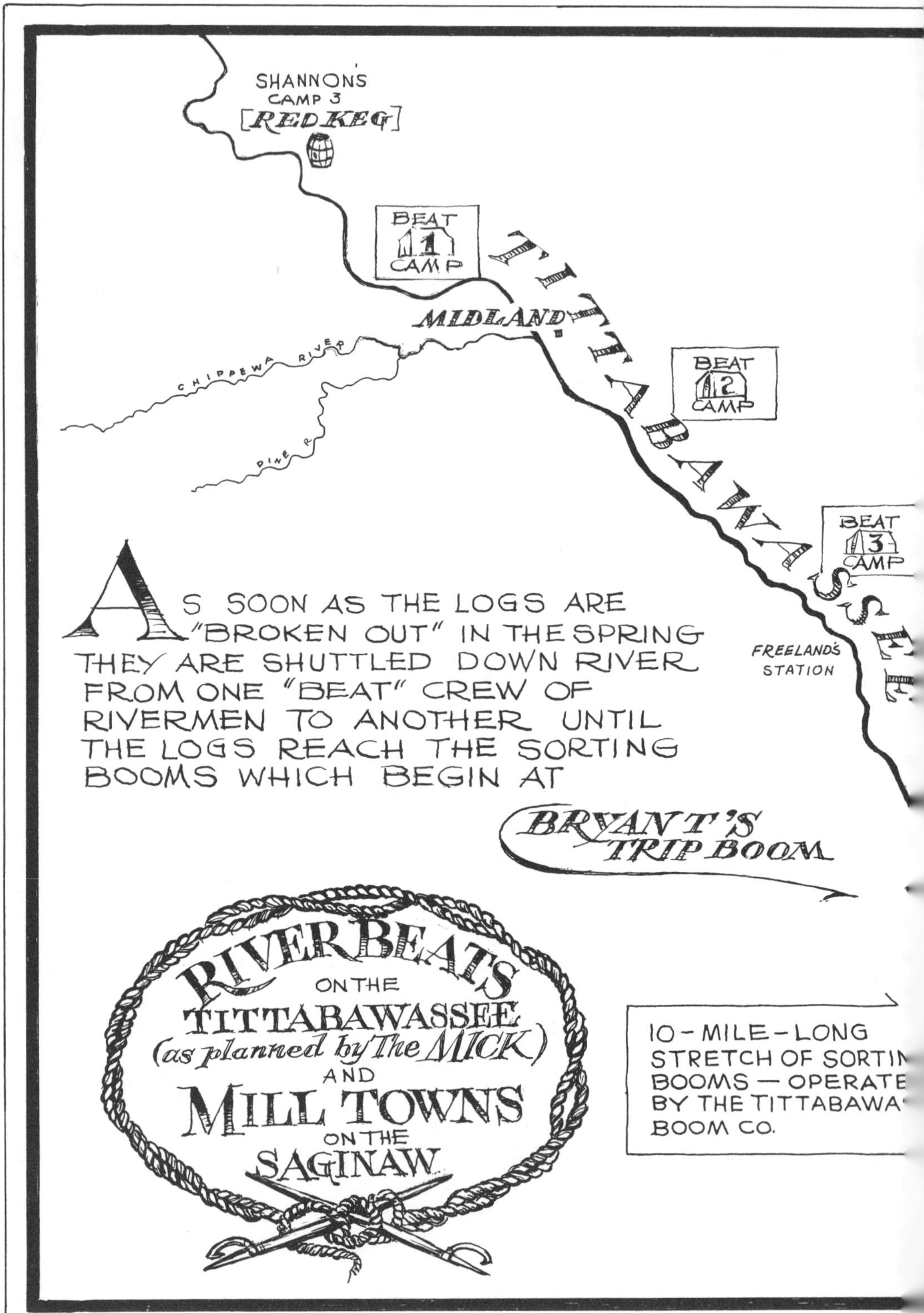
SHANNON'S
CAMP 3
[RED KEG]
BEAT 1 CAMP
TITTABAWASSEE
MIDLAND
CHIPPEWA RIVER
PINE R.
BEAT 2 CAMP
BEAT 3 CAMP
FREELAND'S
STATION
AS SOON AS THE LOGS ARE "BROKEN OUT" IN THE SPRING THEY ARE SHUTTLED DOWN RIVER FROM ONE "BEAT" CREW OF RIVERMEN TO ANOTHER UNTIL THE LOGS REACH THE SORTING BOOMS WHICH BEGIN AT
BRYANT'S
TRIP BOOM
RIVER BEATS
ON THE
TITTABAWASSEE
(as planned by The MICK)
AND
MILL TOWNS
ON THE
SAGINAW
10-MILE-LONG
STRETCH OF SORTIN
BOOMS — OPERATE
BY THE TITTABAWA
BOOM CO.

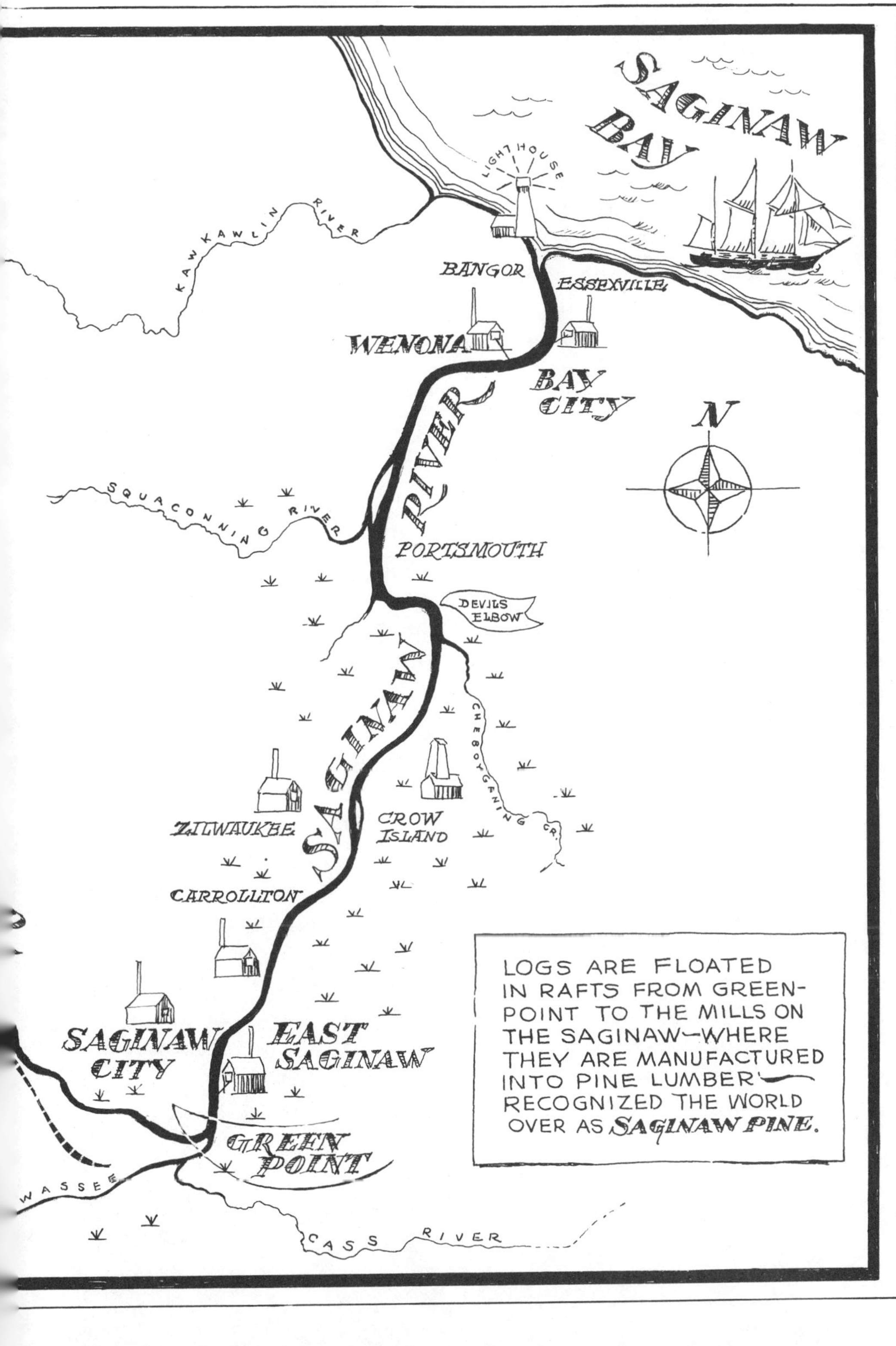
SAGINAW BAY
LIGHTHOUSE
KAWKAWLIN RIVER
BANGOR
ESSEXVILLE
WENONA
BAY CITY
N
SAGINAW RIVER
SQUACONNING RIVER
PORTSMOUTH
DEVILS ELBOW
CHEBOYGANING CR.
CROW ISLAND
ZILWAUKEE
CARROLLTON
SAGINAW CITY
EAST SAGINAW
GREEN POINT
WASSEE
CASS RIVER
LOGS ARE FLOATED IN RAFTS FROM GREEN-POINT TO THE MILLS ON THE SAGINAW—WHERE THEY ARE MANUFACTURED INTO PINE LUMBER—RECOGNIZED THE WORLD OVER AS SAGINAW PINE.

explosives expert, Sam Bean, and others set off several small charges of dynamite up and down the river (to help Mother Nature open up the stream), and when the men went to breakfast before dawn on Monday, they could hear the sounds of the river rushing by. Later on, when the crew assembled on the banks for the big splash, they could see the Tittabawassee running high and swift with jagged pieces of ice whirling amongst quantities of driftwood that floated by in the fast-moving current.

The early morning was cold—but as the sun rose, the day became almost spring-like. "A good beginning," thought McTaggart.

For the momentous event, Duff Kelly and Cap Wilkes had come down from the cook-camp to watch the first logs go into the water. They stood well out of the way of the rivermen. The scene they watched was lively–the air filled with the hoarse shouts of the foreman, who hollered instructions to the men as they hurried here and there getting ready to break out the first pile.

The bottom logs of each gigantic stack were blocked up by short logs that were placed in such a way to start the whole pyramid rolling into the river when they were chopped out. The pine that stubbornly clung to the bank would be attacked by the woodsmen, who would pry and push the logs into the stream with their cant hooks.

Chopping out the key logs was dangerous work and called for skilled choppers who were fleet of foot—for when the tiers of pine began to cascade down the slope, there was no stopping the momentum of the lethal pile. The unlucky chopper who got in the way of the falling logs could very well be ground into something unrecognizable.

Jimmy McHugh and his partner, Frank Cole, had the honors this morning. They were posted at the foot of the first heap selected to go into the water.

"We're ready, Jim!" shouted the foreman. "Whack away!"

McHugh and Cole began to chop. The chips flew until the whole mass of logs began to groan—and then settle. The choppers lost no time in making their escape.

"There she goes!" said Cap as the logs began to move. He watched the great pile gradually melt away before his eyes as the logs went rumbling, tumbling down the bank. He heard the thunder of them as they rolled and pitched, one over the other, until they splashed into the cold Tittabawassee.

The logs mushroomed out over the sparkling water and slowly turned to float with the current down stream. To keep them from getting hung up on shore, there were men standing at the water's edge with pike poles in hand ready to shove the mavericks away. To keep the pine straightened out, the two Canadians, Frenchy LaRiviere and Jack Driscoll, leaped onto the back of one log, and then another, and another—prodding the pine toward O'Halloran's first "beat" crew which was stationed downstream ready to receive the floating mass.

"That's far enough!" McTaggart shouted.

Hearing the command, the two rivermen leaped ashore and walked back to wait for more logs to come down the banks.

Six log piles went into the river that day—about 500,000 feet of pine—and downstream, Shannon's Camps 1 and 2 dumped in as many. O'Halloran's riverhogs were kept busy.

By nightfall, The Mick's crews were exhausted and wet to the skin. They were miserable with cold and hunger. Many of the rivermen had been working in ice water almost twelve hours. But before walking back to their respective camps, the men drew a fly-boom across the river and secured it on shore to stop the pine from wandering farther down stream in the night.

Bright fires and hot food greeted the rivermen when they finally limped into camp—but there were some who were too tired to do more than eat. They staggered into tents and, without even removing their wet boots and socks, they simply rolled up in blankets and fell asleep the instant their heads touched their turkeys.

Back at Red Keg, the men ate Kelly's hot supper in the fair comfort of the cook-camp, and in anticipation of the next day's labor went to bed early.

The next morning while Cap was rousting out McTaggart's crew, The Mick's rivermen were rolling out of their blankets, stiff with aching muscles and chilled to the bone with the damp cold of the March morning. At five o'clock they ate a hurried breakfast huddled around the camp fire and got in line to stuff their "nose bags" with hunks of cold salt pork, johnnycake and thick slices of bread wrapped around the loggers' staple, baked beans. They would eat from their haversacks while on the job, at ten o'clock and again at two—following the old four-meals-a-day Penobscot practice.[13] By sun-up, the men were hiking to their stations on the river, and in the cheerless wintery dawn were at work hustling Shannon's logs toward the sorting booms.

Not all the rivermen worked in the midst of the floating pine. Some worked on paths following the stream and did their best to keep the logs away from the shore. They walked miles each day up and down the banks—splashing in and out of the water, through slush and mud and over tangles of roots, struggling through piles of driftwood and over fallen trees. If logs became hung up on land or on some obstruction and required too much time and effort to work them free, they left the stragglers for McTaggart's sackers—the clean-up crew that would be coming along behind them.

It took a day or two for the rivermen to get into the swing of their work. The third night out, they could joke about the duckings they got—and now they took the time to hang up their wet clothing on lines of haywire that were strung up on a wooden frame built around the camp fire.

At O'Halloran's Beat Camp 2, teamster Jim MacDonald (now riverman MacDonald) pulled off one sodden boot and then another and pushed them both close to the flaring heat to dry.

"Lad, you're puttin' yer boots too close to the fire," commented Alex Gow, who was one of the older hands. "They'll be stiffer 'n my great-grandpa by morning if you leave 'em there."

Taking heed of the wiser head, MacDonald leaned over to pull his boots back a ways from the heat. "Don't know why I bothered greasin' them," he complained. "The water just goes over the tops and down the insides anyways."

The older man laughed and then said kindly, "Mac, you been greasin' the wrong thing. Do like I do—grease yer feet before you put on your socks. Cookee sets out a bucket of lard ever' morning fer the purpose. Just dunk yer feets in the bucket—sqoosh 'em around a little and then clap on yer socks right off. The lard'll keep your toes from shrinkin'."

And so it went that week—in sunshine and in pouring rain—in sleet and in the midst of a snow squall. The rivermen kept the logs moving with hands so numb with cold they could barely hold peavey or pike pole. And when the light failed and they could no longer see to work, the men went back to their camps, bone-weary and wet to the skin.

Upriver, John Shannon's logs kept thundering into the Tittabawassee.

By the following Saturday, at noon, every log cut by the choppers at McTaggart's Red Keg was in the water and floating down the river—including the cork pine logs that had lain behind the island since January.

After dinner the men helped the blacksmith launch, with block and tackle, the two big rafts that made up the wangun. When they were in the water, McCloud chained the rafts together and then tested the anchor (called a "grouser"), which was nothing more than a sharp-pointed stick of timber four inches in diameter, rigged at the stern of the cook-scow and made to drop into the river bottom. The point on the end of the grouser held fast the awkward-looking craft, keeping it from drifting in the current. At the stern of the affair there was a sweep, or a long rudder, which was used for steering.

The instant the scow was launched, Kelly and Cap Wilkes began stowing away provisions and stove wood in the cook shack (a rude log building which was built aft on the forward raft). They stacked pork barrels along the outside walls and shoved sacks of beans, corn meal and potatoes under the cook's work table and into every nook and cranny available. Kelly would perform his rites at the portable iron cook stove that occupied the middle of the shack—its smokestack sticking up through a metal-ringed hole in the canvas roof.

On the second raft, a large tent was erected and fastened down. In the center was a small box stove that rested on a shallow box of sand. When the tent was up, the men brought hay from the barn and spread it on the floor over which they would lay their blankets at night. When turkeys and blanket rolls were stowed in corners, and tools and other equipment along the canvas walls, the place didn't look much like home, but it was where the sacking crew would bed down until they finished their job.

Sunday was the last day the men would spend at Camp 3. The place looked almost deserted. All the horses and oxen had been sent down to Shannon's farm near Freeland's Station for a well-earned rest. Barn-boss, Billy Flynn, and ox driver, Able Fry, accompanied the animals to the farm where the men would work during the summer months. Less than twenty woodsmen remained on the premises—which made the bunk-camp seem quite roomy in comparison to what it was like in the wintertime. Even the cook-camp was bare. Except for the big stove, dry-sink and almost empty shelves, most everything in the way of cooking utensils and provisions had been moved to the wangun where the men ate at noon.

After dinner, as soon as the usual chores were out of the way, Cap and Jack Driscoll went off on a final Sunday afternoon tramp—taking off for Devil's Island in one of the three bateaux hauled up alongside the wangun.

The misty rain that had been falling since early morning had stopped, and the sun was attempting to come out from behind low-lying clouds that were scudding across the sky, hustled along by blustery March winds. No sooner would the cookee and top-loader feel the warmth of the sun's rays on their backs when the sun would disappear behind a cloud, and the men would feel again the chill of the wind as it whipped across the water.

When they reached the island, Jack jumped out of the sharp-ended boat and pulled it onto the low swampy ground. From the shore the two friends went into the woods. They walked in silence—each man feeling sad at the season's passing—knowing with the break up of camp, they might not see one another again. Jack was the first to speak.

"Kinda glad I broke my slats last December and wandered down to Saginaw lookin' for work. Hate to see the end of it here."

Cap, not trusting himself to speak, only nodded his head. The two men had become close companions during the past months. After a few minutes the cookee asked, "You going back to Canady when the drive is over?"

"Don't know yet. After our rears is brought down, I'll go to work for the boom company 'till the end of the season—then I'll make up my mind what I want to do." Driscoll turned to look at his friend. "You comin' back here to stay for the summer?"

Again, the cookee nodded. "I'll be keeping an eye on the place—what with the railroad gang coming this way—."

Once more, silence took precedence over conversation.

Spring, as yet, had not come to the Michigan woods. Even though everything had been soaked by torrents of rain, there was snow in the low places and at the base of some trees. It would be several weeks before the moist soil of the island would stir to produce life—and Cap Wilkes was glad he would be back in time to see again the marsh marigolds, yellow lady slippers, trillium and a hundred other rare plants that were still lying dormant beneath their feet. The day was cold and dark, but Cap's imagination let him smell the sweet fragrance of trailing arbutis, of sumac flowers and the spicy odor of bog orchids. In his mind he could hear again the shrill, clear music of the spring peepers[14], and he wondered if the beavers would try again to build a dam across the north end of the island (the timber thieves had pulled out the last one). A year ago in April he had seen a flock of whistling swans winging overhead. Would they stop again on the marshes near Saginaw Bay to rest before continuing their journey to the Canadian Arctic? And would he ever forget the day when he saw the sky go black and heard the thunder of a million wings as flocks of passenger pigeons flew over Red Keg looking for a nesting place.

The cookee had become lost in his reverie—but the sound of Jack's voice brought him back to the present. He saw his friend drop to his knees on the wet ground.

"Look Cap—look at this. Spring must be closer 'n we thought." As he spoke, he gently pushed back a handful of melting snow and ice to better see a speck of bronze beneath. "What's it called?"

The cookee smiled, and then said, "Skunk cabbage. Ain't a very good name for such a work of nature, is it? Probably got some scientific name that don't sound no better."[15]

Driscoll saw the old woodsman look with reverence at the lovely creation that was still ringed by the winter's snow under the shelter of a tamarack, and heard him say softly as he looked around at the trees and tangles of shrubs, "These is the things I hate to see spoiled by the lumberman. Why can't they take some and leave some?"

Jack stood up. He tried to sound sympathetic. "I guess we're all in the wrong business, Cap." He noticed the sky had darkened—the sun having disappeared altogether. Glad to change the subject, he said, "We better be getting back—looks like a storm is brewin'."

The next morning, in a downpour of rain, George Pepper pulled up the wangun's anchor, and the barge slowly swung away from the bank and into the current—the smoke, rising from the tin smokestack, trailing behind as the scow glided down river. Inside the cook-shack, Cap was cleaning up a last batch of iron skillets, baking tins and dishes left from breakfast—and Kelly was at the stove shoving pans of johnnycake into the oven. The clatter the cookee made washing and stacking up tinware was at times almost drowned out by the dreary sound of the sleety rain as it beat down overhead on the stout canvas.

McTaggart's sacking crew had started to work on the river an hour before. Some of the men were now well ahead of the scow—others were near by, hidden in marshy coves and inlets where the logs lay stranded in the mud and reeds. All were hard at work pulling timber back into the river.

Pepper stuck his head into the doorway of the cook-shack. He said, "The boss give me orders to go down stream about a half mile. I'll tie up there. If the weather clears, mebbe we'll go on farther."

Kelly looked up, but made no comment, and Cap only nodded his head. The tote teamster went back to his post at the sweep to watch for the first tricky bend in the river (there were five between Red Keg and Midland).

The sandy banks of the Tittabawassee were in places quite low, which meant that the river was flooding shore property—making the logs spread over the land. Within minutes after leaving the wangun, the sackers were soaked to the skin. Their hat brims, usually turned back at jaunty angles, now hung limp around their ears—the rain falling in rivulets from the soggy felt to their shoulders and down their backs. At times they could barely see the logs—the needle-like rain was coming down so hard. In spite of it, they pulled and pushed with pike pole and peavey—sometimes literally carrying the great logs back into the stream. The work was slow, strenuous and disheartening—but with McTaggart's steady voice encouraging them, and his strong arms helping them, the rivermen stayed on the job until almost dusk.

All day long, the wangun stayed behind the crew—George trying to keep within sight of the sackers and not run over them. Late in the afternoon, he dropped anchor in the middle of the stream, and before dark he climbed into one of the bateaux, lashed alongside, to round up the men and bring them back to the wangun for the night. As one boat would not accommodate all of the crew, the teamster towed another one, which LaRiviere would man.

Pepper was greeted almost in silence by the rivermen as he collected them one by one. It had stopped raining, and now the early evening was still. Only the sound of the boatmen's oars and the river rushing by broke the stillness as the bateaux made their way against the current. The chilled men were too tired to talk.

When the boats rounded a bend in the stream, the rivermen were heartened by the sight of the dark hulks of the scows resting quietly at anchor. The white canvas top of the cook-shack glowed softly in the dusk, and as the boats neared the rafts, the men saw Cap step out of the open doorway of the shack, his figure silhouetted by a rectangle of yellow lamp light behind him. The sackers watched him lift the gabriel to his lips and blow out a long measure of notes that echoed over the water. It was a sound that warmed their hearts as nothing else could have done at that moment.

At supper, the men—now dressed in dry clothing—sat around the cook-scow on an assortment of chests, bags, barrels and boxes, grateful for the heat radiating from the cook stove. They ate hungrily from heaped up plates that rested on their laps. They drank steaming tea from pannikins and set them down again at their feet. And as the food and drink began to warm their numbed bodies, they felt a new surge of life.

For some reason, the observance of the lumber camp no-talk rule at meal time didn't apply on the wangun; and over the sound of forks scraping on tin plates, there was a drone of conversation and occasionally a burst of laughter here and there. Sam Bean got up to fill his plate again from the food set out on Kelly's work table. And as he was heaping up his dish, he turned his head to look at Omer Fuller, who was his partner on the drive. He said with a laugh, "Omer, did you tell the boys about that settler who come at us with a loaded Springfield today?"

The sawyer-turned-riverhog frowned and said, "That ain't too funny, Sam. You sure got a queer sense of humor."

A chorus of protests followed, "Come on, Sam—don't pay no attention to Omer. Tell us what happened—."

"Well," said Sam, as he settled down on his make-shift seat (a sack of potatoes), "Omer an' me was chasing after a bunch of pine what was floating around some settler's pig pen. I was bendin' over, trying to pull out one of our logs that got stuck inside, when I felt something pushin' at my behind. I turns my head around an' there, standing in back of me, was this old codger wearin' a blue Union blouse 'n cap who stood about eight foot high. He was glowering at me with steely eyes and breathin' fire. He was shakin' so hard I thought the dang gun would go off any minute. Before I had time to open my yap he starts hollering, 'What the Sam Hill you people think yer doin' breaking up my property?' he yells. 'Who do you think you are anyways? I got logs all over my place. They broke up my duck pond and knocked off my front stoop. Next thing I know, they'll be movin' my cabin back into the woods! Git them logs off my property and keep 'em off!"

"That's when Omer comes up—an' seeing the situation, he says real nice, 'Calm down mister. We mostly got our logs back. They ain't hurt nothing—which is more 'n you'll be able to say when the main drives get movin' this way—.'

"Well—I tell you, when that old man heard that, he turned purple and when the Springfield went off—me 'n Omer splashed off his land leaving a wake behind us that'd do credit to Shannon's tug *Fannie S.*"

When the laughter subsided, somebody asked, "Did you get the logs off?"

"Thunder! No!" exclaimed Bean. He shoveled another forkfull of potatoes into his mouth before saying, "Don't know why he was so mad. He's got enough pine there to build him a nice barn—."

Another voice was heard. "What good would buildin' a barn do him. Next year more logs would just knock it down."

Again, laughter was heard in the room, but it broke off when Fuller spoke up. He was mad. "You wouldn't think it was so dang funny if you spent a couple of years pullin' stumps and buildin' up a place, and then have a million feet of pine wipe you out of house and home. I got a homestead up on the Tobacco, and I know how that poor cuss feels—."

McTaggart listened to the story with the rest of the crew and, at the end of the telling, he knew that tomorrow he'd be seeing one very mad settler (with a loaded Springfield) about some logs. When he got to Saginaw, he'd go to the boom company office and recommend to the agent that the company lease the man's land or put in some piers to keep the logs off his place—or they would be paying him damages for years to come.[16]

The next day, at dawn, McTaggart and his sackers were about to push off from the wangun when they heard a voice hail them from nearby. Peering into the early morning mists, they saw a ghost-like figure sitting at the oars of a sharp-ended boat. The bateau was almost surrounded by logs.

"Is Alex McTaggart aboard?" shouted the ghost. "I got a message for him from O'Halloran."

"Right here! I'm McTaggart. What's up?" The foreman leaned from the wangun trying to see who the speaker was.

The voice continued, "I come to tell you high water, ice 'n logs has carried away the bridge at Midland, causin' one hellava jam. The whole place down there is under water and covered solid with logs. Awful bad where the Chippeway enters the Tittabawass'. You can't go on much farther—. Logs is backed up within a mile of here and hung up everywheres!"

"You need help busting up the jam?"

"If you and Sam Bean can come on down—The Mick would sure appreciate it."

The foreman quickly passed on his command to Frenchy LaRiviere, and he and Sam followed O'Halloran's messenger in a boat of their own, rowing until the congestion of pine logs made it impossible to go any farther. The rivermen tramped overland the rest of the way.

It took two days of back-breaking labor to get the jam opened up. To make matters worse, the villagers at Midland treated the whole business like it was a picnic, coming from around the countryside in buggies and wagons to watch the rivermen. They got in the way of the men and their horse teams and oxen who struggled to pull apart the sides of the jam. They offered suggestions until The Mick was ready to commit mayhem or murder.

Dynamite was a remedy most often suggested from the sidelines—but the river-boss didn't want to use explosives unless all else failed. Too many logs would be wasted. So he kept the most experienced of his rivermen straining with their peavies, hunting for key logs that would release the giant pile. The men pried and pulled and cursed at the giant heap that loomed over their heads—scorning the danger that threatened them. The Mick was in the midst of it all. There were times when he thought the jam would give way—only to see the logs behind the pile rush in to fill up the holes the rivermen made.

"We're going to have to blow her out," said the sweating river-boss at noon the second day. We're wastin' too much time here."

Sam Bean, who had already looked over the situation, calmly prepared his charges of explosives. He carefully wound a stout cord around ten sticks of No. 1 dynamite, binding them to a 12-foot-long sapling. He fixed the fuse.

The picnickers and rivermen moved back out of the reach of flying missiles and watched Sam maneuver the long shaft, with its deadly charge, as far as it would go into the middle of the pile-up. He lighted the long fuse, and then stood there watching the little trail of smoke run along the wick. Sam counted to "three" before he beat it for cover. The instant he crouched behind a half submerged tree, he felt the earth move under him. He looked up in time to see logs, flood wood and bark shoot straight into the air in a geyser of white water and then drop with the echo of the blast back into the river. The mountain of pine logs settled into the water. At first, the brown carpet seemed fixed in place—and then Bean saw it move. He heard a cheer from the sidelines.

The Mick roared, "She's unplugged boys! She's runnin' out. Get in there! Keep that pine moving! Keep it moving! O'Halloran leaped onto the back of a log as the mass started down stream. The dog Jupe was behind him clinging to the stick of pine almost under the heels of the riverman. His barking seemed to underscore the urgent commands of his master.

As soon as the log jam was broken out, McTaggart and Sam Bean went back up river to join the sacking crew. On their way they saw the great numbers of logs backed up and those that had been left high and dry as a result of the spring floods and the jam; and the foreman knew his crew would be slowed down considerably in their work. He was right. The sackers did not reach Bryant's Trip until almost a month later. By the time they reached Freeland's Station, which was about five miles north of Bryant's, the rivermen were seeing logs from other camps floating with Shannon's.

Pepper dropped the grouser for the last time at a designated place (about a half mile above the trip-boom) where the barn-boss was instructed to meet them with the tote wagon. The day was Wednesday, the 22nd of April. On that day, as far as McTaggart was concerned, his responsibility for the boss-lumberman's logs was over and done with. From Bryant's on down to Shannon's big sawmill on the Saginaw River, the Tittabawassee Boom Company would be in charge. After eight months of hustle and worry, he felt free at last.

The drive paraphernalia was taken off the wangun and put into the wagon. Billy Flynn and Pepper loosened the ropes and chains that held the barges together—freeing the logs so they would go through the booms like all the others in the river.

The sacking crew spent their last night on the drive rolled up in blankets, sleeping under the stars—and the next morning, without waiting for breakfast, they struck out on foot for Saginaw City to collect their $1.50 a day riverman's pay[17] at Shannon's mill office.

The drive over, the foreman takes leave of his sacking crew to hitch a ride to the booms at Green Point. He goes on to Saginaw City where he tours John Shannon's big steam sawmill.

19

S is for the SORTING BOOMS and SAWMILLS of yesterday

McTaggart lingered behind his crewmen. He stood on the river bank near the mammoth trip-boom watching the sun rise—waiting for some sign of life in the boom keeper's house. It was the foreman's intention to stop long enough to say "howdy" to his old friend John Bryant before moving on to the mill town on the Saginaw.

It was so early in the morning the boom company's rivermen were not yet at their posts on the river. But now the sun was climbing above the horizon, its rays slanting over the green backdrop of forest and flooding Bryant's log house with sunshine. The early morning air was motionless, the quiet broken only by the chitter of birds and the sounds made by acres of pine logs nudging one another in the cold river below. The giant-size boom, which held the logs in check, gently rose and fell with the current of the sun-dappled water.

McTaggart felt at peace with the world(now that Shannon's logs were off his mind) and was lost in the tranquility

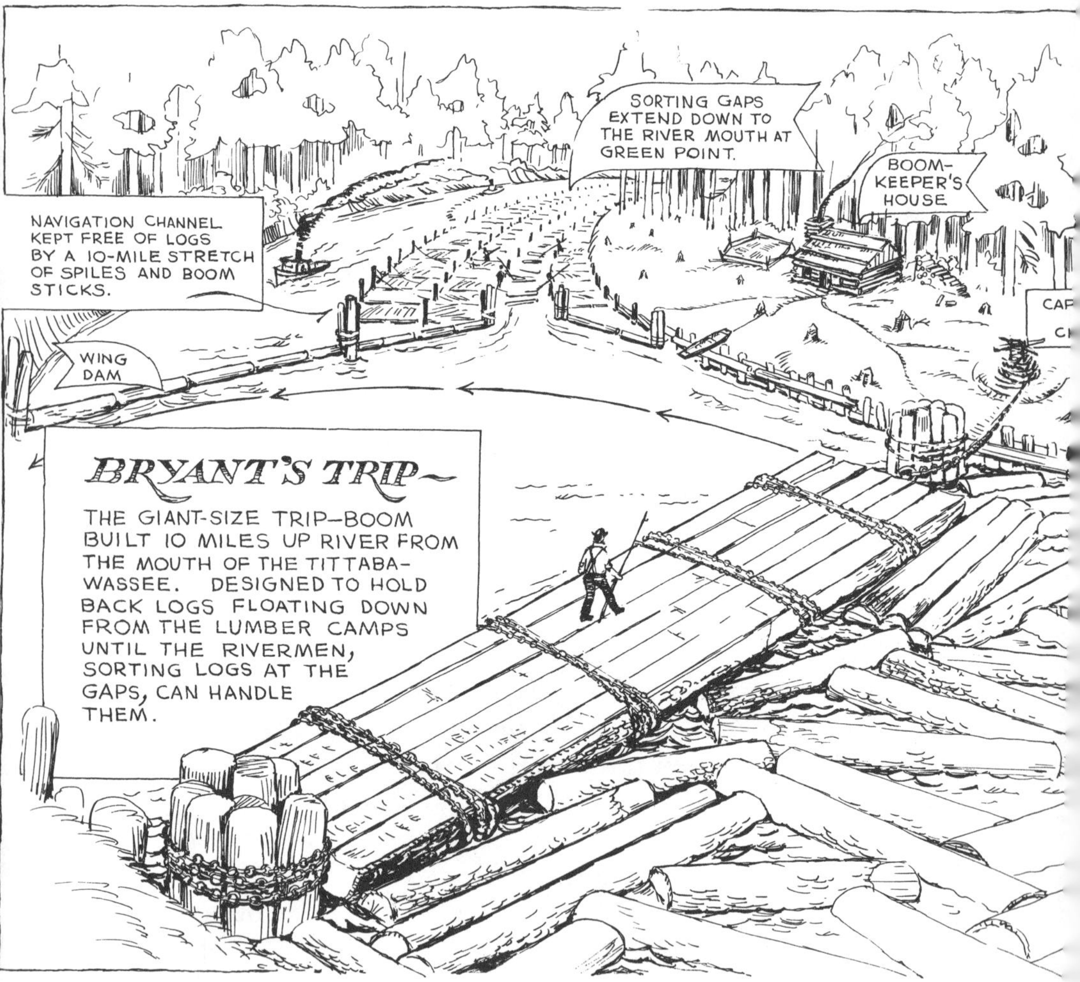

of the scene. Then his ear caught the sound he was waiting to hear—a door opening and closing. He looked toward the house in time to see John Bryant come out yawning, and snapping worn galluses over his broad shoulders.

The boom keeper looked neither to his right nor to his left—but walked to the north of his dwelling place, where the river ended a long curve and created a natural promontory upon which the house was built. The man's interest was on the immense boom that stretched two hundred feet across the river at this point. This was Bryant's first concern each new day. McTaggart watched as the man's eyes swept over the ten-foot wide floating barrier, looking for signs of weakness in the chains that held it in place.

The "trip" was made of pine logs (hewn flat on top for easy footing), bound together by heavy chains. It extended across the water, anchored securely between two clusters of piling. One end opened free. To open the boom, the chain at the free end was unwound from a large capstan. The current in the river did the rest, swinging the massive boom open—allowing the logs behind it to pass on to the sorting gaps that began just below Bryant's.

The place for the trip had been selected with care by the boom company, who built it below a bend in the river—the curve being a natural barrier that helped to hold the logs back. The boom was designed and built to withstand the great pressures created by the logs held in check—miles of pine logs coming down from the upper reaches of the Tittabawassee and her tributaries. All awaited their turn to be released in such quantities as could be handled by the rivermen working in the sorting gaps below.

"Hey there! John!"

The sound of the foreman's voice made the boom man turn with a start. When he saw who was calling his name, he waved a muscular arm that was still covered to the wrist with red flannel (the air would be nippy until well into May). "Where in thunder did you come from?" Bryant shouted.

The two men shook hands with warmth. The foreman said, "Camped over yonder last night. We pulled in the last of our logs and broke up the wangun yesterday—didn't get done 'till late. The crew left this morning before breakfast, they were so anxious to get to town. Billy, Cap 'n Pepper just went off back to camp with a wagon load of tools and stuff from the drive. Thought you might of heard the crew leave—they made enough racket!"

"Didn't hear a thing, Alex. Something must be wrong with my ears." The boom man laughed, and then he said, "I'll wager you ain't had a thing to eat since sundown yesterday. Come on up to the house and have something with us—Dolly is just feedin' the boys—they must be about through eatin' by now."

The foreman hesitated. Dolly Bryant had a reputation for being a good cook. He finally said, "Its mighty nice of you to ask me, John—but I'm anxious to get to Saginaw. I only stopped by to say hello."

But Bryant wasn't to be put off. He said, more persuasively this time, "Come on in, Alex, I ain't had nobody to talk to outside the family in a month of Sundays. The boys can be letting some logs pass down whilst you and me chin a little."

Before the foreman could protest again, the boom man added, "I'm expecting the *Beaver* up this mornin'. O'Neil is bringing us some supplies. You can ride back to Green Point with him. Save you some time and shoe leather." He turned to the house and shouted, "You boys shake a leg in there—."

A sound of protest was heard inside the house, but a

half minute later two lanky boys, about fourteen and sixteen years of age, came out the door wiping sleeves across their still moving jaws. They waved greetings to McTaggart and then went directly to the free end of the trip-boom and began to loosen the big chain that held it in place. And while Bryant went into the log house to tell his wife to set another plate on the table, the foreman watched the boys walk round and round on the beaten path that encircled the capstan, each one pushing on the long beam handles of the rude mechanical device until he saw the heavy chain loosen and sink below the surface of the water. The boom swung lazily out into the stream.

The logs moved slowly from behind the "trip" toward another boom called a wing dam, (also called a sheer dam or side boom). The boom was made of logs, but it didn't cross the river like the big trip-boom did, but slanted diagonally—jutting out from the opposite shore. It was placed there to guide the logs to the right side of the stream. From this point on, another floating boom (made up of "boom-sticks," logs chained end to end) divided the Tittabawassee all the way to the mouth of the river, thus keeping the channel on the left side of the stream free of logs so tugs and other small craft could pass up or down and not run into the congestion of pine—or into the rivermen sorting logs and making rafts in the right channel.

"Grub's on the table, Alex. Come and get it!"

After living for months in the makeshift confines of the lumber camp, it seemed strange to be in a room with rugs on the floor and curtains at the windows. But most of all, it seemed odd to be in the presence of a woman. At first, McTaggart felt uncomfortable—but the graciousness of Dolly Bryant soon put him at ease. After eating his fill of a woodsman's breakfast, he and the boomkeeper were about to settle back for their "chin," when three sharp blasts of a tugboat whistle brought them to their feet.

"Dang it!" exploded Bryant. "Wouldn't you know O'Neil would be on time this week."

The men went outside and watched the tug *Beaver* begin her turn around below the wing dam. Water boiled around her stern, making the booms and logs in the river rise and fall with the undulating water. When the tug was squared away, a chunky figure, wearing a rusty looking black reefer and an old cap, came out of the small pilot house. It was the captain. He shouted to make himself heard over the noise of the idling steam engine, "Mornin' John. I got your groceries and other stuff for you. Are the boys around to help lighter it off—or do ye want McGinty to help?" (McGinty was the tug's fireman-engineer who doubled as cook).

Bryant's sons had been securing the chain on the trip-boom when they saw the *Beaver* come around the bend, and now they came on the run. And while the boys were poling a skiff to the side of the tug, Bryant hollered across to her master, "We got a passenger for you, O'Neil. Alex McTaggart, foreman, Shannon's Camp 3—Red Keg. He's on his way down to Saginaw."

McTaggart waved to the captain, then cupped his hands to his mouth to shout, "Hope you don't mind some extra freight."

"Come aboard—we'll be shovin' off directly."

McTaggart went back into the log house to pick up his turkey and to thank Bryant's wife for his breakfast. When he came out again, the supplies for the boom keeper were already piled on the bank. One of the boys was waiting to pole him to the tug.

The *Beaver* was only one of several river tugs the Tittabawassee Boom Company operated along the miles of waterway that made up the booming grounds. The sturdy little vessels spent their days, during the log driving season, maneuvering boom-sticks, carrying supplies to the boom houses and towing rafts of logs down to the holding booms which were located for the most part at the river mouth.[1]

As the tug pulled away, McTaggart hollered goodby to Bryant and the boys and to Dolly Bryant, who was standing in the early morning sunshine waving her apron at him from the open doorway of the house. The tug whistle sounded, and the *Beaver* went off down the channel leaving a trail of black smoke and sparks behind. The foreman walked forward to stand at the starboard rail where he could observe the activity on the river.

It was only the beginning of the log driving season, but there was evidence that other lumber camps had followed John Shannon's example in breaking out their rollways early. As the tug slowly plowed through the water, the foreman saw a confusion of log marks on logs that were being shoved along by the rivermen. It was obvious, too, that a good-sized crew of rivermen was already on the job—judging by the number of men he saw working and by the amount of washing he saw draped on lines at the occasional riverman's "soup shanty" or boom house the tug passed along the way. The foreman saw menials carrying water from boarding house wells and others stacking wood for cooking stoves. Now and then a drudge came out to toss the contents of a slop bucket into the river, stopping to wave to the *Beaver* as she chugged by. When the tug passed one boom house, McTaggart sympathized with a red-faced woman with sleeves rolled above her elbows, who stood in front of a sagging line of wet wash, shaking her fist at the black smudge of soot that trailed from the tug's smokestack.

In McTaggart's opinion, the work going on at the booming grounds[2] was almost as interesting to watch as the work in the lumber camp. And as the tug slowly progressed down stream, he couldn't help but think how the confusion of cordage, boom sticks, piers and piling would look if one came upon the scene for the first time. The clamor of men shouting and bawling to one another, and the racket made by the pin-whackers, pounding rafting pins into the logs, echoed over the water. The sounds increased in volume as the tug approached a group of rivermen at work—and

drifted off as the vessel passed them by—to swell again as the *Beaver* came upon another crew working at the same occupation in another location.

The tug kept to the left of the channel. The right side was divided into large enclosures called sorting gaps, pocket booms or holding booms—corrals made up of logs chained end to end and hewn flat on top (like the trip-boom) for easy footing.

Watching the rivermen move the logs from one place to another, McTaggart thought how easy they made their job

McTaggart watched the catty-footed rivermen work the logs into the different channels and enclosures with their long pike poles, and thought they looked for all the world like they were standing or walking on the surface of the water.

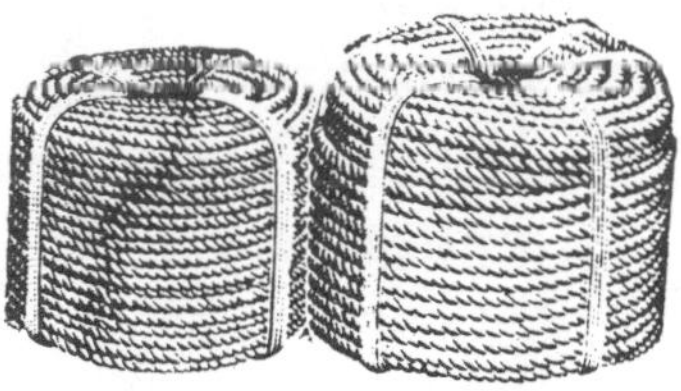

look. At first glance it would seem the men were walking on the surface of the river—the water washing over the boom sticks on which they were standing. The foreman, who in his time had shoved a lot of pine down the Tittabawassee, knew the rivermen were wearing corked boots which enabled them to walk with ease and confidence, balancing on the narrow floating catwalks as sure-footed as if they were walking on a five-foot-wide pine board sidewalk.

The logs, having started through the sorting process when they passed through Bryant's Trip, floated down a long watery lane which was wide enough to admit logs sixteen to twenty feet long. Rivermen, working with pike poles, were stationed on each side of the passage to straighten out the logs so they floated side by side, one log after another down the lane.

Sorting gaps were located at right angles off each side of the long channel. Within each gap there were four or five pens or pockets—partitioned spaces where the logs were assembled and made into rafts for their owners.

Rivermen, called "head gap sorters," were stationed at the entranceway of every gap. As the logs were poled down the channel, the sorters translated the marks stamped on the ends of the logs—calling out the owners' names as the pine floated past them. These men relied almost exclusively on their memory to identify the bewildering number of brands they saw—marks that told them at a glance to which outfit the pine belonged.

Other sorters and their helpers stood by, and when they heard the head man bawl out a particular company name, they reached out with their pike poles and drew the logs from the channel, pulling them into enclosures where there were other logs that had the same brand.

As the *Beaver* plowed down the river, McTaggart heard a

gap sorter sing out, "Sage, one! Rust, one! Shannon, three! And he saw helpers pull three logs that belonged to John Shannon from the channel and send them from one man to another until they were worked into a pen.

"The tail end of Shannon's pine coming down," thought the foreman.

There were other workers posted at the sorting gaps as well as the gap sorters. They were rivermen who watched the logs as keenly as the sorters did. These men were known as "scalers" and "checkers." The foreman could hear them adding to the din, calling out the names of log owners and the estimated number of board feet of lumber their logs contained. Tallymen were on hand as well, to jot down the numbers in their notebooks. Every day these figures were posted at the boom company office where the lumbermen, with only a glance at the sheets, could see how their logs were coming into the booms.

"I was wondering what would happen if O'Neil, after attendin' a three-day wake, would run the *Beaver* through the booms and into that mess of rope yonder—."

McTaggart turned to see the tug's fireman standing at his elbow. "McGinty," he thought to himself.

The man was wiping his hands on an oily rag and looking across the expanse of water at the maze of hemp that extended from large coils resting here and there along the catwalks. The lines were pulled from the center of the coils to the top of the spiling. From there it stretched down to the boom sticks that formed the sorting pens. The ropes moored the sticks in place—yet allowed the floating timbers to be moved out of the way when log rafts were towed from the gaps. In the sorting pens, other ropes extended from the coils over the centers of assembled logs that were being made into rafts. And there were ropes that held ropes out of the water—and ropes to keep ropes from becoming entangled with others. It was indeed a "mess of rope."

"You know anything about raft makin'? Looks kinda like what I seen in Maine. Ain't been here long—so I don't know much about it." McGinty seemed interested.

"Look over there—." McTaggart pointed off the bow toward a slightly-built man who seemed to be standing idly on a floating catwalk alongside a partially finished raft. A length of rope stretched down from the top of a spile and over the center of the assembled logs, (which made them look like a giant-size string of firecrackers). Several more logs had been poled into the enclosure, and they lay loose—side by side, under the line close to the ones that were already fastened together.

"That man there is called a 'pin-whacker,'" explained the foreman. "He's assembling pine for some lumberman. Looks like he's ready to add more logs to the raft he's started."

"There he goes—."

McTaggart and the fireman watched the man hop nimbly off the catwalk and onto the part of the raft already formed. They saw him reach into a keg that rested on the logs.

"What's he doin' now—?" asked McGinty, squinting in the sun to better see the pin-whacker.

The tug engineer's interest seemed to be genuine, so McTaggart launched on a subject he knew well.

"That's a rafting pin[3] he's pulled out of the keg," he explained. "The pin's made of hardwood—probably oak. It's about ten inches long and tapered to a sharp point at one end. The whacker pounds in one of those pins about half its length into the center of every log being made into the raft. Pine being soft, the pin goes in easy. That wooden maul he's using is called a pin mallet," he said as an aside.

The *Beaver* was leaving the whacker behind, but McTaggart and his companion still followed the man's movements with their eyes.

"After he whacks in the pin," continued the foreman, he'll take up the part of the rope that's laying loose across the logs there and make a double half hitch with it round the spike. Then he'll go on to the next log, whack in a pin, make a hitch with the rope, and go on from log to log until the raft is finished—ready to tow to the mill. Its a method brought to Michigan by Penobscot men."

The raft-maker passed out of sight, and, with his disappearance, McGinty's interest. The foreman found himself alone once more.

The tug was now approaching the river mouth, and McTaggart noticed the vessel was traveling at a slower pace. Ahead of them was a long string of rafts being towed to the holding booms. As they inched past the tug and her tow, the foreman saw the marks stamped on the ends of the logs. They were Shannon's. He thought to himself, "There'll be a company tug waiting for those logs—maybe I can hitch a ride down to the mill."

The river was wider now, and McTaggart could see they were coming to the Point where the Tittabawassee merged with the Shiawassee—where the two log-driving rivers flowed into the Saginaw. He looked across the shining waters to the broad expanse of lowlands beyond. The Indians named the place *Shaw-es-sko-skong*. The settlers called it Green Point. When they first saw the land, it was covered with tall blue joint grass, wild pea blossoms and morning glories.[5] This was where early man camped before the birth of Christ—where, centuries later, British traders came to dicker for Indian corn, taking it away in ships to their forts at Detroit and Michilimackinac.[6]. Now—in 1868—Green Point was a meeting place for log dealers and lumbermen—its waters stretching out over the flatlands, its shoreline hemmed in by piling and strings of rafts waiting for tugs to tow them to the sawmills at the Saginaws and the cities near the Bay.

Green Point was the *Beaver's* destination, and McTaggart could see its two-story landmark rising from the flats—a boarding house—a big wood frame affair with broad verandas wrapped around the first and second stories. A collection of other buildings had been put up near by. They included warehouses, a general store, barns, granary, ice house, pin houses and miscellaney of sheds and outbuildings. All were built of pine and had a temporary look about them. To the foreman who had spent the past months in the seclusion of the lumber camp, the place looked like a small city. He could see rivermen, boom men and log dealers; raftsmen, tug men and hands from the boom house—all going about their business, seemingly unconcerned by the racket made by a pile driver, steam engines and a variety of shrieking tug whistles.

McTaggart looked around for one of John Shannon's tugboats. And while the *Beaver* was maneuvering to the side of a well-stocked wood-dock, he caught sight of the lumberman's *Fannie S.*—her fresh green paint gleaming in the sunshine. To attract her captain's attention, the foreman waved his arms in the vessel's direction until at last, he saw the tug head toward them.

The *Fannie S.*, Andrew McNab, Captain, was a small, noisy, low-slung river tug with a high pressure engine that barked and snorted each time the steam in her cylinders discharged into the air. McTaggart watched in admiration the smoke unfold from her stack and puffs of steam shoot from her exhaust. When the tug was close enough to hail, he shouted, "Andy! I'd like a lift down to the mill. When you going back?"

Over the noise of the *Fannie's* pulsating engine, the foreman just managed to hear the tug captain shout, "Be shovin' off in about fifty minutes, Alex—might as well come aboard now—!" With that, he turned the wheel hard over to come in close to the *Beaver*. Without any more ceremony than a quick shake with O'Neil and a wave of his hand to McGinty, the foreman leaped over the side to the *Fannie S.* which picked up speed the minute he hit her deck. The tug was off to pick up her tow.

It was mid-morning before the *Fannie* turned north into the Saginaw River—her 400-foot tow of pine logs wallowing in her wake. McTaggart went forward to see the sights.

The Saginaw was alive with river traffic. Barges, passenger steamboats and small ferryboats were everywhere. Farther down stream, he saw lumber vessels already sidling up to docks for cargoes of pine lumber, laths, shingles, salt and other by-products of the sawmills.

Mill stacks, brick chimneys and salt well drill-towers rose from the banks on both sides of the river. Noticing an absense of smoke rising from some of the stacks, McTaggart could tell that not all of the manufactories were as yet in operation. But the noise of hammers pounding and the sound of sawing and the clank of machinery that drifted out to them from various points along the shore indicated they soon would be. Over the echoing sounds on the river, McTaggart could hear steam-powered pile drivers driving in piling for new docks; and the steady, monotonous thumping noise seemed to him like the heart beat of the Saginaw.

The *Fannie S.* and her log tow traveled down river at a tortoise-like pace and was overtaken time and again by faster-moving vessels. McTaggart was impressed by the apparent increase in traffic over that of the year before. He saw a side-wheel tug approaching, and he was captivated by her walking beam engine see-sawing up and down, and the rolling wash of her paddle wheels churning the water when she went by them.

McNab leaned out of the small pilot house to point to the vessel. In his fog horn voice he shouted, "That's Jesse Hoyt's *Magnet*. Built her in East Saginaw in '55 after a model he seen in Maine. She's a powerhouse! I've seen her towing a dozen schooners out to the Bay at one time—." The captain broke off to utter an oath and shake his fist at another tug—a powerful lake tug that crossed in front of his bow making the *Fannie S.* and her tow bob and tip in her wake. The tug was a long, low, rakish vessel that had brought in a tow of sailing vessels from the Bay and was on her way back to her port in Bay City. When her captain saluted the irate McNab, the sound of her screeching whistle made the foreman's scalp prickle.

There had been a fire at one of the mill docks, and the wind had spread the flames almost a city block down river. Now the gusty breeze was helping to clear away the smoke that still rose from the charred lumber piles. A fire tug was on duty, having steamed to the dock at the first sound of the mill whistle's distress signal. It was still there putting four one-inch streams of water on the conflagraton—the pressure from her hoses lifting the top boards on the lumber piles and scattering them into the river.

As the *Fannie S.* progressed down the Saginaw, McTaggart caught sight of another fire. A mill was burning offal or refuse in an old fashioned open burn called a "hell."[7] It was a hideous sight with black smoke rolling from flames that leaped high overhead. Again, the foreman heard the captain's voice booming in his ear.

"Ought to outlaw them hells. Shannon got rid of his. Built a new brick and iron burner this winter. The stack's forty foot high—a good landmark."

The *Fannie* started around a curve in the river, and McTaggart heard the fog horn again.

"There she is!" shouted McNab.

The foreman saw the new brick burner and watched it until it came into full view.

The tug pulled up opposite Shannon's mill, and when the captain let go with the tug whistle, McTaggart saw two bateaux row out from an opening in the long dock.They pulled up along side. Men, called "runners," climbed from the boats to hop onto the end rafts—and while the boom-boss shouted orders, McTaggart watched the men begin the task of loosening the giant-size ropes that fastened the rafts together.

"I'll be stayin' hard by until all the logs are inside," said the captain. "You might as well jump aboard one of the rafts if you're in a hurry—."

The foreman took the hint. He thanked McNab for the ride, tossed his turkey to one of the runners and followed it over the rail, only stopping long enough to pick up his dunnage before legging it down the long string of logs until he reached the end raft. The workers were already maneuvering it into the storage booms. There were many loose logs floating inside the great mill pond, and McTaggart used them as "stepping stones," leaping from one stick of pine to another until he reached dry land.

It was the foreman's intention to stop at the mill to pay his respects to Shannon's mill superintendent before going off to attend to his own affairs. That is the direction he took.

• • •

John Shannon's sawmill was not yet running at full capacity, but the din of machinery and the whining screech of saws made McTaggart's ears ring as he walked past the big wooden building that was the mill. He dodged slab carts, yard men and other laborers at work, tramping over the short distance up to Water Street where the office was located.

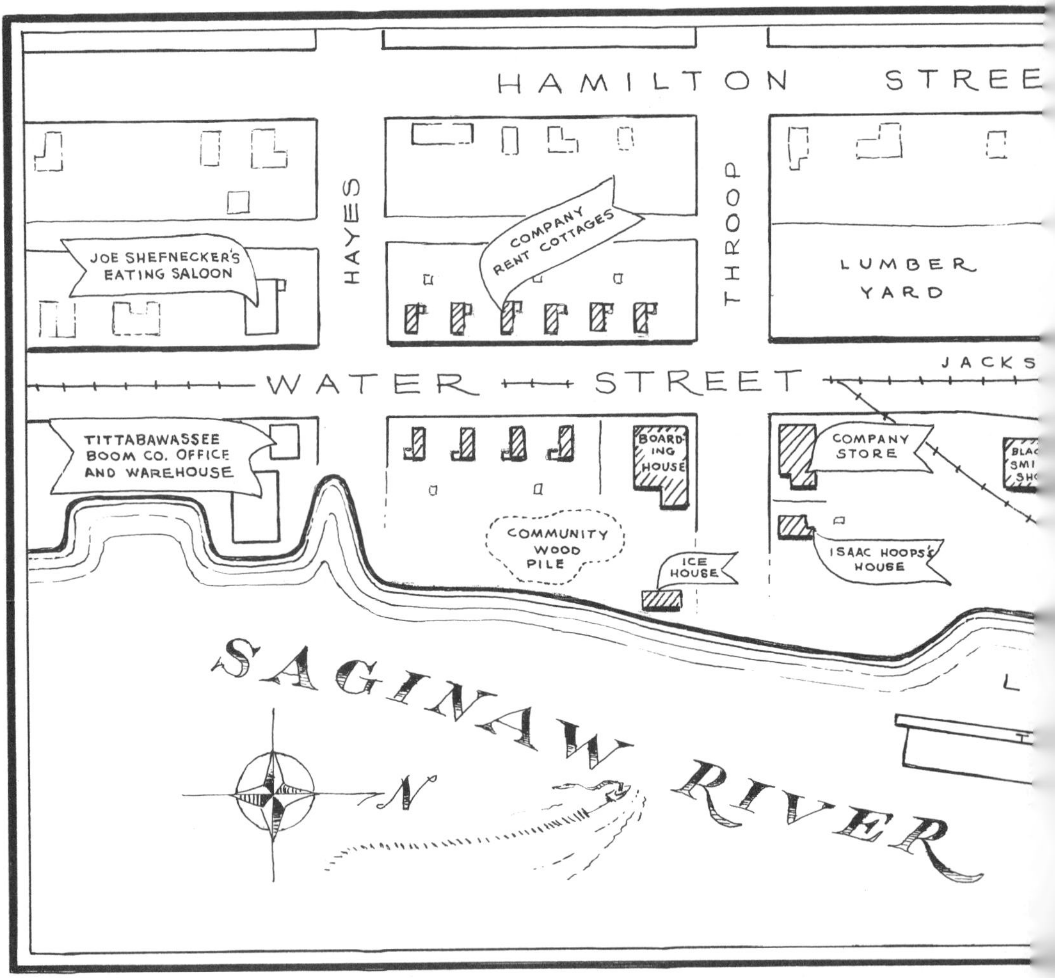

"I'll stop with Charlie just long enough to let him know the drive is over and everything is OK. He might give me a hint who came out ahead in the contest. If I see any of the boys tonight, they'll ask me about it for sure—," mused the foreman as he hustled toward the mill office.

Shannon's property, which was located on the west side of the river in Saginaw City, began at the Bristol Street

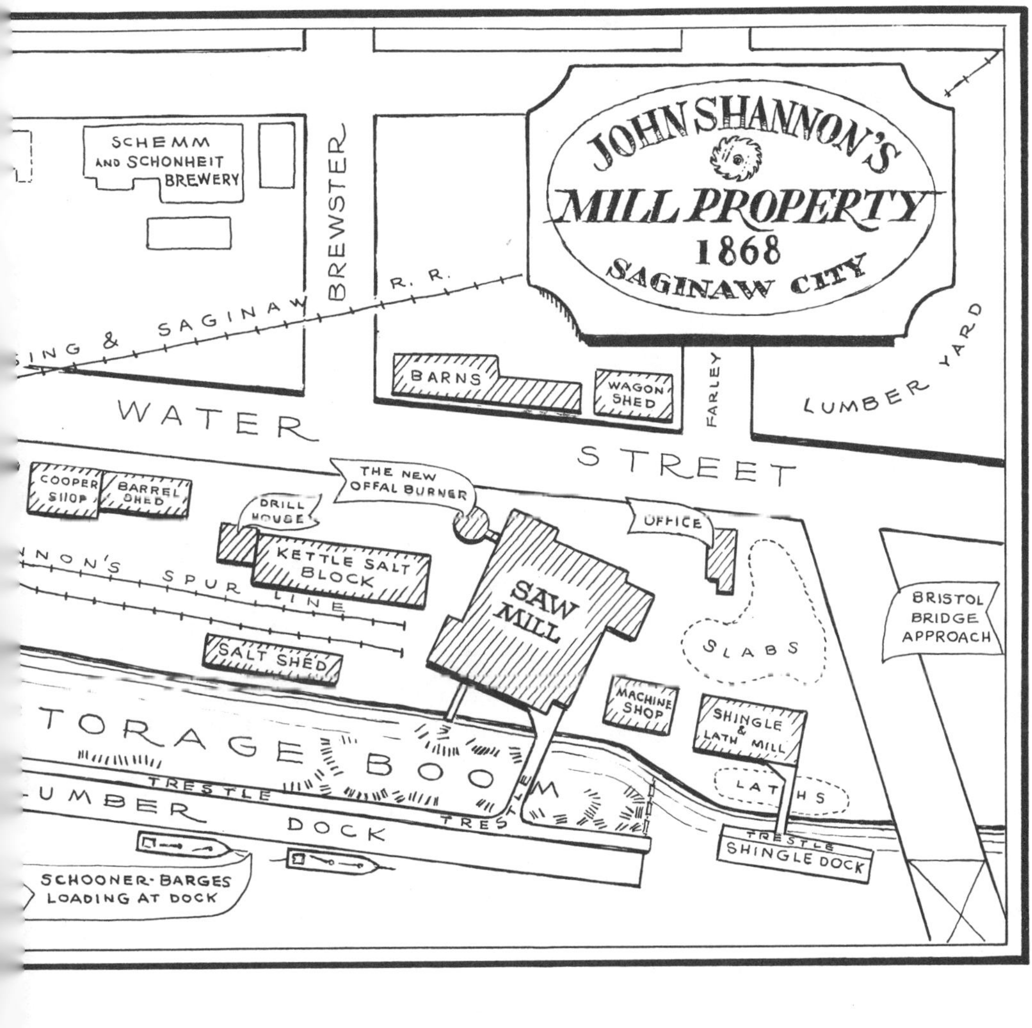

Bridge and ran south four city blocks along a narrow strip of waterfront property to the foot of Hayes Street. In addition to the big mill, the lumberman owned holding booms and docks; a shingle mill, cooper shop and salt block (with attached drill and pump houses); a blacksmith shop, repair shop and horse barns—as well as an assortment of wagon sheds and other outbuildings. Facing each other on both sides of Water Street were narrow, unpainted wood frame cottages—all built alike and set close together. These were the dwelling places the mill owner rented to his workers who had families. Nearby, was a big boarding house for the mill hands who were single. The new company store (managed by the scaler, Isaac Hoops) was the mill owner's newest money-making interest. It also faced Water, located directly across from the boarding house.

The mill office, where McTaggart was heading, was at the extreme north end of the lumbermen's property. It was the only building of all Shannon's real estate holdings in Saginaw City (outside of the mill's boiler house) that was not built of wood. When the foreman came upon the red brick building, he couldn't help thinking how substantial it looked (like John Shannon himself), only separated from the sawdust-covered street by a five-foot wide pine board sidewalk that stretched up and down Water.

To the right of the green-painted door was fastened a small wooden sign that indicated the owner's identity and the kind of business he was in. The message was brief and to the point: "OFFICE—JOHN SHANNON—LOGS & LUMBER." A hand-lettered card with the words, "Walk In," was posted on one door panel. The foreman turned the door knob and walked in.

The quiet inside the building was in sharp contrast to the noise outside, and McTaggart was grateful for the near silence that pervaded the room. But, as he stood waiting for the clerk to look up from his books, he became aware of the recurrent sounds made by a pile driver. Each thump of the big machine, as it drove a spile deeper into the river bottom, made the windows rattle in their frames.

McTaggart looked expectantly at the closed door marked "Private" beyond the clerk's desk. Then he looked at the ticking wall clock. It was almost twelve o'clock. The clerk seemed not to notice his presence. He cleared his throat. "Excuse me—," he said.

The clerk looked up. He seemed surprised to hear the foreman's voice. The man removed his spectacles and blinked near-sightedly at him.

"Is Mr. Shannon or Mr. Burleson in?" asked McTaggart.

The clerk, his mind still on his account books, absent-mindedly cleaned his glasses on his pen wiper. He answered politely, "I'm sorry—but Mr. Shannon won't come in from Albany until late this afternoon or tomorrow—."

Before the clerk could say more, the foreman saw the door of the private office open and the mill superintendent stick his head out.

"I thought I heard voices out here." Charlie Burleson's hand was out-stretched. "Alex—it's good to see you! Come in. Come in!"

McTaggart pushed open the low swing gate in order to step through the room divider, and the two men entered the mill superintendent's office. The foreman noticed the room was as Spartan-like as the outer office—and as blue with pipe smoke as was the bunk-camp any night in the week.

Burleson motioned McTaggart to a chair just as the noon whistles started to blow up and down the river. But he seemed not to notice them, for he at once began to ask the foreman a score of questions concerning the drive.

The mill superintendent was tall and angular, and even though he slouched comfortably in his chair, there was an air of authority about him. He was a man who was as much at ease helping the head-sawyer in the mill adjust a belt-pulley for driving a saw, as he was talking "lumber" with his employer John Shannon in the lumberman's paneled library in Albany. Burleson was an Easterner, like many others in the Saginaw region—a man who had gotten his start in the lumber business on the docks in Buffalo. Off the job, he had a reputation for being something of a "natty" dresser, but on the job he wore durable work clothes—a wool shirt and corduroys that were stuffed into a pair of riverman's boots. His coat, which was hanging limply over the back of a chair—and an old felt hat, laying on his desk where he had tossed it, had sprinklings of sawdust on them—which indicated the mill superintendent hadn't spent the morning in the confines of his office.

When the noon whistles sounded, the foreman, who wanted to make an early escape, rose from his chair. He said, "I don't want to keep you, Charlie. Just came by to check in and to ask about the figures that Isaac brought down to you. The boys are anxious—."

Burleson didn't let him finish. "Sit down, Alex—sit down. Don't be in such a hurry." He looked at his watch. As he closed the lid with a snap he said, "Before you came in, I was going over to the boarding house to get something to eat. Why don't you join me? We can talk over there just as well as here. I want you to take a look around the mill. You might be interested in some improvements we've made this year. And you'll want to see the new store—."

The superintendent got up to put on his coat. He ushered the protesting foreman past the near-sighted clerk, who was still seated at his desk, but was now occupied eating his own lunch, taking sandwiches from a small basket that rested on top of his books and papers.

When the two men stepped out into the street, McTaggart thought how strangely quiet the town was now that the mills were shut down for the noon hour. Mill hands, yard men, salt makers, filers, mill-wrights, boom men and teamsters were streaming from the mill and from the docks, the booms and other parts of the company works. Some men were heading for the boarding house—others, in the direction of the rent houses. Those who lived too far from the mill were going off to sit in groups or alone in the sunshine with their backs against sheds and other mill buildings—or perched on wagons and small hand trucks to eat lunches packed before dawn that morning.

The company boarding house was located several blocks from the mill office, and McTaggart and Burleson made their way with the rest of the hands down the crowded board walk to the barracks-like building.

When they entered the dining room, they made their way to one of the long, oilcloth-covered tables and sat down to share the table with others who were already eating.

There were no signs of comfort in the big room. No curtains were at the windows or rugs on the floor—but the place was clean. McTaggart knew that the second floor was where the transient workers slept. When the mill was running around the clock, the men ate their salt pork and pancakes at the same meal, depending on their coming or going—and it was said the bunks were never cold.[8]

The two late-comers helped themselves to the food on the table and, while they were tucking it away, McTaggart kept glancing at the door, half expecting to see Isaac Hoops come in. Finally he said, "Doesn't Isaac eat here?" Burleson shook his head."He eats most of his meals on the job. Hasn't got much help yet, so he sticks pretty close to the place." The conversation turned to the lumber business—the superintendent telling about the mill, and McTaggart giving him the run-down on the winter at Red Keg.

One by one the mill hands got up from the tables to go back to work, leaving the superintendent and McTaggart to spoon up the last of their pie. A couple of girls came in from the kitchen and began to remove dirty dishes from places left vacant by the hands. They banged crockery and flatware into a big dishpan and took meaningful swipes at the tablecloth with damp rags, eying darkly the two stragglers who were holding up their work.

McTaggart finished his pie and waited for Burleson, who seemed in no hurry. Between mouthfuls, the superintendent said, "I imagine you're anxious to know the results of Shannon's contest."

"I am a little keen on the subject," answered the foreman, glad his dining companion had finally gotten around to the subject. "I told the boys the results would be posted at the mill office tomorrow night. Is that still the plan?"

"It's still the plan," said Burleson, cleaning up the last morsels of his pie. Shannon told me not to make the announcement. I can't even give you a hint—he wants to do it himself—coming all the way from Albany just for the occasion. He wants to look over all the figures again—and see a sampling of the boom scale."

"He's not going to wait until all the logs have gone through the booms, is he? The boom scale isn't as accurate as our camp figures." It sounded like John Shannon still didn't trust the scaler's reports from Red Keg, after that business with Uriah Bell.

Burleson looked to see if his friend had finished eating.

When he saw his plate was empty, he pushed away from the table, saying, "Shannon just wants to see how the camp figures are comparing with those coming down from the booms."

Outside, on the boarding house steps, Burleson stopped to look at his pocket watch. He said, "Business is bound to be a little slow at the store right now. You might go over to see Hoops before it gets busy again. He's real anxious to see you and hear about the drive." The superintendent smiled. "I think he wants to show off the new store. He's been working his tail off pulling it together for the old man."

McTaggart, glad for the opportunity to see the scaler again, said his goodbys to Charlie Burleson. But before he could turn to leave, the superintendent said, "If you've got time later on, come on back to the mill. I've got some improvements I want to show you."

Like the company boarding house, John Shannon's new store was built of pine. The place sat back about five feet from the wood plank sidewalk. Two steps led up to a broad porch on which was displayed a variety of merchandise that included ironstone jugs, butter churns, wooden pails and a display of cant hooks and peaveys, brooms, hay forks and wooden rakes—all bunched together in an empty barrel. Several horse collars and two or three barn lanterns hung from hooks in the porch ceiling.

When McTaggart approached the front of the store, he could see that the usual assembly of loungers had already discovered the place. But today, they had abandoned the stove inside for the warmth of the April sunshine. Two men were sitting comfortably on some bags of grain that were stacked at the side of the door—while a third had pulled up a nail keg and was sitting on that. Under his knees, a yellow dog was lying on his side sound asleep, oblivious to the argument that was waxing hot over his head.

Behind the loungers was a high arched display window. It was covered a quarter of the way down by a dark green shade on which was lettered,

JOHN SHANNON
Co.

A Variety of

Store Goods,

LUMBERING TOOLS AND CAMP HARDWARE

Spring being the season, Hoops had made a display of patent medicines in a corner of the window. They included bottles of blood purifiers and bitters with names like *Radway's Ready Relief* and *Drake's Plantation Bitters*. But a greater portion of the window was taken up with bolts of striped lawns and flowered mulls (unfolded and draped to show the fabric); straw bonnets trimmed with rose sprays and velvet ribbons; gloves artfully laid beside open fans; and bunches of artificial flowers. Several hoop skirts were hung up at the back of the display, and a parasol trimmed with ruffles leaned casually against the side of the window frame to suggest, perhaps, that all was not pine lumber, salt and sawdust in Saginaw City.

"How do you like it?" The voice was Hoops's. He was standing in the open doorway of the store, smiling at the foreman—his steel-rimed spectacles glinting in the sunshine. Isaac's arms were encased in black sateen sleeves from his wrists to his elbows and he wore a blue striped apron of ticking. He was obviously enjoying McTaggart's quiet scrutiny of the new establishment.

"Isaac!" exclaimed McTaggart, taking the two steps at one leap to grasp the hand of his friend. "I swear you look like a real live store-keep."

Hoops led his former boss past the loungers and through the doorway, where the foreman noticed, once he was inside the store, that Isaac had some customers waiting. An Indian woman, who had a sleeping baby strapped to her back, stood behind a small boy who was gravely studying the contents of several glass jars that were arranged along the top of a counter.

Hoops said quietly, "Take a look around while I finish waiting on these people."

While Isaac went to attend to his customers, McTaggart looked over the large, dimly-lighted room, sniffing once again the familiar smells of the general store—a pleasant blending of fresh pine, tobacco, coal oil, harness leather, coffee and hemp—and a hundred other things he couldn't identify.

After making a complete inspection tour, the foreman sat down in one of the chairs provided for the customers near the iron stove, and was astonished to hear Hoops speak to the boy in Chippewa. When at last the child solemnly pointed to the red jawbreakers in one of the jars, Isaac reached into the container and pulled out two candies—putting them into the chubby outstretched hand. The mother said something to the storekeeper and pointed to her basket—then led the child outside past the loafers to disappear down the street.

"Regular customers?" asked McTaggart, after the woman and her children had gone.

"She just brought in some smoked fish and beeswax to

trade for tobacco and pigeon shot for her husband. I do quite a lot of business with the Indians in the area.[9] She'll have wild strawberries and more fish for me in a few weeks."

"The jawbreakers sort of a bonus?" smiled the foreman; and when the scaler nodded his head, he added, "Speaking of bonuses—I just finished talking with Charlie Burleson, and he wouldn't even give me a hint who's got the biggest log cut—one of the other camps or us. You don't suppose they beat us out after all?"

Before Hoops had time to answer, more customers came into the store. The storekeeper went off to scoop up flour from an open barrel, weigh it and tie it up in a package for one shopper. Another wanted to look at some dress goods; so while she was making up her mind as to the kind of fabric she wanted, Isaac went to the back room to draw off some coal oil for a woman who was trying on bonnets; and then came back to count out two dozen eggs and put them into her basket. After that, business became brisk and Hoops finally walked over to where the foreman was and said apologetically, "Alex, I guess I won't be able to do much conversing. People keep coming in. Will you come over to my place tonight? You might as well bunk with me. There's an extra bed for company." He added, "I don't have time to do much cooking, but Shefnecker's Saloon across the way is a good place to eat—or the company boarding house sets a good table."

As the foreman was stepping off the porch, the storekeeper leaned out the doorway and said, "I won't get out of here before ten or eleven tonight. I stay open until everybody's off the streets. Just walk into the house—the door isn't locked."

It was nearly two o'clock when McTaggart left Hoops.

When he walked out of the store, he saw Burleson standing across the street talking with his chief engineer. He waited until the man walked off before going over to join the superintendent, then immediately fell into step with him. The two men made their way back to the mill.

John Shannon's great sawmill faced the lumberman's log holding booms, his lumber docks and the river beyond. It dominated all the other buildings on his property.

The job of supplying enough logs to keep the whole works going was quite a responsibility—but getting ready to saw up the pine into lumber was no small task either. Charlie Burleson had been on the job preparing the mill for an early influx of logs since the end of February. The snows had not melted on the roof tops before workmen were hired on to begin repairing piers and booms. And even now, new docks were being added to the lumberman's extensive water front property. Inside the mill, boilers, engines, pulleys, shafts and belts had been overhauled; new machinery and other equipment were installed, while saws were filed and tested.

Sawmills were always in need of repair—but never more so than at the beginning of a new season. Henry Howland Crapo,[10] a New Bedford man who had investments in whaling as well as in Michigan pine, lamented in a letter to his son, "Mills are like old ships, and like old ships will require a fresh start many times the outlay that was anticipated."[11]

John Shannon couldn't have agreed more with this sentiment when he scrutinized, each week, the letters and reports he received containing itemized accounts of repairs to the mill and money paid out for wages.

As Burleson and McTaggart approached the sawmill, they could hear again the noise of laboring machinery and saws ripping through pine logs.

High, narrow, wooden trestles ran directly from the mill to various points on the docks, and McTaggart could see laborers pushing handcars, loaded with lumber, over the trestle tracks to the place where it was stored to dry and await shipment. Although the season was early, there were great stacks of pine lumber already piled on the docks, and the foreman could see the masts of a schooner alongside being loaded by a handful of sweating dock-wallopers.

Water was still the primary means of transporting logs to the mills (and lumber to the markets), but Shannon, whose eye was on the future, had that year, laid a spur line from the Flint & Pere Marquette railroad into his lumber yard. The road was already reaching a short distance into the pine forests northwest of Saginaw.

Spur tracks ran through the mill yard almost the full

Although the season was early, there were great stacks of pine lumber already piled on the docks.

length of the lumberman's property, and when the superintendent and the boss of Red Keg turned off Water Street, they walked along the tracks that ended at the big mill building.

Like John Shannon, Burleson was enthusiastic over the new rail line and, as he and the foreman walked along the iron tracks, he talked about the future use of the railroad in the lumber industry, not noticing his companion's silence.

To a woodsman like the foreman, the thought of a steam engine thrashing through the pine woods was repugnant, and for the life of him he couldn't think of anything good to say about the subject that was so dear to Charlie Burleson's heart—until he noticed a third rail that ran between the main tracks. He said, "I see she's got a third rail. For logging trains, I suppose?"

The superintendent nodded. "Yes—. We can run in a regular engine and cars as well as those built for a narrow gauge track—the rails being common to both."

As the men walked along, they passed the blacksmith shop, and McTaggart saw the smith replacing an iron tire on a dump cart. He thought of the camp blacksmith and wondered if McCloud had safely reached his old home in Canada where he was going to visit his family.

Two coopers were employed by the mill owner to assemble barrels in which salt (manufactured on the premises) was shipped—and as McTaggart and the superintendent approached the cooper shop, they could see the men working at their trade out in the yard. Staves, hoops and headings were manufactured in the sawmill from waste. The barrel parts were carted from the mill to the stave yard where they were piled up under an open shed to cure and await assembling.

The coopers looked up from their work to nod at the

mill boss and his companion. McTaggart could see their finished work—pine barrels (each one built to hold five bushels of salt), stacked in tiers on a platform near the cooperage awaiting transportation to the salt block.

"Shannon still using kettles for making his salt?"[12] asked the foreman.

"Yes—but this will be the last season for them. We're going to begin manufacturing by the pan method next year. The pans and machinery have been ordered.[13] Also, we'll be piping the exhaust steam from the mill to heat the brine, instead of using fires—which will be a great saving."

They were now passing the drill and pump house which each day brought four hundred barrels of salt brine eight hundred feet up to the earth's surface. Almost every mill owner on the Saginaw had one or more drill houses on his property—salt brine being below the ground just for the taking. Whenever McTaggart saw these drill and pump houses, they always reminded him of small unpainted gable-roofed churches that had square "steeples" towering fifty to seventy-five feet in the air.

When Burleson and McTaggart walked by the pump house, the men could hear the rhythmic throb of steam engine and pump, and see the brine as it flowed into the big settling tank (made of three-inch-thick plank). The great wooden tank rested high on trestlework, so the brine could flow from the tank down to the salt block—a long, wooden structure that housed thirty, 100-gallon cast iron kettles in which the brine was boiled down to make salt.

As the two men detoured around a slab pile (which was handy to the salt block fireplace), they saw steam billowing in white clouds from openings under the low roof of the building. They knew a perspiring salt-maker and his helpers were at work inside tending to the fires and the kettles—

walking up and down the plank walk built between two long masonary platforms on which the brine was heated. In former times, McTaggart had watched the workers dip their ladles into the kettles as the salt crystals formed—observing how they deposited the crystals into draining baskets that hung on the sides of the kettles and then dump the salt into bins. It was an old way of making salt. The method was similar to that used by early Valley settlers when salt-making parties were organized to go out with their kettles to the deer licks, or to the salt springs on the Tittabawassee and Salt rivers, to boil down the brackish water that seeped from the earth.

The railroad track ended at the loading platform of the mill. The two story wooden building was 80 × 120 feet. The engines that powered the saws were on the first story. For safety, the boilers that powered the engines were housed in a separate building that was built of brick and covered with a slate roof.

McTaggart and Burleson turned to go into the sawmill—but, before entering, the superintendent pointed to a smaller two-story structure that was just beyond the mill. "This year our filer and his helpers are located in the loft of the repair shop over there." He shook his head. "There's a bunch that like their privacy. Very jealous of their work."

The workers who "liked their privacy" were kept busy from six to six keeping the mill saws sharp and in condition to cut pine. Their job called for highly skilled workmanship and special techniques that some filers liked to keep to themselves—often working behind closed doors like a secret fraternity.

McTaggart asked what common laborers were being paid this season[14]—no doubt wondering if the mill hands were receiving more than his men for their work in the woods.

"We've hired on about sixty hands so far—that's not counting salt workers, teamsters, filers and the like. We're paying a dollar and twenty-five cents a day to common laborers, and all the slab wood they can carry home for their stoves. The transient workers get a dollar a day and found. We're paying filers, sawyers and other skilled laborers three dollars a day. Of course, if the men prefer store orders instead of cash, Shannon obliges them."

McTaggart looked sideways at the superintendent, but said nothing as they started up the stairs to the work area of the mill. The foreman knew well how many mill owners profited by such methods, since workers were obliged to purchase their orders at the company store.

Before Burleson and the foreman reached the second floor, a two-wheel horse-drawn dump cart was backed up to a slide chute located near the foot of the stairs. At a word from the teamster, the chute opened and pine scantlings poured with a rush from the bin. When the cart was full, the horse automatically walked off toward a wood-lot where the kindling was sold to the townspeople for their stoves. More carts and wagons—some empty—others piled with slabs or staves and laths—rattled by on their way to other yards or the docks. But the foreman could no longer hear them go by because of the deafening sound of mill machinery and screeching saws in his ears. The whole building vibrated with the driving power that was generated by the steam engines below.

As he walked up the flight of steps, McTaggart could see sawdust sifting down from between the rough floor boards—and when he and Burleson reached the work area, he saw it drifting like snow—covering everything in sight—the floor, the machinery, the great supporting beams and the exposed rafters. It was on the workers' clothing, their shoes and on the top of their heads.

A maze of shafts, belts and pulleys criss-crossed the room. They revolved at speeds set by the head sawyer, who adjusted the belts on the pulleys for the different jobs. While some belts labored to drive saws through the pine, others were accelerating—quickening their speed—gathering momentum after saws screeched to the end of a cut.

As soon as they reached the top of the steps, Burleson went off to speak with one of his belt men, leaving McTaggart to look around the great, barn-like room. Six days a week, the mill whistle summoned the hands to work at six o'clock in the morning and released them twelve hours later at six o'clock in the evening.[15] In a few weeks the sawmill would be running full tilt around the clock.

The first thing in the room that arrested the attention of the foreman (other than the deafening roar of machinery and the chorus of wailing saws) was the woodsy fragrance of green, fresh-cut pine. He noticed, too, the barrels and buckets of water that were set here and there around the mill. When he sniffed the pine resin, it reminded him of Red Keg. The water barrels reminded him again how combustible sawmills were.

Fire was a constant threat to mill owners—for sawmills were veritable tinder boxes. Insurance premiums were high,[16] so many owners took their chances, carrying little or no protections against loss by fire.

Sparks from passing tugboats and steamers; exploding firecrackers; lanterns accidentally knocked over; static electricity generated by driving belts improperly installed—these were common causes of mill fires along the river.

At the first cry of alarm, the engineer started the steam whistle blowing, and kept it going as long as the steam lasted or as long as he could stand the heat and smoke. And while the incessant stream of short, rapid blasts of the whistle called for help, water buckets and barrels inside the mill—and along wooden tramways and on top of

A spark from a man's pipe, sparks from a passing tugboat or steamer, a lantern accidentally knocked over, static electricity generated by driving belts improperly installed—these were common causes of mill fires along the river.

buildings—were tipped over by mill hands to wet down as much property as was possible before flames engulfed mill, docks and lumber piles.

The little municipal hose carts and fire trucks of the day were many times useless against such odds—but volunteer firemen bravely did their best to keep fires from spreading to neighboring mills.

The equipment in Shannon's big sawmill included one mulay saw, one sixty-inch diameter circular or rotary saw, and one "gang" of twenty-six saws—plus some smaller circulars that were used as edgers and slabbers. The mill was ready to cut, that year, upwards from twelve million board feet of lumber. Greenwoods' patent stave and heading

Saw Mill Burned Down.

A fire broke out on Wednesday evening, the 6th inst., in the steam saw mill of Messrs. Dolsen & Walker. Being situated in the lower part of the city and at a considerable distance from the engine house, the flames were under too strong headway to be quenched, at the time of the arrival of Protection No. 2. The mill was totally consumed. Loss estimated at $16,000; insured for $6,000. A considerable quantity of lumber was also burned, and the loss would have been much heavier but for the untiring exertions of the firemen and by-standers.

The lack of a sufficient quantity of hose was severely felt, and was the cause of some very valuable time being lost in moving the engine to a position nearer the fire, and also compelling the men to stand in an uncomfortably warm place.

machinery was set up at one side of the mill to make parts for salt barrels. Machinery to make shingles was housed in a separate building.

Like all sawmills operating on the Saginaw, Shannon's mill was powered by steam. Steam engines powered all the machinery from pulleys and belts—the main belts running up from the engine room into the mill where small shafts, pulleys and belts took their power from them. The shafting and belting distributed power in every direction which kept a man, called a "belt-lacer," busy just repairing the leather belts with rawhide lacing.

Most of the movement of logs and lifting of lumber in Valley sawmills was still accomplished by muscle power. A popular saying of the times was, "the only things you need to work in a sawmill are a strong back and a leather apron."

Nor were there safeguards built on mill equipment. A sawyer with all his fingers intact was something of a curiosity. Before the power could be shut off, workers were ofttimes pulled into unguarded, moving machinery—their clothing catching on exposed pulleys or belts. Boilers were not safe either. Consequently, explosions were common—incrustation and stuck safety valves being the greatest sources of boiler destruction. There was nothing done to protect boilers against sudden strains which so disastrously weakened them. They were simply made tight. As to the rest—it was left to chance or good luck. And yet, if steam boilers killed, maimed or carried destruction to all surroundings, the burden was laid, many times, on some mysterious cause, beyond the control of the engineer or mill owner.

Mill accidents were duly reported in the local press, and there was hardly an edition of a daily or weekly newspaper that did not have in them explicit, sometimes shocking,

Spaulding's patent circular saw.

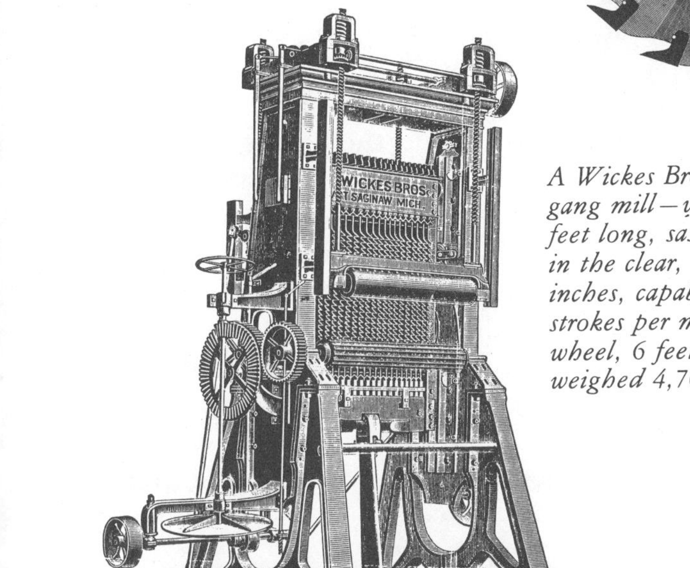

A Wickes Bros. iron frame gang mill—with 26 saws, 4-1/2 feet long, sash 36 inches wide in the clear, and stroke 20 inches, capable of making 230 strokes per minute. The fly wheel, 6 feet in diameter, weighed 4,700 pounds.

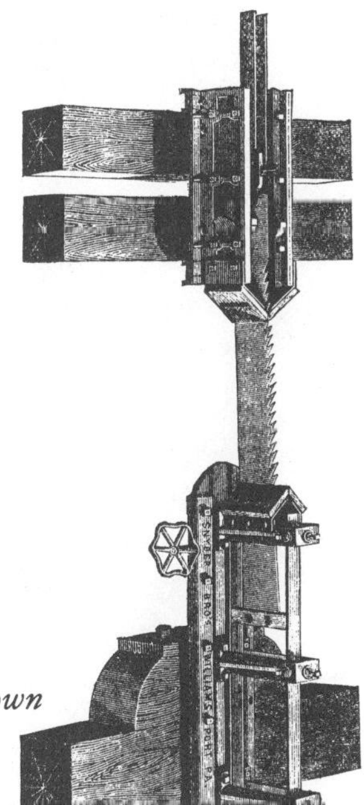

An old up and down mulay saw.

writeups of loss of life or limb in one or more of the mills.

As was true of most other industrialists of the times, lumbermen felt no more responsibility for sawyers hurt in their mills than they did for sawyers hurt in their lumber camps. If time was taken off the job to mend broken bones or heal cuts and bruises, there was no sick pay or other compensation expected from mill owners to pay the doctor (or the undertaker). Unless the hat was passed around for the benefit of the injured man's family, there were no other funds available to tide them over until the breadwinner was well enough to work again. The rent bills piled up, and it wasn't long before the worker was in debt—ofttimes, to his employer.

McTaggart had grown tired of waiting for the superintendent to come back from where he had gone to confer with one of his hands. Not seeing where he could stand out of the way, he wandered over to the big open bay of the mill. He stood looking out over the great holding booms, the docks and the teeming river beyond and was lost in thought when he heard a loud voice call out, "Hey—down there! We're ready for more logs!"

The foreman turned his head to see a wiry-looking mill hand standing next to him on the deck. He had one hand cupped to his mouth, and was waving with his other hand at someone below. The man was evidently signalling to the boom-boss to send up more logs—for the words were no sooner out of his mouth when McTaggart saw the perpetually-moving bull-chain slowly start to haul a dripping pine log from the river. It came wet and shiny, up the slanting V-shape incline—through the great open doorway and into the mill where it was guided by workers over maple rollers to a platform carriage—the conveyance that would carry the log against the sixty-inch Spaulding patent circular saw.[17]

The foreman, his interest aroused, decided to follow the log to the saw. He watched the mill hands fasten it into place on the saw carriage, and after the tail-sawyer had checked to see if the bail clamps were holding the log fast, the head-sawyer pulled a lever, actuating the carrier. The carriage moved, and immediately the sharp teeth of the spinning saw bit into the soft wood with a shriek that sent a shiver running down McTaggart's spine. A spray of sawdust shot from the rapidly revolving teeth, and he heard the saw yowl and sing as it burrowed deeper and deeper into the side of the log until, at last, he saw it tear through to the end.

When the liberated saw spun free, a long slab fell away from the log. A worker pulled the piece from the carriage. (Like others of its kind, the slab would be cut into four-foot lengths by a smaller rotary saw. Parts for barrels and building laths would be made from the slabs. Other slabs would feed the mill boilers). After the carriage was gigged back into position, McTaggart watched helpers with hand hooks and cant hooks struggle to turn the log over. (The mechanical log turner had only been introduced to the industry that year.) Again, the log was clamped to the carriage, and another slab was removed from the opposite side. Thus, the foreman saw the log pass through the circle saw four times. The squared timber was then forwarded to the stock gang—a powerful machine that held a series of twenty-six saws, set upright, about one inch apart, in a heavy frame. The log was fastened down and sent through—whole. Within minutes it was reduced to one-inch boards.

Helpers were on hand to pick up the fresh-cut lumber when it came from the gang. They piled it onto hand trucks and moved it to the trimming tables where a circular saw trimmed off the rough edges and cut the boards to width.

McTaggart watched a sawyer walk several of the boards through the saw (one at a time) and then turn them over, repeating the operation to trim the opposite side. (These were the days before Thomas Munn of Bay City introduced to sawmillers his double edging table with its mechanical feed, which was designed to quickly trim both edges of a board in a single operation).[18]

When the lumber fell from the trimming table, it was moved again—this time to the sizing table, where the boards were cut to equal lengths.[19] The trimmed and sized lumber was then loaded on handcars and pushed out of the mill to the docks where it was sorted and piled in stacks—according to grades.[20]

In spite of himself, McTaggart had become interested in the mill activity and had lost track of time. Now he pulled his stem-winder out of his vest pocket and consulted the dial. He was about to slip it back into its accustomed place when he saw the mill superintendent becken to him across the dusty, pulsating room. Burleson was standing near the sawyer who was operating an old up and down mulay saw.

The foreman dodged and sidestepped moving belts, mill hands, piles of scrap and hand trucks until he reached the superintendent. Burleson pulled him aside—away from the noisy machinery so he could be heard over the din.

"I've been watching the sawyer here work on one of the cork pine logs that came down from Red Keg—thought you'd like to see it under the saw."

McTaggart gave the superintendent a nod to show he had heard him.

"The circle saws may be faster," continued the mill superintendent, "but they still wobble too much for this kind of job. Don't want to take a chance on having the lumber from this beautiful pine look like washboards when we're finished cutting it up. Our old mulay makes about the same width kerf as the rotary saw—a quarter inch—but it cuts straighter."

McTaggart knew all about the mulay[21]—the saw that was presently being replaced by faster-cutting circulars. The mulay saw, with its one perpendicular, six-foot by ten-inch blade (that cut only on the down stroke), resembled its near relative—the cumbersome old-fashioned "sash" or "gate" saw that the first mills on the Saginaw were equipped with back in the thirties. But unlike the sash saw, the mulay wasn't held rigid in a big wooden frame. It was fastened to a cross-head which slid up and down between cast iron guides (to keep it from buckling). The pitman, or connecting rod, was attached directly to the motivating power below.

The slabbed log that was being cut looked enormous resting on the saw carriage. The timber weighed well over three tons. It was moved to the saw on a heavy truck that ran on stout iron wheels. In order to transfer the monster from the truck and onto the saw carriage, a power windlass and chain were used. The hoisting machine drew the car and its burden up a short, inclined track and when it was even with the carriage, helpers manipulated the log (using cant dogs and cant hooks) to roll it onto the conveyance—where it was made fast.

The superintendent and his guest watched the saw move steadily, relentlessly, through the pine. At intervals, the sawyer would pull a lever which brought the feed to a standstill. The long blade continued to jiggle up and down in the saw cut.

McTaggart heard Burleson shout in his ear, "A log as big as that one can't be sawed without clogging up the saw, since a large portion of it is continuously in the kerf. Todd stops the carriage to clear the blade."

Burleson went on with his talk.

"Used to be, if a mill contained two mulay saws, it was considered a first class sawmill. But times have changed. Shannon has decided to get rid of the old upright next season. This is the last time you'll see this old girl in action. The circulars are getting better all the time. Stronger, thinner—and they're so fast, the edger can't keep up with them."

The mill superintendent paused to pull McTaggart farther away from the saw. His voice was beginning to give out.

"Several years ago," he continued, "Bill Taylor, a mill man down river, made a wager with another sawmiller that he could cut double a thousand feet in an hour using a circle saw.[22] The skeptics within earshot thought he was talking through his hat—so a couple of days later Bill set up a demonstration to prove his point. He invited a group of lumbermen to witness the proceedings. Before the sawing began, bets were taken all around—the odds being about two to one that the circular couldn't cut the mustard. After about ten minutes of sawing, there were some pretty surprised spectators watching the Spaulding rip through the pine, and in spite of most of them losing their wagers, there were some happy mill men watching that circular cut over four thousand feet of pine in the hour. The edger couldn't keep up, and the whole mill was covered with boards. They only quit sawing because there wasn't any more room to stack the pine when it came off the saw.[23] The circular had cut in an hour more than twice the amount the old Williams mill cut in twelve hours with their single gate saw back in 1835."

McTaggart, looking impressed, nodded his head to encourage the mill superintendent to continue.

"So—we've seen it demonstrated here on the Saginaw; the capacity of the rotary is only limited by the edger—and the mill hands' ability to move the lumber from the mill and to the docks."

"What's become of the band saw?"[24] asked McTaggart.

"It still isn't perfected for heavy mill work. Jim McCormick[25] tried one at his place down river at Portsmouth about ten years ago. I think he was the first lumberman to use one. That was back in 1858. Imported the saw from France. But after he got it all set up, he cut only one good-size board with it before it broke. Couldn't get proper tension on the blade."

Burleson laughed and then said, "He was so mad and disappointed at the thing breaking, he wove the saw like a ribbon in and out of the picket fence in front of his house. It's still there."[26]

The sawyer Todd finally brought the mulay to a halt. It had sliced up the whole log. But the fresh sawn boards were still attached to the butt end. The clamps that fastened the log to the carriage had not allowed the saw to clear—to cut all the way through. The boards would have to be chopped from the stub.

McTaggart watched a mill hand pick up a broadax and deftly cut away and separate the pine until all of the great, wide, creamy white boards were severed from the log end.

Burleson said, "There's another reason the sawmillers are moving away from the mulay—."

McTaggart understood what the superintendent meant—the circular and gang saws cut all the way through the pine. No time was wasted chopping the boards from the log ends.

The men walked back to the gang to watch it gobble up another log—and as he saw the pine go through the oscillating saws, McTaggart couldn't help but wonder how long the pine industry could last at the rate the lumbermen were accelerating the business.

His thoughts were underscored when he heard Burleson say, "Keeps me busy just reading up on all the advancements they're making on mill equipment. The other day I got a pamphlet from the Wickes Brothers' plant down river.[27] They're making improvements on the old Scotch gangs which will make all the teeth of the saws cut more smoothly and evenly. Hired on a young engineer by the name of Frank Kirby[28] who they say has overcome the difficulty of the lower teeth of the saws doing all the work. Just built a gang for Hackley's over on the Muskegon.[29] We'll be getting reports from them on how it's working out."

The superintendent waved an arm in a wide arc to take in the mill and all its equipment. "In ten years," he said, "I can see this whole plant fully mechanized. No lifting. Conveyors or live rollers will carry the lumber to and from the saws. Sawdust and other offal will be conveyed or blown directly to the refuse burner or to the boilers. I've got it figured out so a couple of slabbing gangs and two stock gangs will cut upward of thirty or forty million feet a year.[30]

"With the railroads going into the woods—we'll be needing new improvements just to keep up with the logs coming in. The demand for lumber is something fierce!"

McTaggart didn't know why he felt so low. After all, he told himself, he was in the same business—and for the same reason as John Shannon and Charlie Burleson—for the profit the pine would bring him. It didn't make any sense—but somehow, the business seemed different in the woods.

All afternoon the sun had been shining through the windows set high under the eaves of the mill—the long shafts of sunlight slanting down through the dusty air to illuminate the mill hands' work. But now, McTaggart noticed the light was failing; and he saw, here and there, lanterns being lighted. He felt a bone-chilling draft pass through the open bay—the current of air rushing in to scatter sawdust and small pieces of scrap at his feet. Peering outside, the foreman saw that the wind had shifted and was whipping up little whitecaps on the surface of the water. The sky looked dark and threatening. A storm was brewing—and it looked like it was going to be a bad one.

Burleson had also noticed the change in the weather.

"I'd better get back to the office before a downpour begins. I want to clean up a few things on my desk before six.

"You're staying with Hoops tonight—is that right, Alex? You know you're welcome to sleep over at my place—."

McTaggart stuck out his hand to pump the superintendent's fist. "Thanks, Charlie—I'm all set. Isaac invited me this noon. We've got a lot to talk over—. It's been a real pleasure seeing you again—."

Reassured that his friend had a place to bed down for the night, Burleson headed for the stairs with the foreman beside him. They were almost at the bottom step when McTaggart said, "We'll be seeing you tomorrow night at the mill office—."

"You bet, Alex—I'll see you and the boys there." And off he went, hunched against the wind as he made his way up the hill to the office.

The foreman trotted across the mill yard in the opposite direction with whirlpools of sawdust flying around him. He hurried past the salt block, the stave yard and the cooper shop—reaching the scaler's little house in time to see a gray wall of rain come pouring out of the heavens.

It was only eight o'clock when Hoops let himself into the house. The instant he walked in, the wind pulled the door out of his hand, slamming it back against the wall. It took him a second or two to close it, and when he turned around, he was grateful to see lighted lamps and feel the warmth of the fire that was burning in the parlor stove. McTaggart (in his stocking feet) was sitting in a wooden rocker with one leg crossed over the other. An open book was on his lap.

"Made myself at home," he said, watching Isaac put down some packages in order to hang up his drenched coat on the rack that was fastened to the back of the door.

The scaler, pleased to see his friend so comfortable, nodded happily, "Glad you did."

"Made myself at home," he said.

He walked past McTaggart to carry his bundles out to the small kitchen, saying over his shoulder, "Haven't seen a storm like this since that bugger we had in camp last January. Not a living soul on the streets—so I decided to close up early."

"The Lakes will take their toll this night," commented the foreman. "I saw a tow going out past the mill early this afternoon—decks loaded over their rails with lumber. There'll be some anxious people waiting out this storm."

From where he sat, McTaggart could hear sounds in the kitchen that indicated Hoops might be preparing something in the way or a repast.

After waiting only a few minutes, he saw the scaler come back into the parlor carrying a tray. There was a loaf of bread on it, as well as a tea cannister, a saucepan (that contained six brown eggs), a hunk of butter in a deep dish and a crock of honey. A teapot, some boarding house crockery, a salt dish and an assortment of knives and spoons completed the tray setting.

"Thought you'd like something to eat," said Hoops. "Brought the stuff from the store—don't keep much food in the house. I'm too busy to eat at home—usually go to the boarding house when I have time to eat a full meal."

He smiled. "I never thought I'd admit it, but I miss Kelly's cooking."

The scaler-turned-storekeeper set his small burden on a table he had pulled up near the stove, and the foreman watched him pour steaming water from a tea kettle into the saucepan of eggs. Hoops returned the kettle to the top plate of the oven that projected from the back of the woodburner—making room for the saucepan.

"Hope you like your eggs hard-boiled," he said.

McTaggart, who hadn't seen an egg in seven months,

said he'd take them any way the scaler chose to serve them.

Isaac nodded his head and picked up a bread knife. He proceeded to cut the loaf into thick slices. Having done this, he opened the door of the stove with one hand while, with the other, he neatly stabbed a slice of the bread with a long-handled fork and held it to the radiating heat. As he toasted the bread, Hoops handed the browned slices to McTaggart who generously buttered them one by one (the foreman hadn't seen butter in seven months either) until an eight-inch stack of toast rested on the granite pie pan (which the scaler had thoughtfully put into the little oven to keep warm).

The next order of business was making tea, and Hoops handily scooped green hyson from the tin cannister, unceremoniously dumping it into the teapot without measuring it. Boiling water was poured over the leaves and up to the brim of the pot. After giving the tea a vigorous stir with the toasting fork, Hoops turned to the stove to slide the saucepan of eggs off the top of the oven and onto the table. He pulled up a chair and sat down before pouring the steaming liquid from the teapot into two big cups.

As McTaggart sipped at his tea, he examined his friend through the fog of steam rising from his cup and thought that Hoops looked tired (trying not to notice his friend looked happy as well—for the foreman still had hopes the scaler would come back to work at Red Keg next winter).

"You look a little beat," he said as he put his cup down to let the tea cool.

"Funny—I always worked hard at scaling—climbing over log piles and out in the cold all day—but now, I'm more tired at night from keeping the store than I ever was after a day in the woods. It's going to take some getting used to."

Isaac got up to take the toast from the stove. "I'm going to give it a try though—. I feel it's time I settled down."

While the two friends ate their small feast (their feet propped comfortably on the fender of the stove), they could feel the wind shake the little house and hear it rattle the windows in their frames—while the rain, in sudden furies, lashed at the roof and ran down the downspouts to flood a swamped rain barrel and flow over a yard that was fast becoming a lake.

Within the house, the lamp light reflected softly in the nickel trim on the wood-burner, and the fire glowed red through the mica windows in the stove door. The little parlor seemed like an island of security. In his heart, McTaggart understood why Hoops would not want to trade what he had now for more hard winters in the isolation of the lumber camp. And he envied him a little.

The scaler motioned with a buttery hand at the wall behind them. It was covered from the floor to the ceiling with home-made pine shelving, on which there were books of every description. The foreman, while he was waiting for Hoops to return from the store, had read some of the titles—Dickens's *Pickwick Papers*, *Oliver Twist*, and other books by the same author, as well as Scott's *Ivanhoe* and *The Scottish Chiefs*—books by Longfellow, Tennyson, Byron, Burns, Holmes, Shakespeare, Bulwer-Lytton—and many more authors that McTaggart didn't recognize.

"Had my books sent out from St. John. My mother has been keeping them for me—." The scaler hesitated (McTaggart would swear he saw his friend's face turn pink), then he went on to say, "I've met someone whose interests are the same as mine—." His voice trailed off in embarrassment.

McTaggart grinned. "Why you old so and so! You mean you've found a lady friend."

Isaac, who seemed greatly relieved to get his announcement out in the open, answered, "We met at a Lyceum two weeks ago—and I'm thinking of asking her to become my wife." He hurried on. "Shannon's business is really growing—and I'm certain he'll be enlarging the store—even building a better one in time. This place is where I can put down some roots—raise a family—maybe invest a little in the business—."

The scaler got up to shove more wood into the stove, saying over his shoulder, "And what about you, Alex—what are your plans?"

The foreman hesitated. How could he tell Hoops that his future might depend on the outcome of some uncertain business he had to attend to the next day? So, he only said, "After tomorrow, I'll be going off land looking for Shannon—and then be back in the fall to hire on another crew to take out the last of the pine at Red Keg. Beyond that—we'll wait and see."

In town, McTaggart tries to forget Red Keg and its problems, but finds trouble has followed him into Saginaw City.

Shannon announces the winners of the contest—the prize bonus being more than the men bargained for.

Saginaw Town, with sawdust-covered streets and pine board sidewalks.

20

T is for the TOWNS, where the boys went their way

In spite of the lateness of the hour when they retired the night before, McTaggart and Isaac Hoops were out of their beds before dawn the next morning—Isaac hurrying off before breakfast to build a fire in the stove at the company store (the temperature had fallen in the night) and to dust off and sweep out before his first customers appeared.

The foreman had gotten up early out of habit and after his friend left the house, he felt at loose ends. His own business, which had been postponed the day before, would take him to the land office across the river in East Saginaw. But the place wouldn't open up for two hours or more. He looked around the small rooms for something to do and when he saw a little pile of dirty dishes (left from supper), resting on the drip board of the iron sink, he immediately rolled up his sleeves and with characteristic energy went to work. In no time, the collection of cups and saucers and other eating utensils were washed and put away in the cupboard.

McTaggart was hanging up the flour sack dish towel on the line strung up over the stove when the kitchen clock struck six-thirty. He still had an hour and a half to kill.

Again, sounds of domestic industry filled the house as the grates in kitchen and parlor stoves were vigorously shaken, and accumulations of wood ashes were taken out and dumped onto a small ash heap in the back yard. Pushing a wheelbarrow through puddles of water to the mill hands' communal woodpile, the foreman replenished Isaac's stock of stove wood. He stacked it neatly in the small shed that was attached to the house and threw an armload of dry wood into the woodbox behind the kitchen stove.

For the first time since he awoke that morning, McTaggart felt pangs of hunger gnawing at his stomach—but the clock told him it was too late for breakfast at the boarding house. He decided to try the bill of fare at Joe Shefnecker's eating saloon (boarders upstairs) which was located not far from the house.

As the foreman walked down the street, he could see some of the damage the storm caused in Saginaw Town the night before. Tree limbs and bricks from chimneys lay scattered about in yards. Rainwater filled the ditches and other low places, and picket fences (those that hadn't blown over and floated away with the wooden sidewalks at the height of the storm) were leaning drunkenly to one side or the other.

Low-lying Water Street was usually paved with sawdust, but there was hardly any evidence of the fact this morning, it being mostly under water. Passing vehicles splashed down the roadway behind Shannon's and other mill owners' properties, and pedestrians gingerly picked their way, like so many wading birds, through the water to cross the street.

The morning was still blowy. The sun struggled to come out from behind dirty patches of racing, low-hanging clouds. The foreman shivered and pulled his jacket collar up around his ears as he made his way down the street, stepping around holes where the sidewalks had washed away—sinking to his ankles in mud when it was unavoidable.

Before pushing his way into the gas-lit saloon, it started to snow, and he thanked Providence that he and his crew were early in getting all of their pine down from Red Keg—feeling sorry for the rivermen who were out on icy rivers and streams struggling with logs in such wintry weather.

Inside Shefnecker's, it was steaming hot—and crowded, even at half-past seven in the morning. Men had been coming in from the woods for a week or more, and many of them, before disbanding, had sought out the grog shops, as woodsmen had done since the old pine days in Maine. By the size of the milling crowd, it looked and sounded to the foreman like every camp on every river in the Saginaw Valley had congregated at Joe's—although he knew that all the saloons in the towns on both sides of the Saginaw were as crowded as Shefnecker's.

Tables, covered with gravy-spotted red checkered table cloths, were set close together along one side of the room. A long mahogany bar and card tables occupied the other. All dining tables were crowded, so McTaggart waited and when he spotted a chair being vacated by a departing woodsman, he slid onto it before anybody else had a chance.

A waiter—a man with a drooping mustache and wearing a distracted look (his apron resembled Kelly's at the end of the logging season)—came over to take his order. There was no printed menu other than the one scratched in chalk on a big piece of slate hanging on the wall—so, after briefly consulting it, the foreman (who had worked up a sizable appetite) ordered flapjacks, fried potatoes, steak and eggs, dried apple pie, coffee and beer. Then he settled back in his chair to enjoy the animated scene before him.

The room was hazy with smoke and in the wavering, harsh light of flaring gas jets the lively goings-on looked like those in a dream. It all seemed unreal and far, far away from the solitude of the past months in the pine woods. McTaggart's enjoyment began to take hold. He was beginning to relax—feeling the responsibilities of the lumber camp at last slipping from his shoulders.

Presently, the waiter came back with the foreman's beer—telling him his breakfast would be along soon. When he went off again, McTaggart idly watched the man's back until his shoulders and then his head were swallowed up in the crowd.

Much of the noise and confusion in the room was caused by old friends greeting one another after months in the lumber camps. There was a lot of hallooing and signalling going on—men trying to attract the attention of others in the crowd. So, when McTaggart saw a waving hand that was attached to a frayed red sleeve surface over the heads of the saloon's clientele and disappear again into a sea of humanity—and then rise and fall again as it advanced toward the dining area, he didn't pay much attention. He only thought, "Whoever the jumping jack is, the poor devil is having one deuce of a time getting the attention of somebody over here."

The foreman made a half turn in his seat to look behind him. Not seeing anyone who was responding to the hand signals, he turned back—in time to see the arm appear again. He stiffened. This time McTaggart saw a head covered with a wild mop of yellow hair bob up along side the red sleeve. Cully Wills! The foreman ducked. But it was too late. The camp swamper was already pushing his way toward him between the crowded tables—his partner, Bill Bragg, in tow.

"We been lookin' all over for you," panted Cully as he and Bragg elbowed their way close to the foreman. His greeting was accusing—as if McTaggart had promised to send him a notice of his whereabouts and hadn't done it.

The foreman frowned. "What's the matter with you two?" He could read the long faces before him and knew something was amiss—but he was determined to eat his breakfast in peace, no matter who was in jail—or had gotten killed the night before.

The waiter took that moment to come back and set down platters and dishes in front of him, and while McTaggart dug into his fried potatoes, the other occupants of the table got up and left for the bar, which gave him a chance to shovel down several more mouthfulls before Cully and Bill could settle into their places.

As soon as the harassed waiter went off to pick up the swampers' orders, Cully leaned over the table and said, "Toohey's in jail!"

When McTaggart didn't look up, Cully appealed to his partner for help. Bill jumped in. "Pat and the 'Highlander' sto— borrowed some dude's horse and two-seater last night after they left the River House—."

The boys had managed to get McTaggart's attention. He nodded his head but went on eating. He knew by heart the rest of the story—or he thought he did.

It was the old, old tale. Woodsman comes into town; cashes in his time—has dollars burning a hole in his pocket at noon; buys a pair of pants and a three dollar watch at 2 p.m.; and at 9:30 a.m. the following day, pawns pants, watch and boots for $2 to pay fine in Recorder's Court.

And McTaggart knew the Highlander of bagpipe notoriety. Dan Richie, he was by name. Occupation—chopper, Camp 2. But he was best known in the Saginaw country for carrying, wherever he went, a kilt of the Mackintosh clan, a collapsible stovepipe hat and bagpipes in his turkey. As soon as he hit town after a winter in the woods, it was his habit to don his kilt and "lum hat"—wrap a red woolen sash around his ample middle and make tracks for his favorite drinking establishment, which was the River House on south Water Street in Saginaw City.

The hat was doffed and placed bottoms up on the bar (for contributions to pay for quenching the Highlander's fierce thirst). One drink was ordered—and quickly dispensed. And with no further encouragement than three ounces of whiskey burning his gullet, the Scotsman would tuck his pipes under his arm, puff a few times on the instrument (to inflate the bellows), and then proceed to squeeze out savage sounds that recalled, for the piper, tales of old feuds and battles fought on misty moors between the Gordons and the Mackintoshes.

When the first squalling sounds droned from the pipes, the patrons of the River House would gather around the Highlander—and those standing outside, who heard the sound, would crowd into the saloon—to see the wonderful spectacle of Dan Richie in kilt and sash, stalking up and down in front of the bar, playing his pipes with eyes closed—his majestic beard aquiver.

In spite of himself the foreman had become interested. Guardedly he said, "You better start at the beginning," feeling in his bones the stolen horse had either run away, frightened out of its wits by Richie's bagpipes, or the buggy was smashed up against some lamp post, or run into the Saginaw River.

Cully, encouraged by the foreman's interest, launched into his story.

"Bill and me first seen Toohey last night when we was comin' out of a saloon near Court Street. It was darker 'n sin and stormin' something awful. We wouldn't of seen the chickadee at all if it wasn't for that gas jet burnin' at the side of the saloon. But we recognized Pat in the light when him and Richie went lickity-split past us in that fancy buggy—.

"Bill hollers in my ear, 'Let's follow 'em Cully!' an' we takes off right behind 'em leggin' it up Court Street hell bent fer leather. A course they was 'way ahead of us—an' by the time we gets to the town square, we could see 'em in flashes of lightnin', circlin' round and round the square on two wheels—goin' past the Court House, the jail 'n the Lutheran Church. That's where they turns around on that empty lot between the German school and the church an' then barrels it back down to Water again—."

Wills's partner took up the story at this point—not wanting to be left out of the telling.

"—The Highlander was in the back seat of the buggy playin' his pipes," said Bill. "It were a rare sight to see, I can tell you—with barkin' dogs chasin' after the rig—and the wind ahowlin' and it thunderin' and lightnin'—and the rain comin' down like it wouldn't never quit—." The swamper took a deep breath. "—And them bagpipes! Lord—what a sound!" Bill shivered even though the temperature in the saloon must have been a steamy eighty degrees and the swamper was wearing three wool shirts over his flannels. He hurried on with the story before his partner had a chance to jump in.

"—When they gets back down to Water again, we seen ol' Toohey suddenly pull up on the lines and stop that poor horse—almost knockin' over a street lamp. Two wheels of the buggy was up on the sidewalk in front of that jewelry store what's on the corner—."

"Smyth's," injected Cully, whose helpfulness drew a frown from Bill.

"Over the wind we heard Pat holler at the Highlander, 'Look see what time it is, Richie—I promised me wife I'd niver stay out later 'n ten o'clock at night!"

"A course the Highlander didn't have no pockets for no time piece—but chancin' to look up, he sees that big wooden sign what looks like a pocket watch hangin' over the front of the store. It was right over his head. Old Richie climbs up on the buggy seat—seemin' to want to get a better look at it. He stands there a second or two—a weavin' back 'n forth—then yells down to Toohey, 'Don't think she's runnin' laddie. Can't hear her tickin'—and then, by damn, if we don't see him reach up and pull the whole contraption off the chains and start shakin' the daylights out of it.

"That's when the sheriff come runnin' around the corner, all out of breath and madder 'n hops. He'd been chasin' after Toohey and Richie—same as Cully an' me—follerin' the sound of the pipes. Well—when the Highlander sees Quackenbush, he heaves the big watch to Toohey and roars, 'I dinna know wha' teem it be fa' ye, Toohey laddie, but fa' Richie, its teem he was on his way oot o' th' toon!' An' with them words, he lights out—over the back seat of the buggy—aclutchin' his hat with one hand an' his pipes with the other and disappears down a alley leavin' old Toohey in the stole buggy holdin' that fool jewelry store watch sign."

The swamper paused again—this time to catch his breath.

The foreman took the opportunity to say, "That the whole story?"

Bill replied, " 'Ceptin' for the part where the sheriff carted Toohey off to the pokey."

McTaggart had finished eating. He looked at the big wall clock behind him. It was after eight o'clock. The camp chickadee must be pretty anxious by this time. He pushed his chair away from the table and stood up. While he fished in his pockets for some money to pay for his breakfast, the two swampers waited for him to make some comment—taking his silence for meaning that he wasn't interested in rescuing their pal.

Now they pleaded, "Boss—you got to get ol' Toohey out of that jail-house. It weren't all his fault. You know how he is—he jest can't stand to be locked up. His wife's expectin' him home sometime next week. He's probably goin' crazy by now."

McTaggart found a fifty cent piece and tossed it to the waiter. He started pushing his way through the crowd and out onto Water Street with the two swampers at his heels—their breakfast forgotten. When they were outside, the foreman said, "All right now—where's he at—the city jail or the county lock-up?"

Relieved, Cully said, "The jail whats at the town square near the Court House."

Without another word, the foreman turned and started walking down the street with Bill Bragg and his partner Cully Wills beside him.

• • •

The county jail was a good eight city blocks from Shefnecker's Saloon—and as McTaggart and the two swampers began their hike up Water Street,[1] the foreman couldn't help but think old Saginaw Town had come a long way since the time when the place was thick with trees and only the fort, five log shanties and a wildcat bank made up the place. That's about the way Cap Wilkes remembered it when he first came to Saginaw City from York State in 1834. Cap once told his friend, Jack Driscoll, that the winter following his family's coming to the Valley thirty-six wolves were killed for which the County paid a bounty of eight dollars a scalp. It was wild country then—and still belonged pretty much to the Indians and the fur traders.

Saginaw (*Kah-bay-shay-way-ning*, the Chippewa called it—meaning the "gathering place"[2]) had in ages past been an Indian center—a base camp—a station on the longest trail in the Michigan Territory.[3] It was where the Chippewa rendezvoused after their winter hunting and spring sugar making. It was where they fished—where they raised grain and held their annual corn feast in the month of October.[4]

The Saginaw country had been known to French and British fur traders since the 17th century. But it was the expansion-minded Yankees who thought in terms of making permanent settlements in the territory. The War of 1812 decided once and for all who would have access to the domain—the British or the Americans.

Only the Chippewa stood in the way of the land promoters. So a treaty was drawn up and an appropriation of $10,000 was made by the United Stated Congress to cover the expenses for a council with the Indians.[5] General Lewis Cass (then the Governor of Michigan Territory) was commissioned as agent to head the treaty negotiations,[6] and one day in "the full moon in the month of September,"[7] in the year 1819, he and a staff of aides, secretaries and interpreters, started out from Detroit on horseback, riding over the old Indian trail to "Sagana." The Revenue Cutter *Porcupine*,[8] loaded with troops, provisions, presents for the Indians and "three kegs of cartridges and arms sufficient for the detachment,"[9] was sent around by way of the Lakes to meet Cass and his party at Kah-bay-shay-wing-ning. The Indians, who had been notified of the meeting by runners, months in advance, were already assembled for the bargaining sessions.

There followed ten or twelve days of elaborate speech-making at the treaty site. Finally, on the 24th day of September—after much palaver and doling out of gifts and whiskey—the chiefs and headmen of the Chippewa (as well as a handfull of Ottawas)—affixed their totems to what would be known thereafter as the Saginaw Treaty.[10] With awkward strokes of the pen, the chiefs—for a token payment of silver and a miscellany of presents—ceded six million acres of their brothers' hunting and fishing paradise to the United States government.[11]

After the native population was thought to be out of the picture, land surveys were begun. It looked like smooth sailing for the Saginaw land promoters. But a few Indians, who were understandably dissatisfied with the Treaty, decided otherwise and began a campaign of harassment that didn't let up until it was decided in Washington to build a fort at Saginaw[12]—the object being "to cultivate a good understanding with the Indians in the neighborhood of Sagana, also to give protection to the Public surveyors in that quarter, and to such inhabitants as may establish themselves in that vicinity."[13]

In 1822, under orders from General Winfield Scott (who would later be a MexicanWar hero), two companies of the 3rd U.S. Infantry, and supplies that included "nails, hinges, locks, glass & etc. necessary for the huts,"[14] were put on board the schooner *Superior*, and on the 18th of June 1822, it sailed into the Saginaw River and "took position there."

A stockade was put up on the west bank of the river about twenty miles from the river mouth, and the garrison, as well as some civilians, moved in.

But after only eighteen months of occupancy, the order was given to abandon the fort. Not because of attacks made by hostile natives, but because of a more sinister onset—malaria and typhoid fever, which caused sickness and death within the stockade. A retreat was made to Detroit, and it wasn't long before grim stories (enlarged upon by the fur traders who wanted to keep the Valley to themselves) began to drift back to the eastern states about Saginaw's "unhealthy, miasmal climate," that tale bearers said was more fit for bullfrogs than it was for humans.

Instead of seeing an inpouring of settlers come into the newly opened territory, the Saginaw land promoters saw only a handful of stalwarts settle there. Easterners, seeking cheap government land, skirted the region like it had the plague—going on to the other parts of the Michigan Territory or farther west to make new homes.

The national panic of 1837 was an added blow to the Saginaw promoters—but by the late 1840's things began to pick up. And by the year 1854, even though there were no streets laid out in Saginaw city—brush heaps and mud holes served as highways and "stumps of trees were everywhere encountered"[15]—there were forty-one steam sawmills along the river within a distance of seventeen miles—and two steamboats were coming in daily.[16] The predictions of the historian Alexis de Tocquevile, who visited the region in 1831, were coming true. "In a few years," he wrote, "these impenetrable forests will have fallen. The sons of civilization and industry will break the silence of the Saginaw; its echoes will cease; the banks will be imprisoned by quays; its current, which flows on unnoticed and tranquil through a nameless waste, will be stemmed by the prows of vessels."[17]

By 1868, the Frenchman's predictions had come true. Where he and his companion, Beaumont, saw Chippewa camp fires winking in the dusk along sandy shores, they would now see sawmills dotting the Saginaw—seventy-one mills whose spinning saws would, that season, turn pine logs into nearly 500 million feet of white pine lumber.[18] "Pine" had become synonymous with the name Saginaw—the pine lumber industry's growth being nothing less than phenomenal; and so, too, the population growth of the mill towns situated on the Saginaw River. They had become boom-towns almost over night.

In 1850, there were but 251 hardy souls living in the place called Buena Vista,[19] which was located across the river from the old town, Saginaw City. In fifteen years, the place (now called East Saginaw) boasted of 8,003 inhabitants.[20] By the year 1868, the town's population had grown to 14,261[21]—having outstripped her older sister, whose citizenry now numbered 7,134.[22]

Twenty miles down river at Bay City (once called Lower Saginaw) boosters bragged about their town's population rising from 810 in 1860[23] to 7,000 in 1868[24]—while across the river on the west bank, where a scant four years before, there wasn't much more than a mill, boarding house and ferryboat landing, the village of Wenona now counted 1200 people.

Before the railroads reached the Saginaws, the communities were almost isolated from the outside world once the river froze over. Now, no matter the season, two railroad lines[25] made it possible for travelers to ride to Detroit in four hours, where connections could be made to the East and West.

There wasn't a bridge built across the Saginaw until the

year 1864. By 1868 there were four new wooden spans—three connecting Saginaw City with East Saginaw, and one uniting Bay City and the village of Wenona.

Although there were the railroads, new bridges, sounds of industry and new buildings going up all over the place—there was still a look of the frontier about the mill towns. Visitors from the East, who were accustomed to seeing neatly laid out city parks, green lawns and cultivated fields, would be struck by the rawness of the towns. Looking across the river toward East Saginaw, one could still see the tops of forest trees beyond church spires and mill chimneys;[26] and tree stumps would be an aggravation to builders for years to come. The woods and bayous were so close, wild animals were known to wander into town by mistake.[27]

Like other wooden, jerry-built communities of the era, the Saginaws were towns where the clang of city fire alarms was as familiar to the inhabitants as were the sounds of mill whistles and church bells. The alarms set into motion volunteer fire fighters like the Red Rover, the Neptune or the Rough and Ready companies, as well as neighborhood bucket brigades who armed themselves with pails, pans, kettles—anything that would hold water—passing the containers from hand to hand from well or cistern to put out a blaze, or to wet down adjoining property to keep it from going up in flames.

It wasn't easy to confine a fire in a tinderbox town, where buildings, sidewalks and even the pavements were made of wood—where every other empty lot was covered with slabwood or sawdust—and wooden docks along the river front were stacked high with lumber piles awaiting shipment.

The careless use of oil and gas for illuminating the streets,[28] dwellings, the mills and other places of business—

and sparks flying from tug and steamboat smokestacks—turned whole sections of the towns and the water front into raging infernos within minutes.

Only three years before, a fire consumed one-third of Bay City's business section—wiping out thirty buildings and damaging the streetcar tracks, which curled out of shape in the intense heat. Within a week, a fire limits ordinance, which prohibited the erection of wooden buildings within the business district, was passed by the City Council in a special meeting.[29] Now, handsome new brick "blocks" were rising from blackened, charred timbers of former wooden business establishments.

But in spite of the fire limits laws, fires would continue to plague the mill towns until well into the "nineties" and volunteer fire companies, resplendent in handsome uniforms, would continue to be a familiar sight,—their members running down streets in the smoky, ruddy glare of burning buildings—pulling hook and ladder carts and hand-operated pumps, or clinging to the backs of horse-drawn trucks. They would be long remembered by those who witnessed or took part in the almost daily ritual of fighting fires.

For all the noise of carpenters' hammers and the sight of new buildings rising every day, the demand for houses to rent continued to be greater than the supply.[30] People moving to the Saginaws were advised to "buy, build or emigrate," or to "bring along the house in which you expect to live."[31] It stands to reason that much of the population "boarded out." Boarding house signs hung between porch posts of every other house on streets like Water, Hamilton, Monroe, Adams and Granger. "As a general thing," one man wrote, while reflecting on his early life living in Saginaw boarding houses, "we have greasy reminiscences of them which even now haunt us at the sight of soup, hash and bread pudding."[32] No doubt the dreary life in cramped back rooms gave more than one young man and his bride the impetus to work hard for success so that they might, one day, own their own home, or live "up on the hill" with the rest of the "high muckedy-mucks" (as Cap called the lumbermen, bankers and lawyers who lived there).

"On the hill," in Saginaw City, meant Washington Street[33]—a broad high and dry, tree-shaded avenue that followed a natural ridge running parallel to Hamilton and Water. The street was in contrast to those located below it, in that there was a genteel, unhurried—yet unfinished—look about it with its two straight rows of nursery-grown trees planted, one after another, down both sides of the wide roadway. The houses on the broad avenue were built far apart—in many instances, a single estate occupying, in imposing grandeur, an entire city block.

The big houses on the hill might have seemed very grand to the mill hands and lumber shovers, who were restricted by their meager earnings to the often wretched living quarters of the mill owner's boarding house or rent cottages. But the spacious houses and carriage barns that were built on Washington Street were far less pretentious than the mansions the so-called lumber "barons" would build in the eighties and nineties—when they would attempt to out-do one another by seeing who could build the biggest house with the most stained glass, projecting cornices, towers and cast-iron finials. In 1868, and for more than a decade afterwards, the rich were content to live in roomy Italianate-style houses, ornamented with shuttered windows, broad verandas and sometimes a cupola on top (known then as a belvedere or observatory), where mill owners could stand with spy glass to watch for their lumber vessels—as anxious as any New Bedford man watching for the return of his whaling ship.

And most houses and property in the sawmill towns were surrounded by wooden fences built to keep out the wandering livestock that roamed the city streets. In Bay City, the town council had but recently passed an ordinance to restrain swine and other animals (except cows) from running at large[34]—and though the city pound did its best to round up the strays, complaints persisted from citizens who were kept awake on hot summer nights by the "ting tong" of bells hanging from the necks of cows grazing under their windows.[35] The summer nights were also filled with the sounds of bullfrogs croaking and mosquitoes humming—annoyances that would continue in East Saginaw until the bayou that stretched the length of the town was filled in by slabs and sawdust fill from the mills.[36] With the coming of warm weather, everybody slapped at mosquitoes and drank bitters for the ague just like the early settlers did before them.

As the foreman and the two swampers hurried down Water Street, they were jostled from side to side by returning woodsmen and townspeople going about their business.

The snow had stopped falling, and the sun was daring to come out. Tree buds were swelling overhead, but it was still nippy under foot. Chickens shivered in doorways, and women clutched woolen shawls around their shoulders as they stopped on their errands to wipe runny noses of the smallest of their offspring.

An assortment of horse and ox-drawn conveyances joggled up and down the street—manure carts on their way to the country and farm wagons coming in from the country—as well as a medley of express wagons, dump carts, omnibusses, drays, and carriages. Finding a hole in the traffic, the foreman and his companions dodged between turning wheels and the hooves of animals to make their way to the opposite side of the busy roadway, stepping over horse droppings and around puddles, and jumping to clear the shallow ditch at the side of the street before continuing on down the thoroughfare.

Every other building on Water Street was a saloon, and from open doorways pungent whiffs of beer escaped to blend with a conglomerate of smells in the street. The men heard noises of revelry—out-of-tune pianos, tinny nickelodeons, punctuated by the boisterous laughter of ladies of questionable virtue—sounding over the deeper outbursts of woodsmen and rivermen on leave.

Not all returning woodsmen were in the saloons. Some congregated on street corners—others leaned idly against buildings and lamp posts. McTaggart saw several familiar faces and acknowledged smiles of recognition with a wave of his hand. He noticed, too, that the lightening rod men were in town and that quantities of furs were being brought in by trappers—certain signs of spring on the Saginaw.

Other evidence of the season were the gaudy displays in shop windows to lure woodsmen inside who had money in their pockets. Jewelry store windows were stocked with quantities of trinkets—cheap pocket watches, watch chains, stick pins and lockets. Clothing stores displayed eye-catching selections of clothing ready-made of "English and French and American cassimeres," as well as paper collars and kid congress gaiters. The number of men wearing corked boots and red sashes, who were coming out of the shops with bundles tucked under their arms, proved that the owners of the establishments knew something about the tastes of woodsmen.

As the men continued down the street, they passed several vacant lots piled with stave bolts, slabs and other refuse from the mills. Another empty lot had a large tent pitched on it where an enterprising photographer had set up shop. On each side of the entrance to the canvas gallery were large portable signs to attract passing woodsmen—"Tintypes—2 for 10¢! Finished in 5 minutes!"

The photographer was doing a brisk business—even better than the jewelry and clothing store operators—for the ferrotype (tintype photograph) was cheap and quick and easy to make in sizes that would fit into the family photo album or into lockets, brooches and rings—making ideal gifts for wives and sweethearts.

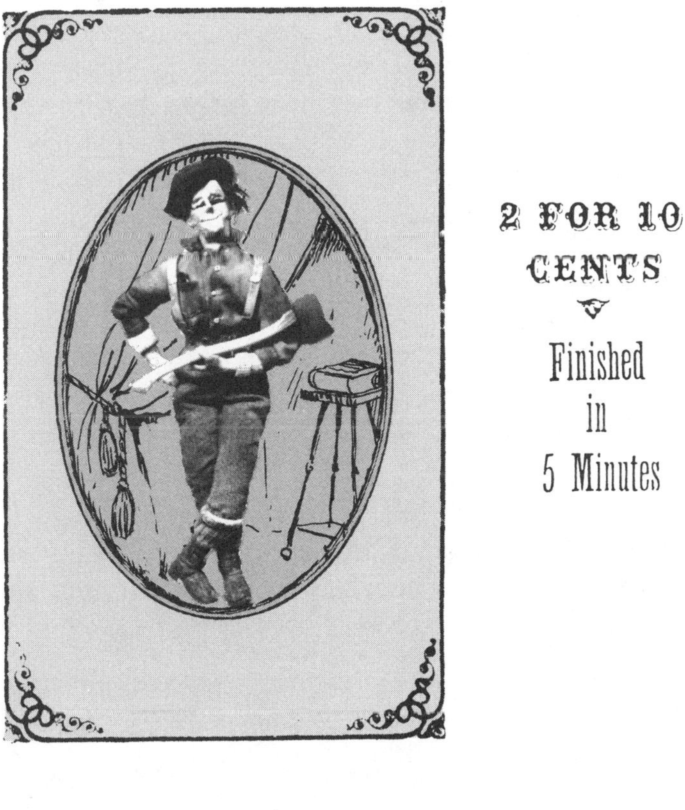

The line-up in front of the tent was a long one, and McTaggart smiled when he saw woodsmen—some of them fresh from the public bath and barber shop—adjusting sashes and taking off their hats to slick down stubborn, newly-trimmed locks—careful to fold back felt brims at jaunty angles when hats were put on again.

McTaggart was still absorbed in the sights at the photo "studio" when he felt a tug at his sleeve and heard Cully exclaim, "Boss! Look at them poor hay-burners!"

The swamper was pointing down the muddy street where a drove of used up lumber camp horses were approaching—prodded on by several handlers.

Traffic was pulling to the side of the road to make room for the odd-looking parade. A boy led the procession shouting, "Auction! Auction!" at the top of his lungs.

The horses looked in bad shape—their coats unkempt and mud-specked. Some were limping as they made their way through the mire to the empty lot where they would be auctioned off to the highest bidders.

Bill said, "I'm glad Shannon don't allow that to happen to his horses." And McTaggart, who was saddened by the sight, said, "Amen to that!"—knowing that the animals would meet their end pulling a broken-down slab cart, a horsecar or plow. Most of them would be shamefully neglected and even cruelly treated by their new masters. It was the ultimate fate of many camp horses.

The three watched the animals and their handlers until they were out of sight and then proceeded on their way—turning onto Court Street at the end of the block.

When they rounded the corner, they heard another street crier bawling, "Extry! Extry! *Albatross* goes aground in gale! First vessel stranded this season! More ships on Carrollton bar!"

Hearing the word, "Albatross," McTaggart hailed the newsboy. The schooner was one of John Shannon's lumber vessels, and the foreman was interested in knowing if she was a total loss. He exchanged two cents for a copy of the newspaper and quickly turned the pages to the local news. Bill and Cully waited impatiently near by.

The outer bar at the mouth of the Saginaw River had plagued vessel owners for years—until last year, when dredging had deepened the channel to twelve feet.[37] But, now, McTaggart read that yesterday's southwesterly gale winds had lowered the water to a scant eight feet at the outer bar and a little over four feet at the treacherous bar north of East Saginaw. The steam-barge *Estabrook*, which had attempted to enter the river during the storm, had successfully come in from the Bay, only to get stuck on the sand bar at Carrollton. And there were four more boats aground there. The tug *Moiles* and several lighters had been engaged all night lightering and discharging cargoes—while another big tug had been sent out into the teeth of the gale up the east shore of the Bay and around the Charity Islands to look for ships in trouble—coming back to report the barge *Albatross* was ashore (her crew rescued by her tug), and lumber scattered everywhere.

While McTaggart was scanning his paper, Bill and Cully, for the want of something to do, walked over to where passersby were gathering to watch an advance man for a circus paste a gaudy poster on a six-foot board fence that enclosed the untidy yard of Charlie Pocket's Saloon. There were numerous other advertisements on the fence (pasted over old ones that were faded and peeling) touting everything from patent medicines and "Pure Monongahela & Old Kentucky Bourbon Whiskey" to moonlight excursions aboard the steamer *Ariel*. But it was the flaming poster picturing bare-back riders; slack wire artists and other circus performers; snarling tigers and mammoth elephants bedecked in red and gold tassels; that made Bill and Cully forget the foreman and even their pal, Toohey.

As soon as the poster man slapped on the last panel and picked up his paste bucket and brush to stow them in his buggy, the two swampers pushed in between a couple of small boys to examine the big poster up close.

They were thus occupied, absorbed in spelling out the flamboyant advertising on the still damp sheets: "Adam Forepaugh's Grand Menagerie and Gigantic Circus—the only Legitimate, Undivided, Unepitomized and Comprehensive Wild Beast Show in America!", when a stranger, who was seedy in dress, appeared out of nowhere to stand at the side of Bill Bragg.

While feigning interest in the colorful display, it might have been observed that out of the corner of his eye the man was at the same time sizing up the noticeable bulge in the swamper's back pocket. He stood there for a minute more and then cooly eased a slim forefinger and thumb into Bill's pocket.

McTaggart chose that moment to look up from his paper—in time to see the pickpocket relieve the unsuspecting Bragg of his wallet.

"Hold on there!" shouted the foreman.

Startled, the thief looked around. He saw a tall, red haired man waving a newspaper at him—coming on the run. The pursesnatcher didn't linger. He swiftly pocketed his prize and jumped—catching hold of the top of the fence and vaulting lightly over the wooden palings.

Bill was still not aware of what happened. He watched in disbelief McTaggart running after a complete stranger. When he heard the foreman shout over his shoulder, "A pickpocket stole your money!" and saw Cully following after McTaggart—the swamper's hand went swiftly to his back pocket. Feeling no familiar lump, Bill came alive. He made a dash for the fence—pulling himself up and over in time to be greeted by a hail of empty whiskey bottles. The thief was picking them up from a trash heap in a corner and throwing them at the men who were chasing him.

In spite of the flying missiles, the three woodsmen steadily advanced across the yard until they saw the purse-snatcher throw the last of his ammunition and retreat around the woodshed. Cully yelled, "Head him off at the ice house!" But the nimble-footed thief was already making tracks through the dirty sawdust that was leaking out the door of the building. The next instant he was scrambling back over the fence—the foreman, Bill Bragg and Cully hot on his heels.

Now, the pickpocket seemed to be on familiar ground, no doubt having been in similar circumstances before, for he led his three pursuers on a wild chase into a maze of alleyways, through back yards, and around manure heaps at livery stable doors—between slab piles and over wheelbarrows and rubbish heaped in vacant lots—circling back to Pocket's Saloon. He ran into the barroom and out the back door where Bill, puffing like a steam engine, caught up with the winded pursesnatcher as he sailed around the pump.

The man went limp—his feet sliding in the sawdust that had been washed from hundreds of chunks of river-ice with water pumped from the well.

The chase had attracted some attention—and while the thief was loudly expressing his innocence—at the same time giving up, at the insistence of Bill, the stolen pocketbook—the sheriff and his deputy burst from the back door of the saloon. They were followed by a crowd of curious imbibers.

The sheriff, Quackenbush, walked up to the man now held in a tight grip by McTaggart.

"Harley! (the lawman was evidently acquainted with the thief) Are you at it again?" Quackenbush sounded reproachful.

"I thought I told you to go across the river—or anyplace else besides Saginaw City to practice yer trade." He looked over the man's head at McTaggart and said, "Just as soon as the woodsmen start coming to town, Harley and his friends go into business. What's he stole this time?"

Before the foreman had time to answer, Bill spoke up. "All my winter and drive pay! But I got it back." The last was said with some satisfaction.

The man called Harley looked glum. He was silent while Quackenbush proceeded to turn the thief's pockets inside out.

Finding nothing but lint and tobacco crumbs, the sheriff said to the foreman, "You want to press charges?"

McTaggart answered, "It's up to Bill, there."

The swamper shrugged.

The foreman looked over at Wills. He was at the pump holding a red-stained rag to his nose.

"Well, Cully—what do you say—?"

"Don't care," answered the bloodied swamper, "as long as Bill got all his pay back—."

McTaggart turned back to the law man. "You do with Harley what you want, Sheriff. We'll be out of town by tomorrow. Haven't got time to press charges."

The trio started out the gate, but halted when they heard McTaggart shout, "Wait a minute—!" The foreman and the two swampers, in all the excitement, had forgotten about the camp chickadee. But now, remembering, McTaggart asked, "Do you have a man locked up in your jail by the name of Toohey—Pat Toohey?"

Quackenbush, after instructing the deputy and the prisoner to go on ahead of him, turned back to the foreman.

"You related to him or something?" The sheriff sounded curious.

McTaggart answered, "No—Pat worked for me last season." He stuck out his hand. "My name is Alex McTaggart—Foreman, Camp 3—Red Keg—up on the Tittabawass'. We were heading for the jail a while ago to see about Toohey when we were waylaid by the pickpocket—."

Quackenbush interrupted the foreman to say, "You must be the one he left the note for."

"What note?" asked the foreman. It was his turn to sound curious.

The sheriff answered, "Sometime in the night, your man Toohey broke out of the jail house. Somebody sawed off the window bars and he lit out for some place up in Canady. Left a note for you—."

Motioning McTaggart aside, Quackenbush handed him a soiled envelope with cramped writing on the back. In a low voice he said, "The rig the boys stole belongs to one of our big lumbermen what lives up on the hill. He's one of the directors of the boom company and is on the board of a couple of banks here—." He lowered his voice to a whisper. "He was visiting his lady friend who works at the River House, when Pat Toohey and the Highlander swiped his horse and buggy."

The foreman smiled. He got the message. No charges would be made against the chickadee—the lumberman-banker wouldn't want to risk the notoriety of the arrest.

The sheriff went on to say (this time in his natural sounding voice), "I was going to let your man out after he sobered up—but he left of his own accord—with some help from Richie. Found a torn piece of wool plaid under the window they broke out of." This last was said by way of an explanation.

He added. "If you have any ider where this Toohey is at, you can tell him he's off the hook."

McTaggart thanked the sheriff, and before the man was out of the yard he was reading the note—so laboriously penciled by the chickadee on the scrap of envelope.

Der McTagart

I leve my best respecks at you all.
dont think hard of me for gitting out
of this plas McTagert. I cant ber the
thotes of bein confine whin i ante to
blame and they wont let me out.
All i want is my liberty till trial.
If they want to bale me they son
kin find me. I will rite you

Patrick Toohey

P.S. I am agooing home to Canady
to see my famly i cant stop
here no longer. I shall son rite
to you.

21

U is for the UNWINDING, they all had their reasons

McTaggart finished reading Toohey's note and unconsciously folded it neatly in the middle—creasing the paper between his thumb and index finger. He took a deep breath. So, it had begun—the "unwinding." How many more of his crew would "go off the deep end" before they met at John Shannon's mill office after supper? He looked in Cully's and Bill's direction. They looked, expectantly, back at him—but his gaze went past them—somewhere beyond the saloon keeper's yard.

"They've got to blow off steam," McTaggart was thinking, "but for once, I wish they'd come into town, get a bath and a haircut, lift a glass or two with their chums and then call it quits."

In spite of himself, the foreman had to smile at this bit of fantasy. It wasn't an easy thing to do—"call it quits" in Saginaw Town. He knew that from past experience.

It was a tradition among liberated woodsmen, dating back to the early lumbering days in Maine and in New Brunswick, for the boys to seek out a "blowing-in" spot in the spring time—places like The Forks on the Kennebec or Haymarket Square in Bangor.[1] McTaggart recalled hearing Kelly (when he was in one of his mellower moods) tell about his old Dad, who was a woodsman up on the St. Croix. How he and forty other bearded men, at the end of the logging season, would parade into town. They first stopped on the outskirts of the village to wind red scarves around their waists and decorate their hats, the yokes of their oxen and the sleighs with handkerchiefs and long streamers made of red flannel torn from old shirts.[2] The woodsmen, picturesque in their impromptu finery, drove their caravans down the main street—causing more excitement than the town had seen in months.

Like the mill towns on the Saginaw, the logging towns in Maine depended upon the woodsmen for their livelihood, and the tradesmen welcomed them with open arms. Grog shops were almost as numerous in Bangor in the 1830's as they were in the Saginaws in the 1860's. One spring in 1832, thirty-five hundred gallons of "ardent spirits" were consumed on the St. Croix upon the arrival of the woodsmen.[3] There is no record available of the number of gallons of spirits that were dissipated during the logging days on the Saginaw—but the figures would be impressive.

To returning woodsmen with cash in their pockets, the towns were paradise where they could feel free again after living for six or seven months the secluded, disciplined life of monks—isolated from strong drink and female companionship.[4] The mill towns were where they could feel the firey contents of the whiskey bottle running down their throats and the sublime, sweet relaxation that followed (before the liquor befogged their brains). Whiskey is what they dreamed about, shivering under blankets on cold winter nights in the lumber camp—that and the glorious romps with the painted "frail ones," the noise of the brass band and the rowdy frolics at Joe Mills's Music Hall down in Bay City.

To many of the men the splendid blow-out in Saginaw or in the river towns near the Bay, was worth every hard-earned cent they had—and they happily squandered their winter's wages with no regrets.

It is understandable that the "good people" of the mill towns only took notice of the antics of the hell-raisers who kept the towns lively from early spring until fall. The townspeople did not give much thought to the majority of the woodsmen who, when spring came, headed straight for home—which was likely as not a pine stump "farm" in the midst of the slashings of a former pine forest. They handed over their pay to waiting helpmates before pulling on overalls to go out and plant a crop of potatoes among the stumps. The wide open towns had only a passing interest for these men, who set their sights on the day far off when they would have enough money put away under the corn-husk mattress to pay off "the place" or to plunk down for a patent stump-puller.

But for the rest! Only recently, outraged by the numerous fights occurring on the streets in front of the Shades Saloon and other "evil dens," seventy-one angry citizens in Bay City got together and signed a petition asking the town council to pass an ordinance "enabling such places to be prosecuted as nuisances and the peace of the city be promoted by the closing of them."[5]

The petition was referred to a "special committee" and was never heard of again. Perhaps the Council thought that the town's blue law (which closed the saloons and hotel barrooms on Sunday) was enough of a concession to make to the petitioners. Business was business.

The "dens" continued to have it mostly their own way until the 1870's when the new decade brought with it a big temperance revival that rolled across the country, gathering momentum as it went along.[6]

Like other temperance people in the nation, those in the Saginaw Valley were waiting for the trumpet call; and when they heard it, they got busy and organized into red ribbon societies and other temperance organizations—like the Reform Club and the Royal Templers of Temperance[7]—recruiting members who took up the crusade to save the inebriates and make life miserable for people like Joe Shefnecker in Saginaw and Mattie Morgan at the Domino down in Bay city. The crusaders kept things stirred up—marching in parades with banners flying—and storming into the "gin palaces" with axes swinging.

Mattie confessed to her colleague, Ross Robb, that the zealots were more trouble than a November gale was to a tow of logs—and did as much damage.

Rivermen and loggers in from the drives—and sailors waiting in town for their vessels to take on an another load of pine—were the prime targets of the temperance people. Certainly these men were their biggest challenge—and the Royal Templers and the red ribbon brigades steadfastly (but without much reality) sought them out on the docks, on street corners and in the saloons. They passed out temperance tracts and exhorted candidates to attend their meetings which were held in public halls and in the towns' churches.

The woodsmen attended the meetings (it's not clear

whether they went for an evening's lark or to be "saved"). They listened to the recitations, the chalk talks and the lectures—and they joined in singing songs like "Throw out the Life Line,"[8] and the more spirited "Wear the Badge Proudly."[9] The meetings gathered momentum when the aspirants, aroused by impassioned speeches and songs, summoned the courage to stand up and talk about their past experiences with the "demon rum." The impromptu confessions, being on the emotional side, often brought the evening gatherings to a dramatic climax. With the "Temperance Fire Song" still ringing in their ears ("Fire! Firé! The rumfiends fire / Sweeps t'ward our hearth-stone nigher and nigher;")[10]—weeping, remorseful bibbers stumbled forward to take up the pen and sign their names (or their X's) to temperance pledges, vowing to "neither make, buy or sell or use as a beverage any spirituous liquors, wine or cider."[11]

Ribbon badges—the symbols of the movement—were given to the signers to wear. With red (or white or blue) ribbons pinned to lapels and on sleeves, the teary-eyed wearers,still buoyed up by their emotions, went out into the night, knowing for certain that they at last, had been brought up "from the depths."

But, in spite of the well-meaning efforts of the temperance people—in spite of their rallying songs, prayers and exhortations (and broken plate glass mirrors and smashed whiskey bottles), the saloons did not disappear from the sawmill towns. If anything, they multiplied in number—keeping up with the enormous thirsts of the ever-increasing numbers of men who took part in the pine lumber business—the gigantic industry that ruled over everything in the Saginaw Valley.

In 1873, which was about the time when the temperance movement began in the mill towns, there were sixty-two saloons divided between Bay City and the village of Wenona (most of them were in Bay City).[12] By the time the year 1887 rolled around, there were 156 saloons in Bay City alone. That year, there were 737 arrests made by the town's police department—559 being charged to the tipplers who raised one glass too many.[13]

Only when the woodsmen said goodby forever to the Saginaw towns, did most of the hangouts on Water and Lapeer streets go out of business. Some saloon keepers followed the boys west, where they would continue to "blow 'er in" in places like Eureka on Humbolt Bay or in the City of Aberdeen on Washington's west coast.[14]

The foreman had folded and creased Toohey's note until it was almost in shreds. Ordinarily, it was none of his business what the men did away from camp—or how they spent their money. Several of the boys who wanted to enjoy a few hours "on the town," but not trusting themselves to the temptations of the city, had asked McTaggart to hold onto their wages—taking out only a few bills for a little hurrahing—each man rolling his money into a "Michigan bankroll"[15] to make the bills look like a bundle to impress their colleagues.

"If only I had insisted on holding Pat's wages for him," he thought to himself, "he wouldn't be going home to his family empty handed and in disgrace. And tonight—what if the whole crew shows up at Shannon's ceremony drunk as skunks—."

Bill and Cully, waiting impatiently for the foreman to make a move, watched him tuck what was left of the chickadee's note into his coat pocket. They heard him say something that sounded like, "Idleness—mother of mischief—."

McTaggart motioned the men to him. He seemed to

have made up his mind about something. He said, "I've got some important business I have to take care of at the land office over at East Saginaw. It can't wait any longer. While I'm gone, I want you two to round up every man in our crew that you can find in town. You know where to look. Start up at the Steamboat Saloon and work on down Water Street to the River House and on towards the mill. I don't care if the boys are in a bordello or drinking sarsaparilla in an ice cream parlor—haul 'em out and get them over to Shannon's new company store. Tomorrow being Saturday, Hoops will be glad for some extra hands. Maybe if the men are kept busy, they'll stay out of trouble."

Cully, who was still nursing his bruised nose, interrupted the foreman to say, "How about East Saginaw?"

McTaggart answered, "I'll take care of the East Side while I'm over there." He went on with his instructions. "If anybody puts up an argument, tell them if they want a bunk at Camp 3—or any other camp on the Tittabawass', they better be at the store by the time I get back. Got that?"

Both Bill and Cully nodded their heads—at the same time looked doubtful as to their mission. But the look on the foreman's face put an end to any more conversation about it. They turned to leave the bottle-strewn yard to disappear down the street—while McTaggart hurried off in the opposite direction.

• • •

Every half hour or so horsecars ran across town and over the Genesee Street Bridge (a mile farther down river) to East Saginaw. Not seeing a car in sight, the foreman decided to hike it over the "middle bridge" at Bristol Street.

When he got there, McTaggart saw that the wooden span was open. "Another delay," he thought impatiently.

The river traffic was heavy that morning—and, as was the usual custom during peak hours, instead of opening and closing the bridges at every signal of tug or steamboat, the bridge tenders simply left the big spans open to let the parade of boats go through without interruption. The townspeople were obliged to wait to cross the river until there was a lull in the traffic below.

Most of the waiting crowd seemed resigned to the delay in their business and stood huddled in groups—the women's voluminous skirts flapping around them in the chilly April wind. Truants from school chased each other around carts and wagons and under the legs of patient horses—and tried to see who could get closest to the unguarded edge of the open draw without falling into the river. Those engaged in conversation had to speak loud enough to be heard over boat whistles and the everlasting noise of the pile driver.

"How long she been open?"

The man to whom McTaggart directed his question looked down from his perch on a springless seat of a broken-down wagon. The foreman guessed he was a backwoods farmer by the look of the contents of the wagon—a jumble of hand-hewn ax handles and ox yokes—no doubt the results of his winter's work. He was more than likely taking the articles to East Saginaw to sell or to trade for provisions.

The farmer spat a stream of tobacco juice leeward, then wiped a ragged coat sleeve across his stained beard before replying. "Been here 'bout twenty minutes. I'd pull out and go down to the Genesee bridge—but sure as rain at a Sunday School picnic, she'd be open too."

Just then, the bridge's steam whistle sounded its warning to the river traffic, and McTaggart was relieved to hear the creak and rumble of the great swing span as it began to move back into place. He cordially nodded his head at the farmer and turned away to jog on ahead of the slower-moving horse-drawn vehicle. At the toll house he paid his three cents and proceeded down the long wooden approach to East Saginaw's Water Street where he turned north.

McTaggart stretched out his long legs—dodging street traffic and mud holes. As he hurried along, he tried to dismiss the camp crew from his thoughts and concentrate on the business at hand—a mission that had been delayed by one thing and another ever since he got into town yesterday afternoon. The business had to do with some pine the foreman had discovered last summer while cruising on the upper reaches of the Tittabawassee. His discovery gave meaning to his future.

Like the scaler, Isaac Hoops, McTaggart was ambitious to get ahead in the lumber business. Up until the present time, he had always worked for others. Now, he had a chance to strike out on his own. It came about like this:

Several seasons ago, the foreman had been looking over specific pinelands on orders from his employer, John Shannon. Having finished his land looking assignment for the lumberman, he decided to do a little cruising on his own behalf. He had heard rumors about good pine acreage situated somewhere between the Tittabawassee and Rifle rivers—so he started his search by working away from the headwaters of the Tittabawassee in a northeasterly direction.

The first two days McTaggart saw only scattered stands of pine—none of which interested him. On the third day his luck didn't change. It was misting at dawn, and as the morning gave way to afternoon, he was beating his way through a wilderness of underbrush in a blinding rain. Nightfall came early—the steady downpour and darkness drawing a thick curtain around him. He made camp as best he could under a water-proof—ate some ship's bread, rolled up in a blanket and fell asleep heartily discouraged.

The next morning he awoke to find that he had wandered by blind chance into a stand of virgin white pine—the sight of which almost took his breath away. The trees seemed to soar to the blue sky and beyond.

McTaggart knew about where he was—but he would need a point to work from to fix boundary lines. With racing heart, he spread out his maps. Slowly, he ran his finger down the Rifle River following the river's course until he found what he was looking for.

"There she is," he said half aloud, "right smack on the Arenac-Ogemaw county line." The foreman was referring to a section corner located on the bank of the river. It was a point from which he could start locating other corners.

"But first," he thought to himself, "I've got to find the river." He would follow its course south to find the section corner on the county line.

Because he had been traveling northeasterly the last three days, McTaggart thought he was about a mile west of the Rifle River. He struck camp and started going due east.

Three-quarters of a mile and half an hour later, the foreman saw the sparkling river shining through the trees.

The log drives were still going on. Pine logs were drifting down the rippling current, and he saw an occasional riverman working them down to the sorting booms at the river mouth.

McTaggart tried to keep out of sight as he made his way along the edge of the meandering stream. At the same time he kept his eyes open for old blaze marks on nearby trees that might indicate where the survey crew had placed the section post that he was looking for.

It took three days of searching, but he finally found it.

The old post—half rotted from exposure—was about six feet back from the river where the government surveyors had driven it many years before. It marked the SE corner of Section 33, Town 21 North, Range 3 East. Now the foreman knew exactly where he was, and he started to back-track to locate "his" stand of pine.

McTaggart spent another week with compass in hand—pacing off boundaries and making notes as he went along. At the end of his inspection tour, he knew that the pine he was after was on three, one-quarter sections of land—480 acres. It was all within hauling distance to the Rifle River.

Like most land lookers, all the time the foreman was looking over the terrain, he was nagged by the thought that someone else had already discovered the pine; and he kept his eyes open for signs of man—for evidence of camp fires and fresh-cut blazes on trees. And as he worked, he was careful to leave no trace of a camp or a blaze that would be discovered by another party.

Something else bothered him. His bank balance stood at a little over $200. He only had enough ready cash to purchase one of the quarter-sections. Borrowing money was out of the question, for he had no collateral other than his good name. He could only hope that no one would discover the stand until he could afford to purchase all of it.

When McTaggart completed his notes, he rolled up his gear and followed the Rifle down to where the river empties into Saginaw Bay. He arrived at the dock only an hour before the shoreline steamer *Reynolds* hove into view, trailing a smudge of black smoke behind her as she approached to "wood up" and take on passengers.

The foreman arrived in Saginaw dead beat with a two months growth of whiskers on his face. But he didn't stop to rest or to clean up. He went directly to the bank and withdrew all of his savings—taking the money to the land office where he found to his everlasting amazement (and relief) that the pineland he had selected on the Rifle River was still owned by the United States Government. He purchased one-quarter section—160 acres at $1.25 an acre.

Now, almost eight months later, he was on his way back to the land office to see if the remaining two-quarter sections were still unsold. If they were for sale—he was ready to buy. His winter's wages and his drive money were buttoned up in his coat pocket.

"Another winter at Red Keg," he was thinking, "and with my percentages earned while cruising for Shannon—and if we win the bonus—whatever it is—." So went the musings of the foreman as he hustled down the street.

The land office was located on the second floor of a brick building only recently put up on the corner of Washington and Genesee streets. It was across from the elegant new Bancroft House. McTaggart was sweating a little when he ran lightly up the wooden stairway at the side of the building. And when he opened the big tract book and started to turn over the pages, he felt a trickle of moisture run down the middle of his spine. But his new-found luck was with him still. The book revealed that the land he wanted had not been sold.

The foreman gave the clerk instructions to register the pineland in his name and then counted out the money to pay for it. He picked up his receipt and left the office—walking down the steps in a daze—still not believing his good fortune. When he reached the sidewalk, he took out his bandana and wiped his forehead, in relief. He was free now to take care of the next order of business—looking up the Red Keg crew.

• • •

There were over sixty saloons in East Saginaw and even though most of them were located on Genesee, Lapeer and South Water streets (all within walking distance), it would take too much time to look into each one.

"I'll try St. Nick's since it's right here on Washington," McTaggart said to himself.

A brief survey of the crowded saloon netted no results—so after treating himself to a beer (to celebrate his recent good luck), the foreman left Nick's and turned off Washington onto Genesee Street. He was walking east, stepping around the usual hawkers, reeling woodsmen and lamp post loungers, when he heard behind him running footsteps pounding the plank sidewalk—and a familiar voice call out, "Boss! Hey! Wait up—!" It was Jack Driscoll—all out of breath.

McTaggart waited for the top-loader to catch up and then took off with Driscoll at his side.

After the usual pleasantries, Jack said, "Got me a job—bouncer at Zoe Herto's place. Didn't turn in 'till about four this mornin'. Zoe give me a blanket and I got some shut-eye stretched out on some whiskey cases in the back room. The job's something to do till Monday when I go to work for the boom company drivin' logs." He paused. "Where we headed?"

McTaggart explained his mission, and the top-loader, glad for something to do the remainder of the day, immediately fell in with the foreman's plan.

The two men walked on until Driscoll pointed to a weathered frame building with a high false front. "You might find Hilly James and Whisperin' Bill over there. Last winter they was forever braggin' on how good they was playin' pool."

A sign was nailed across the front of the building with

the message, JOHN NERRETER'S NATIONAL SALOON & BILLIARD ROOM. Painted under the saloon keeper's name was a life-like foaming schooner of beer with two billiard cues crossed under it.

The suggestion was a good one. When the foreman and top-loader passed through the bar and into the billiard room, they spotted the two loaders almost at once. The men were standing in the shadows along one wall with other onlookers, who were intently watching a couple of rivermen playing pocket billiards under the wavering light of two hanging oil lamps. One of the men was preparing to make a play. While his opponent cooly chalked the tip of his cue the player crouched over the table and deliberately slid his tapered stick back and forth between his fingers. When he flicked it out, the stick connected—leather tip against ivory—and the spectators narrowly watched the balls explode into a blur of colors as they raced across the smooth green felt of the table top.

There was betting going on among the men, and McTaggart noticed that Hilly and Whispering Bill were clutching folding money in their fists. He was glad he had arrived on the scene while they still had some of their wages left. So while the players studied their next moves, the foreman edged over to the two loaders. He touched each one on the shoulder and motioned them out of hearing distance of the other spectators.

A brief, tense discussion followed. The words of the men were almost drowned out by the usual barroom noise, which was heightened by some woodsmen who were harmonizing off key through a half dozen verses of a popular song—accompanied by the clarion voice of one of Nerreter's girls, who was decked out in ostrich feathers and striped hose.

Whatever was said between McTaggart and the two loaders, the latter seemed to get the message, because at the end of the speech they glumly followed him and Driscoll out of the saloon.

Outside, Bill grumbled, "We was just startin' to win some money—."

The foreman ignored the complaint. He only said "We'll

Two woodsmen were harmonizing off-key through a half dozen verses of a popular song accompanied by the clarion voice of one of Nerreter's girls who was decked out in ostrich feathers and striped hose.

check out the Anchor House next—and there's a couple of other places nearby—." The little procession proceeded down the street.

Like St. Nick's, the Anchor House didn't reveal any of the Red Keg crew. One of the bartenders told McTaggart that he had seen Jimmy McHugh and Frank Cole playing cards in the place about an hour ago—but they had left the saloon to find another game. So the foreman and his followers angled off Genesee Street crossing at Jefferson onto Lapeer. They continued on their way, looking into places that had card games going. None of the boys had been seen in any of them. There was nothing to do but start back toward Pete Nickodemus's saloon.

They found the two choppers there. Frank Cole was at the bar, and McHugh was at a table playing cards. While Hilly and Whispering Bill went to give Frank the foreman's message, McTaggart and Driscoll walked over to the card game.

Jimmy-the-Chopper looked up and smiled in recognition when he saw the foreman. McTaggart curtly nodded his head toward the door, indicating the direction he wanted the chopper to take when he finished out his hand—then stepped back to wait.

The men were playing stud poker, and the dealer had just dealt the first round of cards. Most of the chips were stacked in front of Jim, so it's not surprising when the hand was played out and he made preparations to leave the table, there were some protests heard.

"You ain't goin' no place, chummy—not until me 'n the boys here gits a chance to win back some of them chips."

The speaker was a woodsman of considerable size. He smelled of bay rum and whiskey and had a blue jaw and fists that looked the size of McCloud's anvil. He shot out one long arm and roughly pulled the surprised chopper back into his chair—then looked around the table and said in a voice full of fight, "Whoever's dealin', *deal*!"

The players, anticipating trouble, had started to get up but settled back into their chairs when the dealer nervously started to shuffle the cards.

The big man said, "Come on—hurry it up—. Dealer's choice. What's yer pleasure?"

But before the dealer had a chance to name the game, Jack Driscoll moved in. He rested two fists on the edge of the table and leaned over so his eyes were almost level with those of the obstinant woodsman. "Gentlemen—pay up," he said. The voice did not betray any emotion—it was the top loader's look that gave meaning to his words.

Now, there was a sound of scuffling boots and shifting chairs as the barroom clientele pushed back from tables and craned their necks to get a better look at the man who was challenging the hostile thunder cloud at the card table.

McTaggart, Hilly James and Whispering Bill moved around to get into position to support the top-loader. And Frank Cole, who had started out the door, came back into the saloon when he heard the sudden hush fall over the room. The piano player stopped playing—the girls their singing. The attention of everyone in the barroom was suddenly arrested by the little drama that was being played between the logger with the blue jaw and Silver Jack Driscoll.

"I said PLAY! You deef 'er something!" bellowed Blue Jaw, at the same time taking great pains to ignore the man who was standing over the table. He seemed determined to keep Jimmy McHugh in the game.

Jack was just as determined. Once again he said, "One more time boys—count your chips." The top loader's voice had an edge to it now that neither his fellow crewman nor McTaggart had heard before.

The players responded instantly. There was a sudden click-clicking sound of poker chips being counted. Money began to exchange hands.

The situation had become unbearable for the enraged woodsman. With an oath he thundered, "I'll be choked by a damned saw-log if I cash in *my* chips!" Saying this, he flung Driscoll's fists off the card table, sending the surprised top-loader sprawling over the table. Drinks, poker chips, money and playing cards spilled over the floor as Jack struggled to gain his footing—all the while pushing the table before him into the laps of the astonished poker players, who landed backwards on the floor in a series of crashes.

At last, Jack regained his balance. He straightened up. Seeing the Jaw on the floor with two other players—he yelled over his shoulder, "You got your money, Jim?"

The chopper made a hasty calculation. He answered, "Yeah—most of it, I guess—."

"Well then beat it while I hold this guy off!"

As Jimmy started for the door, the big man on the floor slowly got to his feet. He was hell-bent on murder. Without taking his eyes off Jack, he picked up a chair and with all his strength, hurled it at the head of the top-loader.

That was the signal. The fight was on. The girls scattered and Nickodemus ran to the bar, imploring, "Please boys! Don't hit the mirror!" as a shower of bottles and glasses sailed through the air.

Within seconds there was a roaring melee inside the big room with combatants fighting and yelling for the utter joy of it. And in the midst of the contest were the top-loader, the two pool-playing experts, the chopper and the foreman of Red Keg, who, whether he liked it or not, had been forced into the fracus.

In spite of himself, McTaggart got caught up in the spirit of the battle and landed a few well-aimed hay-makers even while working himself and his men toward the swinging doors of the barroom. He reached the doorway at last, and, pushing Hilly and Bill out onto the sidewalk, he shouted after them, "Take the horsecar to Hoops's store and stay there. We'll be along directly."

The foreman looked back into the saloon in time to see Driscoll land a blow under the blue jaw of his adversary. The man went limp and sagged in a heap. Jack straightened, and, in the midst of the uproar, he calmly slapped at the sawdust on his pants with his battered hat, at the same time looked around the room for his friends. Catching the eye of the foreman, he waved, and then stepped over sprawling bodies, broken glass and furniture to join him at the door. Frank Cole limped behind, grinning from ear to ear.

On the street, the men stopped long enough to get their breath back and then trotted with McTaggart down Genesee Street—catching up with the others at the bridge approach.

The horsecar had halted there to allow a third horse to be hitched ahead of the regular team (to help pull the heavy car up the long incline to the bridge) and Hilly, Whispering

Bill and Jimmy-the-Chopper waved from the car windows, hallooing to their comrades as they approached.

The car was slow in starting up the grade and Jack Driscoll, still buoyed up with high spirits, called out, "Let's give the old Dobbins a hand, boys."

The Red Keg crew jumped to the pavement and put their shoulders to the back of the car—pushing until it was on the bridge. They hopped aboard to ride the remainder of the way to the company store.

When the foreman and his jubilant troupe reached Hoops's place, they found the two swampers, Cully Wills and Bill Bragg, and five other members of the troop at work. Isaac, glad for the extra hands, found jobs for the new arrivals.

It seemed like no time at all before the sun went down behind the big mill smokestack and whistles were blowing up and down the river announcing the end of another work day.

• • •

As expected, the lumberman, John Shannon, arrived in Saginaw City coming in that afternoon by through-ticket from Albany via the Great Western Railway of Canada—taking the F & P M cars up through Flint to East Saginaw where he hired a hack to drive to his new store.

Hoops had been primed for the last two days, ready to show his employer the company books, but when he started to lead him to the office, he was surprised to hear Shannon wave aside store business. "I've brought some boxes from New York and want them picked up from the freight office—right now," he said. "There is one with an X marked on it. Tell your men to deliver it to the mill office. The others—about ten crates more—are to be brought here, to the store, where they are to remain unopened."

John Shannon seemed to be enjoying himself. If he noticed the look of curiosity on his store manager's face, he chose to ignore it. Instead of enlightening him as to the contents of the mystery crates, he only asked the whereabouts of McTaggart. Hoops pointed to the storeroom, and as he watched his employer walk away, he could only think the mystery had something to do with the ceremony at the mill that evening. Whatever the crates contained, they sure put the old man in a good mood.

The boss-lumberman found his camp foreman performing one of the homelier duties of a tradesman—drawing molasses from a barrel—running it into a jug for one of Isaac's customers. While McTaggart looked around for a rag to wipe his hands on, the lumberman asked, "When are you leaving for the North?"

"Tomorrow morning," answered the foreman. "I'm leaving on the first steamer that pulls out of Bay City for Alpena. Got my gear all packed." He wondered what his employer had on his mind.

Shannon pulled a letter out of his pocket—unfolded it and handed the paper to the foreman. "It's about that forty I've been wanting that's adjacent to Camp 3."

McTaggart saw that the communication was short and to the point. It was from Ross Robb. The saloon keeper wanted to sell off the standing timber on the land next to Camp 3, and he was giving the lumberman first chance to purchase it. His asking price—$10.00 an acre. (Three years ago, Robb had paid the old trapper who homesteaded the land, $30 for the forty—or 75 cents an acre).

When the foreman finished reading the letter, he looked up. "I see Robb just wants to sell stumpage—not the land. Looks like he's going to stay in the saloon business up there." He saw the lumberman frown.

"You want me to act as your agent?"

"That's right," answered Shannon. "I want you to make a deal for that pine. You can see there in his letter, Robb will be at his place—the Domino, I think he calls it—down in Bay City, all day tomorrow. Dicker to lower the price, but if he won't come down—pay him what he's asking. *I want those trees.*"

"All right—I'll do what I can," said McTaggart—knowing as well as the lumberman the importance of the sale. The kind of virgin timber that was standing on Robb's property was fast disappearing from the river. But he couldn't help thinking about next winter—his last one at Camp 3. If the deal went through, he would be falling trees smack in the back yard of the Red Keg Saloon. And he thought he had problems with the place last season!

• • •

When the echoes of the last mill whistle died away on the river, the members of the foreman's crew (those who had been helping Hoops), went off to the boarding house for supper with instructions to go directly to the mill office afterwards. McTaggart stayed with Isaac at the store where they shared a scoop of crackers from the cracker barrel and a wedge of store cheese—washing it down with a growler of beer the foreman brought over from Shefnecker's saloon. Both men speculated on the contents of the mystery boxes over their repast.

When the substitute clerk arrived, the foreman and Hoops went off to join the other woodsmen who were gathering in front of the mill office.

The front steps of the red brick building were almost even with the edge of the scarred plank sidewalk, and by the time McTaggart and Isaac arrived, the men already assembled were spilling out onto the street and into the slab yard at the side of the mill office.

Except for the rise and fall of the men's voices, the scuffling of their boots on the board walk and the occasional rattle and jingle of a horse-drawn vehicle rolling down Water Street, the town seemed strangely still. The mills were not yet working their night shifts—and the pile drivers and river traffic had been quieted by the approaching darkness. A few stragglers from Shannon's mill and salt works hurried by on their way home. They looked with curiosity at the milling crowd of woodsmen standing restlessly in front of Shannon's office. A team hauling a Schemm & Schonheit beer truck (loaded with empty kegs) rumbled off the long Bristol Street Bridge approach onto Water—passing down the street behind the mill. It turned onto Brewster and rattled across the railroad tracks to disappear into the big brewery yard opposite Shannon's salt block. In the daytime, the noise made by the heavy truck would hardly have been noticed.

McTaggart and Hoops stood looking over the crowd.

"Most of the men look sober enough," thought the foreman, although there were several who looked a little flushed and unsteady on their feet. He recognized most of the men from Camps 1 and 2. Lanky Zip Somers, who figured in last winter's bee tree episode, was among them. His long nose looked purple in the twilight. McTaggart wondered idly if the stings he got from the bees had anything to do with its color.

There were a few faces missing from the Red Keg crew. For something to do, McTaggart mentally ticked off the men who would not be present at the ceremonies that evening—beginning with Cap Wilkes. He had gone back to camp with George Pepper and the drive paraphernalia. It would take more inducement than the possibility of winning a prize to get the cookee into town. The barn-boss, Billy Flynn, and a couple of the teamsters, as well as ox driver, Adam Oliver, were at Shannon's farm near Freeland's Station. And Alex McCloud, the blacksmith—and of course, Toohey, were well on their way to Canada. Several of the crew, who had farms in Saginaw and other nearby counties, had set out for home the instant the drive reached the booms—and the foreman knew that the choppers, John Tree, and his pal, Jack Potter, had gone down to Bangor to look for jobs at Crosthwaite's shipyard.

McTaggart saw Duff Kelly slouched against a slab cart talking with the cook at Camp 2. He was still wearing his battered, flour-specked bowler. He looked strange without his apron on. "I'll probably see him on the boat going down to Bay City tomorrow," thought the foreman. (Hoops had told him Kelly was shipping out from Wenona as cook on one of Henry Sage's lumber vessels).

Ben Sweeney, Ed Flatt, Omer Fuller and Sam Bean joined the group. Each was carrying a dinner bucket—having come over the "middle bridge" from East Saginaw where they were employed at the big Sears and Holland mill.

Daylight was fast disappearing, and the men idly watched the lamp lighter come down the street on horseback to light the lamp on the corner. But the pale illumination did nothing to light the yard and street where the woodsmen were congregated.

Presently, McTaggart saw the mill superintendent walk out onto the narrow porch of the office. He was followed by the near-sighted clerk, who was carrying a couple of lighted barn lanterns. Burleson looked over the crowd and then went back inside. The bookkeeper hung the lanterns up on hooks between the porch posts.

In the lantern light, the foreman observed, for the first time, Mick O'Halloran who was standing at the edge of the gathering. He was talking with Frenchy LaRiviere. The boss-riverman's dog, Jupe, still aloof, was at his side.

Hoops's voice broke into the foreman's train of thought. "Wonder what's holding up the show," he said. The scaler-turned-store-keeper was anxious to get back to his job. He looked at his watch, peering at the dial in the dim light. "Just seven o'clock. Seems like we've been here longer than that."

"There's Burleson again," said the foreman. He and Isaac watched the superintendent come out and beckon to two men standing near the steps. The woodsmen mounted the porch and followed him into the office. When they came out again, they were bearing a large wooden packing crate.

"Now, what," thought McTaggart. He heard Isaac say, "Looks like one of the crates that was delivered to the store this afternoon." There was an X marked on the side of the box.

There was some pushing and shoving in the crowd, as the men tried to see what was going on. The foreman wished Shannon would begin the ceremonies. They had been waiting around long enough.

John Shannon took that moment to walk out the green-paneled door.

In honor of the occasion, the boss-lumberman was wearing a high silk hat and a cutaway coat. The lantern light created gleaming highlights on the hat and made the clean-shaven cheeks of the mill owner look polished—as if he had recently come from the hotel barber shop.

Shannon stepped up on the big box (so all could see him). He held up his hands for the men to quiet down.

The visiting stopped and everyone's eyes turned to the boss-lumberman.

The mill owner made a few preliminary remarks, and then he said, "—And so we come to this night—the night when I promised awards to the camp that put the most logs into the river—." He paused, and the men waited expectantly. But instead of hearing the lumberman name the winning camp, they heard him say, "However—there has been a change in plans—."

McTaggart heard groans of disappointment in the crowd, and before the mill owner could go on with his speech, the foreman saw several men begin to turn away.

To stop the men from leaving, the mill superintendent came forward and shouted, "Hold on a minute—."

Speaking more loudly this time, the lumberman continued, "In addition—." He paused and then, cupping his hands to his mouth, he said, "*In addition* to the awards for putting the most pine into the Tittabawassee, I have prizes here for those who brought the logs down to Bryant's so successfully."

The woodsmen, who had begun to leave, stopped in their tracks.

McTaggart was as surprised as anyone to hear this piece of news, and he turned to look at Hoops—his eyebrows raised. The scaler shrugged his shoulders as if to say, "Don't look at me, I don't know anything about it."

Now, the lumberman had the full attention of all the men, for there was hardly anyone there who hadn't participated in the log drive. He said, "This past week I received in Albany reports from Mr. Burleson who told me that in spite of heavy snowfalls this past winter, the waters on the upper reaches of the Tittabawassee and the Cass are already remarkably low. Logs are laid up at some camps—and their owners are despairing—knowing that they will not be gotten out this season unless unwarranted rains fall—or expensive dams are built.

"You men know, as well as I do, that low water means closed mills and rusty saws. So I am grateful for the early planning and the efforts that brought down all of John Shannon's logs while the water was up.

"To show my gratitude for having all my logs safe at the booms and coming nicely to the mill—*additional* awards will be presented this evening."

A lusty cheer went up from the crowd.

Shannon held up his hand, saying, "However—." The lumberman smiled good-humoredly when a laugh went up from the men at hearing the word that nearly broke up the meeting. He continued speaking, "First things first—." The boss-lumberman paused to take a slip of paper from his vest pocket. He looked it over, and then in a loud voice announced, "The award for putting the greatest number of board feet of pine into the river this season goes to Camp 3. The cork pine retrieved from the log thieves put the camp over the top with 4,853,000 feet."

Pandemonium broke loose. The Red Keg crew shouted and hugged and beat each other on the back—and while the lumberman waited for the tumult to quiet down, the foreman of Camp 3 couldn't help but look over at Hoops. The man's eyes were beaming. McTaggart thought about the scaler, who in spite of the displeasure of the whole camp, had doggedly gone ahead and re-scaled all the logs that Uriah Bell measured; and how, in his quiet way, Isaac had gained the respect of the woodsmen with his skill in treating Sam Bean's injuries.

Shannon waited patiently until he could be heard over the noise. Then he called out, "Alex McTaggart—. Where's

McTaggart?" He peered into the dimly lighted yard, and when he saw the foreman raise his hand, he said, "Come up here, will you please?"

When he went forward to stand before the lumberman, McTaggart was clapped on the back again and again as he walked past the members of his crew. Shannon gave him an envelope. "In token of my appreciation," he said, "please accept this bonus of one hundred dollars."

The foreman heard someone in the crowd yell, "Good for you, Boss!" He was about to express his thanks to his employer, when the lumberman said, "Wait." He turned to his mill superintendent, who put into his hand a small green plush-covered box. Shannon opened the box and took from it a gold pocket watch, snapping it open to read aloud what was inscribed on the inside of the cover.

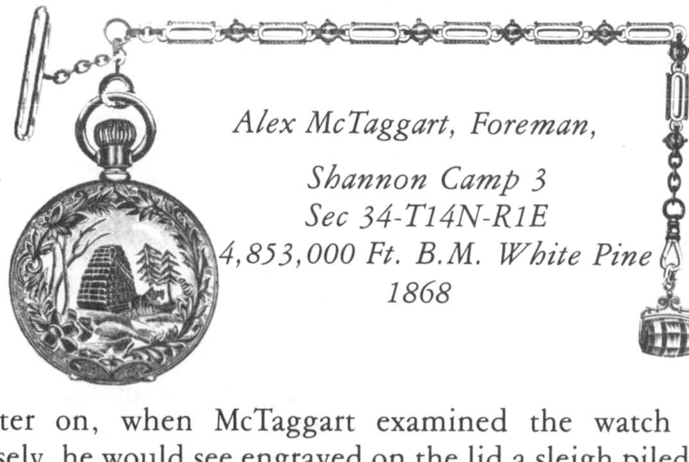

(Later on, when McTaggart examined the watch more closely, he would see engraved on the lid a sleigh piled high with a "brag load" of logs).

Another cheer went up from the Red Keg crew. The foreman tried to speak. They wouldn't let him—but he managed a loud "Thank you, boys!" before turning to the lumberman to express his gratitude and to shake his hand. Then he walked back to where the scaler was standing, pocketing his bonus on the way as he pushed through the admiring crowd.

An expectant hush fell over the gathering when the men saw Shannon motion to Burleson and say something to him on the aside. The superintendent nodded his head and went inside the office. When he came out again, he was carrying two pine saw horses which he set on the porch. He said to a couple of men who were standing close by, "Give me a hand, will you? Lift the box there onto the horses." (The lumberman had stepped from his improvised platform).

While the men were picking up the crate and placing it on the wooden stand, Shannon took hold of a crowbar that the superintendent handed to him.

The men crowded forward. There was some pushing—a few elbows finding their way into neighboring ribs. And the foreman heard, here and there, voices saying, "Quit yer shovin' " and "Take off yer hat—your blockin' the view," as the men twisted their necks and stood on tip-toe to see.

The crowbar was inserted under a slat in the box, and the men heard sharp, protesting sounds as the wooden top was raised. When the lid was taken off and placed against the wall of the porch, the lumberman put his hand into the crate and pulled out an article that made the man standing closest to the box shout, "Who-o-o-e-e! Will you look at that!"

The object the boss-lumberman held for all to see was a fancy-figured, silk-brocaded waistcoat. A double-breasted, scarlet-lined waistcoat that was buttoned in front with eight gold buttons.

Shannon held the vest up so all could get a good look at it—turning back the lapels to show off the bright lining.

The sheen of the brocaded fabric glowed in the lantern light and, as the lumberman turned the vest, he said, "The crewmen who worked at Camp 3 last winter will receive a waistcoat like this one—a bowler hat and a coat and matching trousers. And those men at Camp 3, who worked on the drive, will also receive a pair of box toe gaiters to match their new wardrobes."

An explosion of yelling, whistling and stomping feet broke out in the audience that kept the lumberman from continuing his speech—even though he held up his hands for silence.

The articles of clothing were magnanimous gifts in the eyes of most of the men present, who owned nothing more in the way of a wardrobe but worn, down-at-the-heels boots, shabby felt hats and shirts with patches sewed on them. Some of the men might have preferred a gift of cash—but that wasn't the lumberman's way of doing business. He could purchase the clothing at wholesale prices in Albany. The men, could not have bought the articles at the company store—or anyplace else in Saginaw City—for the amount in cash the boss-lumberman might have given them for a bonus. In the eyes of most of the woodsmen, Shannon's gift looked far more generous than a gift of cash—a consideration the lumberman hadn't overlooked.

The shouting continued until The Mick, who was standing to one side of the group, shouted, "Quiet! Dang it—. Shut up and let the man finish!"

The hub-bub of voices finally died down and Shannon continued. "All of the other rivermen who helped bring down my logs will receive a vest exactly like this one—a new bowler hat and gaiters like those the other men received."

The voices began to rise again—but this time Shannon stopped them before they got out of control. He quickly pointed to O'Halloran and said, "Micah—come forward." And while the riverman pushed through the crowd, the boss-lumberman continued with his speech, saying, "Micah—besides the articles of clothing you have earned for your prize, I want you to have this Waltham stem-wind Boss watch for the splendid job you did for me on the log drive."

The Mick was overcome—and showed it as he thanked his employer. He held up his watch so all could see it. It was the first time he had ever received anything in recognition for a job well done. A job was a job to him, and getting paid his wages was all he ever expected in return.

As far as he could see, McTaggart thought that all of the men present had participated in the drive—all, that is, except scowling Zip Somers, who was known to hate water like poison. The foreman could see that Zip was one unhappy man, and he knew, in spite of his loathing, the cookee of Camp 2 would have swum the length of the Tittabawassee and back again to possess one of those vests. McTaggart didn't blame Somers much if he was as mad as he looked.

Shannon was still speaking, and McTaggart turned his attention back to the lumberman in time to hear him say, "Mr. Hoops, will you please distribute the clothing to the men? Ten more crates are in the back room at the store. All sizes of clothing are represented. Those who are not here this evening to accept their prizes, will receive them by mail."

The lumberman wound up the ceremonies with a few more words—and while O'Halloran was leading the boys in a hearty "Hip, hip, hurrah for Mr. Shannon!" McTaggart heard Isaac say over the uproar, "I better be getting back to the store before the mob comes along. If you can help me out this evening, I'd appreciate it." When he saw the foreman nod his head, he continued, "Bring along several others to keep order. Maybe Driscoll and O'Halloran would be willing." The scaler turned to go, shouting over his shoulder, "I'll send Joseph (the substitute store keeper) over with the delivery wagon for the box that's on the porch. Keep your eye on it until he arrives."

By midnight, all the clothing in the crates was distributed to the men—and it can be said that after that chilly April night in 1868, the pine plank streets in Saginaw City and in East Saginaw would never again see such a splendid array of bowler hats, floral waistcoats and box toe gaiters as were displayed on the persons of John Shannon's woodsmen.

McTaggart had done his best to keep the men in line that day—but there was no holding them back after they were decked out in their "bonuses." So, it was with resignation the foreman went back to Isaac's little house that night—glad that his camp had won out over the others—but even happier in the knowledge that he would be off and away from the town the next day.

A brief encounter with a young woman very nearly makes McTaggart forget his business at the Domino in Bay City.

At the saloon the talk shifts from pine to lumber vessels—ending when a dispute in the barroom almost causes the foreman to miss his boat to the North Country.

A. G. Maxwell

22

V is for the lumber VESSELS, built stout for all seasons

At seven o'clock the next morning, McTaggart was in East Saginaw, his lumber camp and log driving responsibilities—as well as his goodbys to Hoops—behind him. He was pacing restlessly up and down the scarred planks of the steamboat dock with his pocket watch in his hand, waiting for the paddle wheeler *Evening Star*[1] to pull out. He was anxious to get to the town at the lower end of the Saginaw so he could take care of his business with Ross Robb and get on to the North Country.

"If the boat gets to Bay City by ten, I can take my time dickering with Robb for his stumpage and still have time to catch the *Huron* for Alpena," he thought, as he slid his watch back into his vest pocket again.

McTaggart's mind was on his meeting with the saloon keeper. He took no notice of the passenger boat and how pretty she looked in her new spring paint, rocking gently in the water, her bumper posts nudging the dock spiles.

It had been a cold night (the window glass in the scaler's house had frosted over as if in winter), and when the foreman purchased his boat ticket, the agent had told him the *Star* had a "half dozen inches" of snow on her decks when he came to work that morning. But now, like the town's sidewalks and the dock, the boat had been swept clean of snow. But the top of her pilot house and wheel boxes were still wet—glistening in the early morning sunshine.

The *Star* was a wood-burning, two-decker steamboat of 381 tons, powered by a walking-beam engine that projected above her upper cabin. Like her rival, the *L. G. Mason*, she

The "Evening Star" was a wood-burning, two-decker, side-wheel steamboat of 381 tons, powered by a walking-beam engine.

acted as a taxi for people traveling up and down river. Unlike the big shoreline steamboats *Huron* and *Metropolis*, the *Evening Star* did not have overnight accommodations for passengers—only for her crew.

The *Star's* pilot house was octagonal in shape and sat forward of her big stack. Her main deck was housed from her stern to within about thirty feet of her bow—the open deck forward being covered with a canvas awning. The half dozen fire buckets, lined up behind the wheel house, testified to the wooden vessel's vulnerability to fire.

The only other adornments on the boat (besides the awning) were a graceful rack of elk antlers, fastened on top of a flag pole forward of the pilot house, and a slender iron weather vane, in the shape of an arrow, which was turning lazily on its pivot on top of the black smokestack as the wind changed from north to northeast.

McTaggart glanced now and then at the crowd of woodsmen jostling for position at the ticket window at the side of the steamboat office. He recognized several of his crew and others from Shannon's camps 1 and 2. They were still decked out in their capering clothes—their turkeys lumped at their feet. All were waiting to buy tickets for the trip down to Bay City.

The foreman thought he never saw so many bowler hats in one place. Some were balanced on the backs of newly cropped heads, others were tipped over an ear, an eye or a nose. Ed Flatt, the teamster, was there and Frank Cole. He saw Cully Wills and Sweeney too—and Bill Bragg. All looked a little unsteady on their feet—their clothing disheveled. Paper collars were hiked under one ear, and shiny buttons on flowered brocaded vests were twisted into the wrong button holes as if departures had been hasty. But the boys looked happy.

"I'll bet they haven't got two bits among them," thought the foreman.

Another bowler hove into view—a battered, flour-specked hat resting on the back of a ragged head of black hair that strayed down over the owner's eyes in spite of frequent brushing back. McTaggart was idly speculating on what Kelly had done with his new bowler when the camp cook caught his eye. He walked over to where the foreman was standing—neatly side-stepping the lineup at the ticket window and an assortment of deck hands who were loading freight on board the vessel—and cordwood from the adjoining wood yard, stacking it on the main deck near the engine room. Kelly looked unhappy.

"This is going to be some trip," he said glumly. He nodded his head in the direction of the woodsmen. "Half of 'em is already three sheets to the wind and we ain't even started yet. Most of them are heading fer Mattie's place an' acting like the *Star* is the only boat on the river and the last one out this season."

When the foreman didn't make any comment, one way or the other, the cook spat into the river, balefully eyeing the crowd. "They must of heard about the free whiskey that's being offered to ride on the *Star*. Free whiskey or a free newspaper—take yer choice. Competition between the river line boats is getting to be something fierce," he added.[2]

"I hear you signed up with Sage to sail on one of his boats this summer."

Kelly nodded his head, and, in spite of his seeming reluctance to let go of the steamboat situation, he answered, "That's where I'm headed."

"What vessel you sailing on?"

"Don't know yet. Only that she's a big schooner and her capt'n lodges at the Montreal House in Bay City, when he's not sailin'. Frenchy LaRiviere's cousin, Rolly Rabideaux, runs the place. When he found out the boat was lookin' fer a good cook, he told the capt'n about me workin' in the same camp as Frenchy. I run into LaRiviere yesterday an' he give me the dope about the job."

The cook decided to enlarge upon his story. "Seems, last season, the capt'n hired on some miserable cook who'd been workin' fer a haywire lumber camp up on the AuSable somewheres. He signs on this bird without findin' out whether he knows a fry pan from a pot holder. Turns out he couldn't cook nothin' but pork and potatoes—which, Frenchy told me, wouldn't of been so bad, 'cept'n he burnt 'em on a regular basis. Now, the ship's got a reputation fer bein' a poor feeder and the capt'n can't get a decent crew together until he can prove he's hired on a good cook."

The foreman, who had some experience with punk cooks, nodded his head in sympathy. "When are you shipping out?" he asked.

"On Tuesday—which'll give the capt'n a little time to advertise the fact he's got hisself a good cook aboard." (Kelly wasn't one to hide his talents under a basket).

By this time the mob at the agent's window was thinning out, and the cook said, "Are you comin' aboard now—or are you goin' to wait until they pull up the gangplank?"

McTaggart decided he would just as soon keep to himself going down to Bay City. He made the excuse, "You go on ahead, Kelly. I think I'll get some tobacco before we pull out."

The cook, taking the hint, turned toward the gangway, while the foreman ducked into the waiting room.

McTaggart took his time making his selection, giving Kelly plenty of time to board the waiting vessel. When he came out again, he saw Zip Somers standing to one side of some stragglers at the ticket window. He was talking confidentially with a man who had his back to the foreman. Somers's companion looked familiar to McTaggart, even from the back, but he couldn't pin down in his mind where he had seen the man before.

"Whoever he is," the foreman thought, "he seems to have put Zip in a good humor," remembering the look on the cookee's face the night before when he was the only one who didn't receive one of Shannon's prizes. It wasn't natural for old Zip to get over a mad so fast.

McTaggart's curiosity was aroused. Wanting to size up the two men without their noticing him, he pulled from his pocket a folded newspaper he had been saving for the trip down river. He buried his nose in it—looking up every so often to observe the cookee and his companion.

Zip's usual sour expression was twisted into a smile, which was enough to make anyone who knew him suspicious. But when the foreman heard him laugh out loud and saw him playfully jerk off his cap and clap it on top of his friend's hat, and then turn to nod his head in the direction of the boat, McTaggart was certain the two were up to no good. When the mystery man turned around, the foreman's suspicions were confirmed. Zip's confidant was none other than Christy Dodge, the log thief.

"Now, how the Sam Hill did Christy manage to get out of jail so soon?" was McTaggart's first thought. As he watched the two men walk over the gangway with their heads together, he asked himself, "What's that darn fool of a Somers mixed up with the likes of Dodge for?" Zip didn't have enough ambition to get into the log stealing business. It must be the two were planning some devilment on board the *Star*.

The foreman watched the conspirators until they went

on deck, then jammed his paper into his pocket—mad at himself for being concerned about Zip Somers and his pal.

The ticket agent slammed down his window—which seemed to be the signal for the deck hands to pull up the fender posts and fasten them to the side of the boat. The captain of the *Star* stuck his head out of the pilot house and nodded at the mate, who in turn shouted to the hands below, "Let 'er go!"

McTaggart legged it over the passenger gangway just before the deck hands hauled it in.

Bells clanged in the engine room. The boat whistle sounded its shrill warning to the river traffic, and the paddle wheels began their revolutions—causing the water to boil and stir up the sawdust on the surface of the river. The walking-beam had begun its up-and-down movement, and beneath the foreman's feet, he could feel the power—the throb and vibrations of the steamboat engine when the *Star* slowly pulled away from the dock. Flags and pennants at her bow, midships and stern stiffened in the breeze, and a roll of black smoke uncurled from her stack. They were off.

McTaggart leaned comfortably on the rail—looking back at the dock which grew smaller as the sidewheeler began her journey down river. He made a decision. Whatever Zip Somers and Christy Dodge were hatching up between them, he was going to stay clear of it by remaining out on deck. Most of the woodsmen on board would be in the bar soaking up the free whiskey. At least he would have a little peace and quiet going down to Bay City.

But the boat had only passed Jesse Hoyt's flour mill, before the foreman heard a commotion above him, and he turned to see several woodsmen being forcedly ejected from the top deck down a steep, narrow flight of steps by a husky man wearing a dark blue uniform. The steward, stepping half way down the stairs, hollered after the men, "If you want to come up here, take off your boots and leave them below. You know the rules." After delivering this ultimatum, he turned to go back to his station midst the jeers and catcalls of the woodsmen.

When the steward reached the top of the stairway, he squeezed past another blue uniform—the purser—a jowly individual who slowly inched his way down the iron steps and over to the men. He politely asked them for their tickets.

As the portly official collected the bits of red pasteboard, he tore them in two—handing the ticket halves back to their owners. The remaining stubs were added to the pack he held clamped in his fist.

The woodsmen scowled at the purser's back as they stuck their ticket remnants into hat bands. (Only a glance would tell the ticket collector that the men had paid their passage if they moved to another part of the boat).

"What's the beef?" asked McTaggart when the purser came over to him.

The man laughed. "They're wearin' corks—." he said.

The foreman waved away any more explanation, knowing about the rules in regard to wearing calked boots on board the steamboats. The needle-like steel calks in the heels and soles of the men's boots were death on canvas-covered and varnished decks; and, for this reason, the woodsmen and rivermen, who wore them aboard the vessels, were kept to the freight deck.[3]

The purser seemed amused by something; and, after he picked up McTaggart's ticket, he lingered to tell him about a comical shipboard incident that made the foreman laugh, too—although for a different reason than the ticket collector.

"All this week," said the purser, "the management's been advertisin' free whiskey as a bonus to ride on the *Star*—knowing full well that just about every man what comes aboard will be coming direct from the log drives and wearin' rivermen's boots. 'Course, once they get 'em on board, they tell the boys they're sorry, but its the rules—they got to stay out here.

"I swear (the purser raised his hand heavenward) all week long, I ain't seen one woodsman or riverman in the bar. Not one got as much as a sniff of the cork, let alone a drink of red-eye."

The man paused again, chuckling in amusement as he went on with his story, "Well, just about a minute ago, a half a hundred men rushes on board an' heads right fer the saloon. I watched the steward hot-footin' it over to tell them about the regulations—when all of a sudden he puts on the brakes—noticing that just about all of 'em—except that bunch over there—was wearin' new fancy gaiters! Not a calk among 'em."

McTaggart couldn't help but smile at the joke on the boat's management. But he didn't bother to enlighten the ticket man about the shoes. He was wondering what the teetotaller, John Shannon, would say if he knew he was a party to his lumber crew drinking up the steamboat's stock of whiskey (presently being watered down by the steward).

The purser turned to walk back to the second deck, pausing at the bottom of the ladder to say to the banished woodsmen, "If you take off yer boots, you're welcome to come into the bar." He looked at the foreman and winked.

When the purser went off, McTaggart found himself alone again—sharing the deck with the half-dozen noisy woodsmen, who immediately began to complain loudly about the cold.

The foreman decided to stand at the port side of the boat where the wind would be at his back. The breezy gusts were whipping across the open deck, making the striped awning belly and flap over his head. He turned up his collar and then searched in his pockets for the tobacco he had purchased at the steamboat dock—ducking his head to light his pipe. Getting it going to his satisfaction, he directed his attention to the river traffic as the *Star* moved into the midst of it.

McTaggart's noisy companions settled down too. After deciding it was going to be a dull trip down to Bay City, they gathered up all the chairs on deck and piled them up to make a windbreak. Someone broke out a flask—another, a greasy deck of cards, and within minutes they were engrossed in a game of Hi-Low-Jack, paying no attention to the foreman—or the young woman who walked out onto the freight deck.

She, at first, hesitated, taking note of the card players and the lone man at the ship's rail. Making up her mind that the tall red-headed man was the lesser of two evils, she went to stand at the rail a short distance away.

At first, McTaggart was not aware of the young lady—but then he felt her presence. Out of the corner of his eye, he cautiously looked in her direction, and what the foreman saw made him suddenly conscious of his rough clothing—his worn rivermen's boots, smelly pipe and a hat that had served him well but now he wished at the bottom of the Saginaw River. Without realizing what he was doing, he tapped his pipe on his hand and slipped it into a pocket. He straightened his shoulders.

She was tall and slender (he liked that), and she was wearing a plain, but fashionable, blue traveling suit. The snug fitting bodice and full skirt were trimmed with braid.

And she wore a new spring hat—a small straw hat that dipped over one eye. It was decorated with rosebuds. A large cashmere shawl, with a pattern woven into it resembling the tartan of the Highlander's kilt, was loosely draped across her shoulders. But the wind was brisk, and she drew it up—wrapping it snugly around her body. And now, her face was almost hidden—except for the tip of her nose—a tantalizing feature that made the foreman wish to see all of it. He could not now see her hair—which was dark brown and pulled back and fashioned into a kind of roll or bun at the nape of her neck—and McTaggart found himself wishing the shawl would drop down again, so he could see her hair shining in the sun.

The girl seemed completely absorbed in the passing river scene, and the foreman grew bolder. He turned his head so he could better observe her. She must have come aboard before he arrived at the steamboat dock, he thought. He would have remembered seeing her at the ticket window.

The shawl slipped down, and the young woman adjusted it again. As she was pulling it around her shoulders, she turned her head toward the red-headed man—and smiled.

McTaggart started—then swiftly looked away, feeling his face flush scarlet—the red welling up from his collar to disappear under the despised hat. He hadn't blushed like that since he was a kid in the old Chumung Valley camp, he thought angrily. He knew she was looking at him, and he felt like a chump for not turning to make some polite civility about the weather—or anything—anything at all. Why was he always so tongue-tied when it came to women? He felt his heart banging away in his chest cavity. And he felt hot, even though the temperature was about fifty degrees, and the wind was whipping across the deck at thirty knots.

Was she laughing at him? Blast it all—that wouldn't do at all.

Impulsively, the foreman turned toward the girl and burst out, "My name's McTaggart—Alex McTaggart." He stuck out his hand as he moved to her side. "Are you riding down to Bay City?" he asked.

Again, his face turned crimson. What the Sam Hill did he say that for, he thought, smarting to himself—where else was the damn boat going—except to Bay City.

"Well, Mr. McTaggart—yes, I am," the girl answered. Her eyes were teasing.

"They're blue—," McTaggart thought, even in his distress. "Her eyes are bluer than deep wood's violets." The finely-shaped features were animated and cheery.

"My name is Margaret Carr. Miss Carr—Meg for short." The young woman rushed on—evidently trying to put the foreman at his ease, "I've been in East Saginaw visiting my father who is in the lumber business there. I live with my Aunt Julia—my father's sister—in Bay City—. My mother died when I was a baby—." The girl hesitated—knowing she was running on too much.

But McTaggart, whose only interest in life, up to that moment, had been *Pine* with a capital "P," had not noticed Miss Carr's impulsive self-introduction. He was absorbed by little wisps of her hair that were trying to escape from under the little hat with the rosebuds on it—imagining the shining tresses loosened and streaming in the wind.

"Are you going to Bay City also?" She laughed at her little joke.

"I have some business to attend to in Bay City and then I'm off to Thunder Bay Country and another summer timber looking." McTaggart found his tongue at last.

"So, you're a cruiser—a land looker."

"That and a few other things," smiled the foreman. "I've been working the past few years for the lumberman, John Shannon—."

"From Albany?" asked Miss Carr. When she saw McTaggart nod his head, she said, "My father and Mr. Shannon have had business dealings—." She stopped short. "I'm sorry—please go on Mr. McTaggart."

The foreman waved away her apology and continued, "I've done considerable 'looking' for Shannon and other parties, and the last two seasons I've been managing his Camp 3 on the Tittabawassee—."

Again, the girl interrupted him, "Red Keg," she said. "I've heard my father talk about the splendid pine that comes from there." She stopped abruptly. "I'm sorry—please go on."

McTaggart laughed outright. He hardly knew why—he just felt like it. The girl's obvious interest almost made him feel giddy. What would Hoops think if he could see him now? He said, "There's not really much more to tell. I've got another year's work at Camp 3—and then I'm off on my own. Bought a piece of pine up on the Rifle River—."

Miss Carr turned to stand with her back to the rail. When she spoke again, her mood had changed. "I really envy you, Mr. McTaggart. I wish I could go off on my own—go into the woods—build something—be useful—." She shook her head. "Do you know what I'll be doing while you are off timber looking this summer?"

McTaggart confessed he had no idea.

"I'm being packed off to Europe—'to round out my education' (that's what my aunt calls it). I'm going to do nothing for two whole months but travel and visit while you'll be living in the beautiful woods, tramping over bogs and windfalls and cooking over camp fires—."

It was McTaggart's turn to interrupt, "—and slap at mosquitoes and black flies 'till I'm crazy—get chased by bears and go without washing, except in a stream—and eat salt pork and hardtack until it comes out of my ears."

In spite of his trying to make his life in the woods sound as miserable and unattractive as possible, the foreman couldn't quite put his heart into it—the reason being, perhaps, that's all he had been thinking about the last two days.

"Ah—," the girl said softly, smiling in triumph, "do I hear a note of enthusiasm?"

The conversation was getting out of hand. It was being spoiled and McTaggart wanted to keep things on an even keel—so to speak. He was relieved when the young woman changed the subject.

"Mr. McTaggart, do you know we're almost down to Carrollton and we haven't heard a sound coming from the bar? The last time I went to Bay City, I traveled on the *Mason* and two fights broke out between East Saginaw and Crow Island. Some rivermen nearly killed each other over cards before we got down to the salt works."

McTaggart had forgotten all about the men in the bar.

"Several of my old crew are on board and I've been expecting to hear some disturbance or other before the trip is over. As long as it's quiet, we can be thankful," he said.

McTaggart saw Miss Carr's eyes light up at the mention of a little excitement happening. "Do you think—?"

"No—no, of course not. I hope not." He looked around for a couple of chairs.

"Let's not sit down, Mr. McTaggart. I'd just as soon stand at the rail. It's such a beautiful day!" The young woman turned again to look out over the water.

The foreman pretended interest in the activity on the

Saginaw—but he was, in fact, weighing in his mind his chances of seeing his companion again. Obviously, she was out of his class, he was thinking. She seemed well-off—well-bred—her manner and speech were clues in that direction—and she had mentioned traveling abroad—. "Hold your horses," he admonished himself. "Take your time. She's only been talking to you same as she would to anyone else." Still—countered the foreman—he had seen something in her eyes that told him she was interested in him as well. He'd take one step at a time—but he vowed he *would* see her again—he made up his mind about that.

The *Star* had been churning down river past the mills and docks of the Saginaws—and now, after curving gently to the right, the steamboat approached the dreaded flats at Carrollton. Normally, only four feet of water covered the treacherous sand bars, but, according to the weather vane on top of the *Star's* smokestack, the wind was still holding its northeast course—blowing water in from the Bay. The paddlewheeler eased through the shoal water with no trouble, sounding her whistle to salute the mills that rose from the marshy shore like cranes on stilt-like legs.

On the left bank, farther down river, the steamboat thrashed by another mill settlement called Zilwaukee. The big salt works at Crow Island was situated on the shore opposite. Willow Island—still another mill outpost with a satellite boarding house, company store and school house—was the last pine manufacturing site the foreman and Miss Carr would see until the *Evening Star* passed around Devil's Elbow—a tricky bend in the river near Bay City.

The steamboat paid her respects to island mill hands with a couple of toots on her whistle and passed on down the river to overtake a river tug towing a long string of logs bound for the mills below. Other tugs were chugging up river in search of their tows, having received their orders by telegraph at their home port in Bay City. And they passed a powerful lake tug that was inching her way down river towing in tandem four wooden schooners. Their holds and decks were loaded with pine. (The tug would haul the vessels out to the Bay where the ships would hoist sail and continue on their way, using only the wind for power).

McTaggart could see that the schooners in tow were not supporting capacity loads. Like most lumber carriers loaded

They passed a powerful tug that was inching her way down river towing in tandem, four schooners—their holds and decks loaded with pine.

at the Saginaws up river, their masters had orders to stop off at a lumber dock in Bay City for "topping out."[4] It was a precautionary measure that enabled the ships to sail past Carrollton without getting stuck on the flats.

The Saginaw River—formed by the confluence of the Tittabawassee and Shiawassee rivers—flowed contrarywise, in a northerly direction, for a distance of twenty-two miles to empty into the south end of Saginaw Bay. At Saginaw City, the river was about five hundred feet wide. It gradually widened as it flowed toward its mouth, where the stream was a quarter of a mile wide from shore to shore. The shining surface looked very broad this morning, even at the upper reaches of the river, for the land on either side of the stream, from East Saginaw down to the south end of Bay City, was for the most part, low-lying—the river spreading out to flood the rich watery moors that were called the Saginaw marshes.[5] Many creeks and water courses had found their way into the lowlands. Hidden among the rushes and wild grasses were a multitude of lagoons, water holes, duck ponds and bogs, making the place inaccessible, except by small boats or canoes.

For centuries, the marshes had been the feeding grounds and habitat for a multitude of wild fowl, as well as for the muskrat, the otter, the beaver and mink. And where the ground was high enough for trees to grow, or where a crop could be sown, it was a place where the deer, bear and fox came. Because it was springtime, much of the land was covered with water and would be so until well into the summer months.

There were wild rice beds in the marshes. The beds at Crow Island, Cheboyganing and Squaconning creeks covered hundreds of acres. Thousands upon thousands of ducks were attracted by the rice and the dense cover it provided—

coming every year to feed and to nest. The marsh was the "haunt of coot and hern," of small marsh wrens, and red-winged black birds. The loon came too, as did the great blue heron, the plover, the spotted sandpiper and the fish hawk. Most of the wild fowl came early, only leaving the marshes when blustery gales and snow squawls blew in from the Bay to drive the birds south for the winter.

Sometimes, Mother Nature seems reluctant to perform her rites of spring in Michigan. Even though it was just six days to the first of May, the marshes were still a brown and yellow sea of dead cattails, reeds and grasses that had been nipped and bent by November frosts and flattened by wintery winds. Here and there muskrat houses dotted the moors—adding a dimension to an otherwise flat landscape.

As he looked across the waters, McTaggart could see skims of ice around the shore rushes and noticed that the trees in the forest, far off along the rim of the marshes, were still barren of leaves. The pines and white birches stood out amongst tree trunks that looked somber and gray against the bright blue of the cloudless sky.

He felt a light touch on his arm, and he heard Meg's voice over the thrashing noise of the paddle wheels.

"My father sometimes seeks the solitude of the marshes," she said, "to get away from the responsibilities of business. He has taken me with him as his companion many times—ever since I was a child.

"We paddle down river from East Saginaw in a canoe made out of a pine log—like the Indians do. Because they're so tippy, he used to warn me not to move around."

She paused—remembering. And McTaggart painted a picture in his mind of a little girl with brown hair flying in the wind and bright cheeks that told about high spirits and the difficulty of sitting still in a "tippy" dugout canoe.

"Before we would start out for the marshes, Father would say very solemnly, 'Maggy, I hope you have parted your hair *exactly* in the *middle* today. We must keep the boat BALANCED.' "

The girl laughed at the recollection—the foreman joining in.

McTaggart had heard Cap Wilkes tell about the dugouts used for duck boats on the Saginaw marshes and recounted what he remembered, ending, "—Cap and two fur traders, by the names of Jacque and Louis DuPraw, used to paddle down below Crow Island where the salt works bayou is. They would build a fire and tip their canoes on edge as wind breaks and stay there over night for the early morning duck shooting."

The girl shook her head. "I know the DuPraws—they're market hunters. They take too much from the marshes—."

She suddenly broke off to point toward the horizon. An iron locomotive was chugging along the top of a narrow dike that followed the course of the river down to Wenona.

"It won't be the DuPraws or their kind who will spoil the marshes." The young woman's voice sounded sad. "It will be dredges and pile drivers and railroad tracks—."

The steam engine was hauling several cars behind it, traveling about twenty miles an hour, and McTaggart and his companion watched, in silence, black smoke roll from its big diamond stack as the wood-burning locomotive huffed over the iron rails.

There was another railroad track running to Bay City on the east side of the river and both the man and his companion knew that it was inevitable that the marshes would be drained and more roads built to connect the sawmill towns. But it would be a while yet before Mother Nature would give up her claim, the foreman thought with some satisfaction.

He heard the young woman speaking again.

"Have you seen them in the summertime? The marshes, I mean—." And when McTaggart shook his head, Meg Carr began to tell him about the watery moors, as she knew them.

"In late summer, when the rice is high, it is so ripe it drops into the canoe when you paddle or push your boat through it." She grinned, reaching with her hand over her shoulder to make a scratching motion. "The rice kernels have little barbs on them and if you get one of the pesky little things down your back or into your boot, you have to stop to get it out before you can move on.

"The ponds are covered with water lilies—great, sweet-scented white flowers and yellow buds."

Miss Carr waved a gloved hand toward the land. "It looks so desolate now—it's hard to imagine wild flowers in bloom and small wading birds picking their way across the lily pads as they search for food.

"And the bull frogs hidden in the reeds—. What a noisy lot they are! On summer nights we can hear their booming and croaking all the way in town.

"The carp and pickerel sun themselves in the shallows and when you reach out with your paddle to touch them—they are gone. You look and see only their shadows—.

"One time, Father and I heard a boat whistle sound on the river. It startled hundreds of rails and we could hear them all around us abusing it with their harsh cries."

The girl turned her head to look at the foreman, smiling softly as she continued her account.

"The marshes are beautiful in the spring and summer months—but I think late autumn is the time I love them most of all—when there's a smoky haze over the land. The muskrat is building his home of sticks and mud, and when you look up, you see geese and ducks flying south—." She stopped short when she saw McTaggart suddenly point into the sky.

"Look there!" he said. "They're coming back—."

Over the throbbing of the *Star's* engine and the wash of her paddle wheels, they heard the muted din of hundreds of Canada geese honking, and they saw great flocks of the birds flying in "V" formations high overhead in a northerly direction.

McTaggart and Meg Carr watched the birds in silence until they were out of sight.

Now, the foreman was beginning to recognize certain landmarks that were associated with the towns at the lower end of the Saginaw. Panic gripped him. The trip was nearly over and if he didn't shake a leg and say something to the girl about wanting to see her again, she would be gone out of his life forever.

He saw her pointing to a long stretch of land that was only separated from the mainland by a narrow channel.

"There's Skull Island," she said. "We've only two more miles to go before we see John McGraw's mill."

"Yes—we'll be pulling in soon—." The foreman's words were almost drowned out by a noisy river tug passing.

The girl waited.

McTaggart gritted his teeth and plowed ahead. "Miss Carr—I would very much like to see you again—."

Miss Carr smiled.

Encouraged, the foreman continued his speech, trying to sound at ease and not so stiff and formal. "I would like to meet your aunt," he said, "and with her permission call on you when I get back from the North—sometime late in August." He finally got it out.

"We'll be back home by that time," answered the girl,

seeming not to notice her companion's discomfort. She hesitated and then said shyly, "If we're going to be friends, you must call me Meg."

McTaggart felt a little giddy. He couldn't believe his good fortune. For the second time that week, he congratulated himself.

The young woman held out her hand, saying, "Come—you must meet my aunt before we dock—."

• • •

The sidewheeler *Evening Star* left the Saginaw marshes behind and passed John McGraw's big smokestack—but neither McTaggart nor Miss Carr noticed, for the foreman and the girl with the violet eyes were in the stuffy, stove-heated ladies saloon. Feeling wretched—but determined, McTaggart was putting his best foot forward.

After the formal introductions were over and a few polite pleasantries made, the foreman's question was posed to the aunt. And the elder Miss Carr, like the sensible lady she was, looked beyond the rough woodsman's clothing to look into the direct gaze of two blue eyes to determine whether she approved of the man who was standing so stiff and straight before her.

It seemed to the foreman like an interminable amount of time went by before she made her answer.

"If you and my niece still wish to see one another in late summer," said the lady, "you are welcome to call."

More polite exchanges were made—conversation that resulted in an unexpected invitation to tea later that afternoon.

A flustered foreman was explaining the urgency of his business in Bay City and about having to board the *Huron* shortly afterwards—there being no other boat going north until Monday—when the trio heard the *Star's* whistle blow, signalling the Third Street Bridge to open. They had arrived at their destination.

The sidewheeler passed through the open draw and slowly steamed toward her dock which was located just beyond the wooden span.

Perspiring freely by this time, McTaggart formally shook the aunt's hand and then gripped the gloved hand of the other Miss Carr. He tried in vain to think of something to say to her that she would remember all summer long. He was about to speak when he heard over his shoulder a slightly alcoholic, out-of-breath baritone voice, which sounded both incredulous (at the sight of the foreman clasping the hand of a beautiful young woman) and urgent.

To the amusement of Meg Carr, the foreman let go of her hand like it was a hot poker. He whirled around to confront Bill Bragg.

The swamper, awed by the female presence, pulled the embarrassed McTaggart aside. He said in a loud whisper, "Boss! Ol' Zip won all our bowlers an' vests in a poker game. I swear on a stack of Bibles that log-stealin' friend of his and him was playin' with marked cards—."

The foreman had forgotten all about the Camp 2 cookee and the log thief and was not a little indignant at Bill for breaking in on his goodby. He saw Meg out of the corner of his eye pretending not to notice him and the swamper. But Bill's whispered message was loud enough to be heard in East Saginaw.

McTaggart couldn't tell how the aunt was viewing the intrusion, but he made up his mind then and there he would take no chances on incurring her displeasure. Between clenched teeth, he said to Bill, "You know Somers's reputation as well as I do. If you lost your bowlers to Zip, its just too bad. Go tell the boys to pick up their turkeys and get off the boat without making a ruckus."

Those were the words he said, but he was thinking to himself in disgust, "I sound just like some parlor stiff." He hated it when he saw Bill turn and glumly go out of the cabin. And Meg—she didn't look too happy at his sending the swamper away without helping him. But he was certain he saw the aunt frown at the interloper.

The foreman felt miserable—and looked it. He was caught between his loyalty to his men and feelings for a girl that made his throat tight and himself hot all over.

Julia Carr came to his rescue.

"Mr. McTaggart," she said briskly, "We'll be happy to receive you in August. That's all settled. Now—please—won't you go and see to your men? Margaret and I will stay here until the woodsmen have gotten off the boat." Saying this, she gathered up her newspaper and began looking for her spectacles case. The foreman had been dismissed.

McTaggart never caught up with Bill Bragg. Half way up the iron steps he felt the *Star's* fenders bump against the dock and heard the boat whistle rip off another blast. When the sound died away, he could hear a bedlam of female voices hallooing—calling to the woodsmen on board. And over his head, a male chorus (thick with whiskey) was answering, "Hey girls—you been waitin' long fer us? Where's Mattie—tell her we come back—."

The *Evening Star* had pulled up to her dock, and the Domino Saloon girls were out in full regalia to greet the male passengers.

The water front saloon was one of the first frame buildings erected in Lower Saginaw. It was built of pine lumber, whip-sawed on the site by a Lakes captain, whose enterprising nature had prompted him to invest in ventures other than sailing. In the old days the building was a store which served as a trading post catering to Indians and settlers—and what Lakes' trade existed. But when the town got too crowded, the captain moved across the river to the west side, and Ross Robb and Mattie Morgan moved in—but not before they cleared out the Indian goods and groceries. A bar was installed, and a second story was added on the building with a narrow wooden balcony built across the front (where the girls were greeting the men). All four sides of the building were decorated with the name of the establishment—painted in three-foot-high letters. No matter which side of the building it was viewed from, there was no mistaking the Domino for some other place in the same line of business. The only other decoration on the saloon was the Stars and Stripes which flew from a standard fastened on the peak of the roof.

The Domino Saloon had one big advantage over all the other saloons in town. It was close to the steamboat dock. In fact, it was so close, when a boat docked, its stern extended almost across the front of the building.

McTaggart stopped climbing the stairs when he heard the hullabaloo overhead. He decided while the boys' interests were taken up with the girls, he would get off the boat and be on the welcoming committee when Zip and Christy walked over the gangway.

The foreman backed down the iron steps—and even while the deck hands were letting down the passenger gangplank, he was picking up his dunnage to head for shore.

When McTaggart reached the dock, he looked back at the *Star* and saw that most of the woodsmen were bunched at the stern of the vessel. They were leaning over the taffrail, reaching for the girls who were waving gaudy handkerchiefs and shouting and laughing at the boys' antics.

A woodsman's inventiveness for skylarking after a half dozen drinks was well known to the foreman, but when he saw Zip Somers standing on top of the rail with one arm wrapped around a deck support (to keep from falling off), he knew he was looking at a prize-winner for mischief. The upper part of the cookee's bean pole frame bulged with layers of flowered vests, and on his head were a half dozen bowler hats, stacked one on top of the other. Zip looked like an overstuffed bright-colored bird perched on the rail trying to make up its mind whether or not to fly off. Some of the sore losers were in evidence, for every so often the foreman saw one of them leap up and try to dislodge the hats. Every time this happened, the cookee let fly at somebody's skull with a well-aimed foot, and then everybody would duck.

Christy was there too. The foreman saw him holding a bulging sack over his head out of the way of grabbing hands. And it was evident that some of the woodsmen had taken sides, for they were helping to hold off the men who were trying to retrieve their property.

Every time the cookee swung out from the rail, the hats would wobble over his brow, and he would stretch out a bony hand and set them straight.

"Look at Zip, darlin's," he hollered. "I got all dressed up jest fer you! Where's Mattie at? Tell her Zip Somers is back—!"

McTaggart was thinking, "He'll never get off the boat with those hats," when he saw Ben Sweeney's head and shoulders rise from the crowd and with one long arm knock the bowlers clean off Somers's head—the hats sailing over the rail to fall, one by one, into the Saginaw River.

The girls screamed. And in the brief, tense moment that followed, a dozen pairs of eyes riveted on the bowlers, watching them turn with the current and disappear around the stern of the boat to go bobbing down river, looking like so many brown turtles swimming out to the bay.

Another scream splintered the air. The foreman whirled around to look where the girls were pointing.

At first, he thought the cookee had fallen overboard—but he saw instead, a blur of arms and legs behind the taffrail and saw the human mass moving in the direction of the iron steps to the lower level of the boat. The whirlwind momentarily disappeared from view and then reappeared as it came tumbling down the steps to the main deck. It separated there and went streaking off the gangway and onto the dock where it passed him like a tornado.

McTaggart turned to watch the boys leg it up to Water Street in a cloud of sawdust—relieved that the situation had been taken out of his hands. Turning back to the *Star*, he waved at Meg (who had been watching the chase from the freight deck) and then walked toward the Domino to keep his appointment with Ross Robb.

• • •

When the foreman stepped from the bright sunlight into the lamp-lighted saloon, he stood for a moment inside the doors until his eyes became accustomed to the dimness of the place. Over the noise in the barroom, he could hear tug and steamboat whistles as well as ferryboats competing for attention. From across the river at Bangor (a "suburb" of Wenona), he could hear faint sounds of ship's carpenters and "corkers" at work overhauling and building vessels for the lumber trade. McTaggart smiled to himself. He was in Bay City all right—the sounds were unmistakable.

The place, like the Saginaws up river, was a sawmill town—but with a difference. It was the headquarters for Saginaw River shipping—the port from which two-thirds of all lumber manufactured in the Valley was shipped.[6] On

Water Street, mingling with mill hands, dock wallopers and woodsmen on leave, Lakes' sailors, who had only recently come out of their winter's lodgings, were hanging around waiting for their ships to load up. Others were busy painting the sides of vessels or patching up decks in preparation for another season.

It was springtime, but it mattered not to the Lakes' sailors whether the latest style bonnets were displayed in milliners' windows, or if the country roads were deep in mud—whether stump pulling, tree trimming, fence repairing and sidewalk mending was going on all over town; or that the new velocipede craze was all the rage. It mattered not to them. Their interests were confined to Water Street, which might have been Buffalo's Canal Street or any water-front street on the Great Lakes. To these men and to those on board the thirty or more wind-bound vessels lying that morning between the Third Street Bridge and the Keystone Mill (waiting for the storm to blow itself out), the town was just another port where one picked up or discharged cargoes—a place where a ship could run back to if the weather turned into a howler out on the Bay.

Bay City was lively enough any day in the week—but the grand slam was on Saturday—payday at the mills and shipyards. Along the town's "hell's half mile,"[7] the thump and scuffle of boots on scarred plank walks could be heard until well into the morning as sailors, raftsmen, lumber shovers, mill hands and woodsmen made their way from one saloon or bordello to another. At nightfall, the good citizens, shaking their heads at such goings-on, would call in their offspring, lock their doors, close their blinds and try to ignore the noise made by the rowdy, drunken revelers who would whoop it up until church bells rang up and down Washington Avenue on Sunday morning.

The "law" did its best to keep order. But in a place where it was not unusual to see signs like "Fresh Bear Meat" on butcher shop windows—where the only qualifications to be a policeman[8] were being able to run faster and to fight harder than the brawlers on Water Street—it wasn't surprising that law and order were primitive in Lower Saginaw.

The foreman knew there was a crowd of hangers-on loafing on the bridge approach above him. When the middle span was left open (like in Saginaw), woodsmen and other vagabonds congregated there talking, arguing, fighting. He heard an organ grinder entertaining—grinding out on a hand organ a plaintive tune that mingled with the singsong cry of a fish peddler who was alternately blowing on a tin horn and bawling "Fresh-caught-Saginaw-beef! Six-cents-a-pound!"[9]

"Are you comin' in or goin' out?" The husky female voice was brisk and business-like—but friendly.

McTaggart swung around to face a handsome woman with a full-blown figure. Mattie Morgan herself!

The mistress of the Domino Saloon was done up in emerald green taffeta. The foreman couldn't help noticing that the woman's elaborately-dressed hair was only a little less brilliant in color than his own and that her lustrous skin shone fair where the close-fitting bodice of her gown didn't cover it. The skirt of the lady's dress was without hoops and caught up in the back where it was fashioned into a small bustle. She was wearing fancy jet ear bobs and a quantity of jewelry around her neck—adornments which flashed in the light that ventured into the saloon above the swinging doors.

The foreman touched his hat politely. "Is Robb around?" he asked.

"You must be McTaggart—John Shannon's foreman,"

said Mattie. She pointed to a door near the bar. It had the word "Private" lettered on it.

"Ross is waiting for you in there. Just knock and walk on in. He's expecting you."

McTaggart hadn't seen Ross Robb since the night the log thieves were apprehended in the Red Keg Saloon. When he took a seat opposite him, he wondered if Robb would mention the incident. But the saloon keeper had either forgotten about the episode or he chose to avoid the subject that might strain his relationship with the man facing him.

He offered the foreman a cigar from a box on top of the roll top desk and after the two men lighted up, Robb eased back in his chair and said, "You might be wondering why I decided to sell off the timber on our forty up on the Tittabawass'."

McTaggart hadn't given the subject much thought, but before he had a chance to say so, Robb went on with his speech, seeming anxious to tell about his plans.

"I'm going to use the money to invest in a slice of a schooner-barge," he said.

The announcement came as no surprise to the foreman. Individuals and groups of people who had capital to invest were known to lay it out to build a ship or to buy shares in one. The Lakes' trade was growing, and investors could expect a good return if the circumstances were right.

Two years before, grain freight rates were the highest ever recorded, and other commodity rates were proportionately high—which led to a "sudden craze to secure anything afloat." Men went so far as to mortgage their homes to purchase Lakes' vessels—sometimes at exorbitant prices.[10] Shipping rates slipped a little in 1868, but investors were still as keen as last year, or the year before, to build or buy ships.

McTaggart knew that the going rate this spring for carrying pine lumber down the Lakes from the Saginaws to Buffalo was $4.50 per thousand feet.[11] Counting out disaster, Robb would get back his investment within a short time.

"She's going to be built down at Weed's yard," continued the saloon keeper, "and strictly for the lumber trade—to tow behind a tug. She'll be no converted schooner that's stripped of her planking and top masts."

Robb was referring to the indignity and fate that many sailing vessels on the Lakes were now suffering at the hands of men who looked for regular sailing schedules and greater cargo space for shipping lumber.

Sailing vessels on the Great Lakes went back in time to the explorer LaSalle, when he built for the fur trade the ill-fated, high-pooped *Griffin* in the Niagara River in 1679. Since that time, hundreds of sloops, brigs, brigantines, barques and schooners were built for trade around the Lakes in inlets and in coves and at river mouths in places like Bay City; Vermillion, Ohio; Erie, Pennsylvania; and at Shipman's Corner, St. Catharines, Ontario.[12] Wherever there was oak growing for keels and breastworks, and white pine and fir for masts and planking (and enough water to launch a ship), there would be shipbuilders like Louis Shickluna, Hosea Rogers and Sam Ward at work with adz and saw and caulking tools, fashioning vessels for the Lakes' trade.

In Manitowoc, in 1851–52, ship designer, William Bates, set the style for Great Lakes sailing vessels when he built the two master *Challenge*—a sharp-ended schooner designed to enter narrow channels and shoal waters.[13] His rakish, clipper-type schooners, with center boards hauled up, could be maneuvered into the shallowest harbors. They carried good-sized cargoes—and they were fast. Under a fair wind with sails set, center boards down, they could beat down the Lakes at thirteen knots. It was a sight that Lake men from Buffalo to Duluth fondly reflected on while sitting around ship chandlers' stoves in ice-locked harbors in January.

Those ships, whose only power was the wind, might have drawn a sentimental tear from the eyes of the fresh water sailors who sailed them—and from those who had seen the beauty of one or two hundred white-winged vessels at the head of Lake Erie, lingering there until the wind changed to a favoring breeze before running up the Detroit

Lake vessels, like the schooner "J. B. Newland," whose only power was the wind, couldn't be depended upon for regular sailing schedules. Notice the typical triangular raffee sail on the "Newland's" foremast.

River.[14] But they only drew the ire of lumbermen (who weren't known for their sentimentality), who fumed while waiting for the arrival of a ship that was holed up near Amhurstburg or Bay City—ships that dallied while a storm blew itself out—or were becalmed miles from the lumber market, waiting for the wind to get up.

It stands to reason it was a lumberman who thought of a better way to ship pine. His was an idea that never put the sailing vessels out of business while there was still lumber to carry—but it did, rather, put a crimp in their style.[15]

The palmy days of the big passenger steamboats—the boats that had carried thousands of immigrants from Buffalo to Detroit and points west—had recently come to an end. They were put to rest by the railroads and by the slower, more economical screw-propeller steamers. By the late 1850's, these floating palaces could be seen in any major port, rotting in shoal water or at dock side.

Buffaloian, John Noyes, was one who observed the old hulks. He was operating out of East Saginaw as a lumber forwarding agent, and he saw the derelicts when his business took him to Detroit. They gave him an idea—one that would, in a short time, revolutionize the Lakes' forwarding business.

To carry out his idea, John Noyes purchased two old steamboats—the *Empire*[16] and the *Sultana* (lying useless at the foot of Belle Isle in the Detroit River). He stripped them down to their hulls, removing everything that was removable, including their staterooms, engines and boilers—for the boats would no longer carry paying passengers—nor would they need motive power. Noyes overhauled and strengthened them to carry pine lumber—to tow behind a tug. The powerhouse *Reindeer*,[17] a well-known lakes character, with Captain "Sim" Keeler at the helm, was engaged for the towing job.

In the spring of 1862, the *Empire* set off from the Saginaw River behind the lake tug—bound for Buffalo. She was loaded with 850,000 to 900,000 feet of pine lumber. The *Sultana* remained behind to be loaded and to await the return of the *Reindeer* (one crew, in charge of Capt. Jack Meyers, was hired to go from one barge to the other).

Sailing time for the tow averaged from seven to nine days. The time for vessels that depended upon the wind was (with luck) from eighteen to twenty days—their average carrying capacity, from 150,000 to 200,000 feet of pine.[18]

As soon as the insurance companies were convinced that the towing method of transporting lumber was relatively safe, shipyards began to convert old sailing vessels (which proved to be more practical for conversion). Foremasts and aftermost masts were shortened (others were removed) and decks were cleared to carry lumber.

Shippers of other bulk commodities weren't slow to take up the idea. The era of towing had begun.

About 1870, the steam barge (the forerunner of the present-day Great Lakes freighter) replaced the tug for towing. These propellers were loaded with lumber just like their consorts.[19] Before long, there were so many converted schooners towing behind steamers, the sight of them moved the Master of the schooner *LaPetite* to write in his journal (after observing them in Bay City in 1878), "old worn out vessels converted to barges, towed by old propellers are as thick here as hasty pudding."[20]

As soon as the converted vessels wore out, shipyards replaced them by building steamships and schooner-barges that were designed strictly for the carrying trade.

The early wooden steamships looked very much like the vessels they towed behind them. Their smokestacks were the only things that set them apart from their consorts. Like the schooner-barges, they had big holds, cleared decks, with only a deck house for their crews. Although they were driven by steam, the propellers were built with one or more short masts. When the weather was favorable, sails were set to conserve fuel. The tow barges also carried fore-and-aft sails on their shortened masts, which aided the ships pulling them—and in rough weather, helped to keep hemp tow cables intact and the vessels steady.[21]

Lumber vessels, whether powered by steam or wind, were loaded through their hatches—the lumber being

Schooner-barges, like the "A. W. Wright," carried fore and aft sails on their shortened masts, which helped to keep tow cables over bow and stern intact and the vessels steady. Notice the size of the "Wright's" load—her tow rope and anchors. Also in view, are the captstan and donkey engine, to take up and pay out the hawser.

passed down from the dock to the lumber shovers in the hold. They stowed the boards from port side to starboard (so the cargo would not shift). When the lumber reached the plank sheer, wedges were driven in to support the decks where more pine was piled in layers—sometimes twice as high as the deck house.[22]

The method for testing whether the lumber piles on

deck were reaching the top-heavy limit was unique—if rudimentary. The mate would sing out to some eighteen or twenty dock wallopers to stand in a row on top of the deck load and, on signal, instructed them to walk together (shoulder to shoulder) to the opposite side of the vessel. If the trim of the ship was affected by this action, the mate took it as a warning to stop shoving lumber![23]

Sch. "Jamaica" in distress.
In 1867, 931 Lakes disasters were reported, with 211 lives lost.

Lake men navigated their ships "by guess and by God" on waters that were still uncharted—and schooner-barges strung out behind a steamer or tug, made theLakes even more dangerous to navigate. Shown here is the steamer "Charles H. Bradley" with consorts "Brightie," "Mary Woolson" and "Goshawk" in tow.

Contrary to what one might think, when underway, the barges with their holds filled and their decks loaded as high as fourteen feet with pine, did not roll very much. Their masts acted as stablizers. But—if the towing vessel and consort became separated or were dismasted in a sudden blow, the ships stood little chance of survival.

Lake men navigated their ships "by guess and by God" on waters that were still uncharted[24]; and barges strung out behind a steamer or tug made the Lakes even more dangerous to navigate. At night, if a vessel came upon a steamer with consort wallowing behind, her watchman could only determine the number of vessels she was towing by the number of lanterns (called towing lights) that were hung on her foremast. When visibility was good, the lights might be seen by ships sailing nearby, but in a fog or a blinding snow storm, when visibility was reduced to the pilot house window, it was not uncommon for vessels to become fouled in the lines of steamer and tow.

Imagine coming upon a tow at night that extended over one mile in length—like the steamer *Antelope* with eight schooners strung out behind her carrying over six million feet of lumber![25]

Sudden, violent storms, that have been known to open the seams of Clyde-built ships and sink them without a trace, were a great hazard to the lumber vessels. Countless sailing ships, as well as wooden steamers and barges with tow ropes parted, were stranded on shore or piled up on reefs and on rocky ledges where they were pounded to bits by the same freakish winds and green water that have sheared the rivets and twisted the frames of steel freighters. In the navigation year of 1867, nine hundred and thirty-one disasters were reported, with two hundred and eleven lives lost.[26] Still, the men who sailed on the wooden ships did so with no more thought for their own safety than a landlubber sitting in a rocking chair on his own front porch.

Many lumber vessels were hastily built so they could begin earning money for their investors. The barge *John McDougall*, built in 1873 at Crosthwait's yard in Bangor, was one of these. A tug captain passed the boatyard on his way out of the Saginaw River one day and happened to notice that there was not a vessel abuilding at the yard. When he returned, nineteen days later, he would swear he went by the shipyard and saw the *McDougall* just off the ways and loaded with 600,000 feet of pine![27]

When competition in the business increased, forwarders lengthened their ships to carry more lumber—and tried to

Vessels, like the "Parana" were often left stranded by November gales—or by ice—trying to make "one more trip."

lengthen the sailing season as well. They sent their vessels out in the spring while there were still ice flows on the Lakes and tried to extend the season into late fall for "one more trip," taking chances on having their ships caught out in November gales or frozen in the middle of a lake far from their home ports.

The carrying business was all hustle and hurry. There were ship masters of leaking lumber vessels who did not even take the time to go into drydock for caulking. Every other trip they simply ran their boats into shoal water near a sawmill, and the crew would fill up the seams with manure from the barnyard—or sawdust from the mill—and take off again down the Lakes with another load of pine.[28]

When Ross Robb launched into the subject of the lumber forwarding business, his talk made McTaggart think that it might not be a bad idea to look into the business himself. It's funny, he thought, how a pair of eyes the color of violets can change a man's life and make him even more conscious of his future.

The two men got down to business, coming to terms on the price of Robb's trees—the saloon keeper standing firm on his original asking price, knowing as well as McTaggart that he could get as much from any other lumberman in the Valley.

Mattie and one of the bartenders were called in to witness the signing of the contract. The last signature was officially affixed on the paper, and a draft on a Saginaw City bank was being placed in the outstretched hand of Robb when the signers and witnesses heard a commotion outside the room.

To see what the disturbance was about, the mistress of the Domino stalked over to the door and pulled it open. Just as quickly, she slammed it shut again—but not before a whiskey bottle sailed into the office, narrowly missing McTaggart's head to smash against the big iron wall safe behind Robb's chair.

A barrage followed, and the quartette inside the office heard the tinkle of broken glass and the muffled sounds of angry voices outside the door panels.

"Now what!" said Mattie, her back pressed against the door.

Robb, who was inspecting the bank draft, only said, "Don't get your dander up, Mat. Some of the boys are more than likely having a little altercation over one of the girls."

"Altercation, my foot!" Miss Morgan turned again and opened the door a crack to peer into the barroom. She said (with aversion), "Zip Somers is back." She turned to McTaggart and motioned him to take a look. "Ain't that Bill Bragg chasin' him?"

Hearing the names "Somers," and "Bragg," McTaggart got up from his chair. He joined Mattie, who stood aside to let the foreman see for himself what was going on in the barroom. He cautiously opened the door a crack wider and saw what looked like a free-for-all just beginning.

Mattie said impatiently, "We won't have a blessed thing left in one piece. I'm going out there and settle it." She started to push past the foreman.

But McTaggart was already out the door, making a flying leap over the bar just as Bill ran past him. He caught hold of the swamper by his suspenders and made a grab for Zip with his other hand. "Shut the doors and lock them so nobody can come in or go out," he shouted. "We're going to settle this once and for all!"

The doors slammed shut and the foreman continued, "If there's going to be a fight, it's going to be a fair one. Everybody move against the walls. Give these men room."

Mattie, happy that the place wasn't going to be wrecked, became a regular whirlwind of orders. She directed tables to be pushed out of the way and chairs turned upside down on table tops while shooing her girls and the clientele up against the walls. Others stood on chairs so they could better view the mayhem.

When all was ready, the only sound in the big room was a nervous giggle or two and a light scuffling of boots in the sawdust as the patrons in the saloon jostled one another to find a better place to stand.

Zip was still padded out in the flowered vests, and before the foreman began giving his instructions to the fighters, he told him to take off his purloined finery and put each garment on the bar.

He heard Bill mutter darkly, "Dodge has got some of our bowlers in that turkey there—."

So all the hats hadn't fallen into the river after all, thought McTaggart. He said to Christie, "Give that bag to Miss Morgan." And after the log thief reluctantly handed over the turkey, the foreman turned back to the two warriors.

His instructions were simple. The men were to go to opposite "corners" of the "ring" and come out fighting when he said, "Go!" The fight was to be rough and tumble—no holds barred. They were to fight to the finish until one of them had enough. The prize would be the vests and bowlers.

At the word, the two men squared off and McTaggart was relieved to see that the fighters were pretty well matched. One man wiry and quick with fists and feet; and the other man, smaller in stature, but bullish and strong. Neither was wearing calked boots, and the foreman was glad of it; for in his time, he had seen some ugly cases of "logger's smallpox"—the results of brawls between woodsmen and rivermen.

Bill took the first clout on the nose and his blood spurted ruby red. At the sight, the room rocked with noise. The swamper wiped away the gore with the back of his sleeve and lunged at the lanky Somers like a maddened bear. But the cookee was quick on his feet, and just when Bill thought he had him, Zip ducked and like lightning, lashed out with foot and fist to catch the swamper on an ear, in the pit of his stomach or on the side of his head.

The room was in an uproar—it rocked with shouts and foot stomping. Poor Bragg looked like he was almost crazy trying to catch hold of the cookee. He landed only an occasional punch that did nothing to faze the confidence of Somers. Zip skipped around the swamper, feinting with dainty jabs—taunting and laughing at him in a way that angered the foreman to a point where he found himself shouting over again, "Catch hold of him, Bill—and hang on!"

But yelling advice to the swamper didn't seem to do any good. He couldn't hear the foreman over the tumult; Bill looked winded and almost used up. In spite of it, McTaggart kept up his steady command, "Catch hold Bill and hang on!"—until finally, the sound of his voice got through—piercing the din to find its way to the swamper's ear.

Bill stopped running. And the cookee, whose overconfidence would be the ruin of him, stopped too. Surprised—he stood off a ways, eyeing his adversary. And the swamper, his big chest heaving, glowered back. Bill reminded the foreman of a bloodied rock standing impregnable in the middle of the frenzied gathering.

The cookee made a slight move. But Bragg stood his

ground. Zip leaned toward the swamper—edging nearer—his feet making sliding marks in the sawdust. The rock didn't budge. The cookee moved in again—this time cooly taking a swipe at the swamper's nose—missing by a hair's breadth.

Somers never finished his swing. Before McTaggart's unbelieving eyes, he saw Bragg reach out and catch hold of the fist. He drew the cookee to him, wrapping his arms, like a boa constrictor, around the astonished Zip—and squeezed.

When Bill released him, Zip slid onto the floor with a moan. The swamper—his nose dripping crimson and with both eyes almost closed—stood over the fallen Somers with clenched fists.

"You had enough?" he panted.

The cookee groaned, "You broke my ribs, Bragg. You wait—."

But before he could finish his threat, the foreman stepped in. He took hold of Bill's hand and held it aloft. "Bragg's the winner," he said. Shouting, whistling and stomping broke over their heads.

McTaggart looked around the circle of woodsmen and girls until he located Cully Wills. He called out to him, "Pick up those duds lying on the bar there and hang onto them until Bill is fit to pass them around." And then he turned and said in a loud voice, "Don't anybody open the doors until the place is straightened up."

While the saloon was being put back in order, McTaggart collared Ed Flatt, who with others was congratulating the victor.

The foreman pulled the teamster aside. "After Bill gets his breath back and has a chance to pass around the clothes, I want you to take him up to the German's drug store on

Third Street. Tell Walthausen to put some leeches on his eyes. I got a boat to catch or I'd do it myself." He pulled some money from his pocket and handed it to the teamster.

While Flatt was pocketing the money, McTaggart looked around the barroom. He decided it wouldn't look any worse for wear once the sawdust settled. His business at the Domino was finished. Catching Mattie's eye, he tipped his hat and then turned on his heel and made tracks for the door.

"Open up," he said to the man on guard.

The sunshine almost blinded the foreman when he stepped outside the saloon. He stood there for a moment looking at the *Evening Star*. The gentle rocking of the boat gave him a sense of peace—the same feeling he felt the morning after they broke up the wangun at Bryant's. He thought of Meg Carr. A warm glow crept over him—thinking about their chance meeting—and the future—.

But his moment of reflection was broken by the sonorous message of a steamboat whistle. He pulled out his watch. One o'clock on the dot. The *Huron* was ready to pull out. Shoving his ticker back into his vest pocket, the foreman picked up his gear and took off—hoofing it up to Water Street and over the crowded plank sidewalk to the foot of Fourth where the big shoreline paddlewheeler was waiting at her dock.

McTaggart didn't have time to purchase a ticket—deck hands were already pulling up the gangplank. He'd pay his passage direct to the purser. Tossing his dunnage aboard the boat, he made a running jump and landed on deck just as the *Huron's* paddle wheels began their gyrations.

"I made it," he thought. "I finally got away from the whole dang crew."

The foreman heaved a great sigh of relief and turned toward the bar. He'd earned a beer. He'd get his lager and then do nothing the rest of the day but relax—and think about Meg. And he would have time to read the newspaper he had tucked in his pocket. It seemed like an age since he had read a current issue from the front to the back.

McTaggart elbowed his way to the crowded bar and ordered a beer. Waiting for the bartender to fill his order, he reached into his jacket pocket for his paper. It was gone. Probably fell out during the fracus in the Domino, he thought. The foreman paid for his beer, picked up his glass, and then turned to look over the room, thinking he could find a newspaper that somebody had read and thrown down. But there was only one in sight. A copy of the *Bay City Chronicle* covered the face of a sleeping woodsman who was stretched out on several chairs shoved together to make a couch. The man's snores were regular and loud enough to be heard over the noise in the bar. The paper gently rose and fell with each breath he took.

"I'll read it through and then put it back where I found it," the foreman said to himself.

He carefully lifted one corner of the newspaper, not wanting to disturb the slumbering man. The countenance beneath the paper was reposed. It looked tranquil and without a care in the world. It belonged to none other than Toohey—Pat Toohey—on his way home to Canada.

Without disturbing a whisker, McTaggart carefully lowered the *Chronicle* over its owner's face and turned to make his way to the opposite end of the boat, mentally calculating the odds on the chickadee's staying locked in the arms of Morpheus until the *Huron* docked at Thunder Bay.

23

WXYZ

EPILOGUE

The railroad came too soon to Red Keg—hastening the depredation of the pine forests in the Saginaw Valley. By the fall of 1868, the Flint & Pere Marquette Railroad had completed its 6.5 mile line from the village of Midland to a station[1] they built on the banks of the Tittabawassee. Workers at the railroad station, in keeping with tradition—or perhaps to distinguish the place from other stations on the line—placed a row of red-painted kegs on top of the depot. The region around it was gradually cleared of stumps and made into a banking grounds for lumbermen, who brought trainloads of pine out of the woods—stacking it on the banks to await the spring log drives.[2]

In November 1870, the F & PM line, on its steady northwest trek, was extended 24.4 miles farther into virgin pine country that had formerly been inaccessible to the lumbermen. By the end of 1871, another 37.6 miles of track was laid to form a junction with the Grand Rapids and Indiana Railroad,[3] thus linking the great eastern and western lumber districts of Michigan.

In January 1868, the Secretary of the Tittabawassee Boom Company wrote in his Annual Report, "It is probable that the amount of logs passed through the boom last season (over 235 million feet) will never be exceeded."[4] In writing this statement, Mr. Whittier didn't reckon with the railroads—for in 1882 (the year when the lumber era reached its zenith in the Saginaw Valley), almost three times that amount would go through the company's sorting gaps. Much of this pine was hauled out of the pineries by steam locomotives.

Well over half of the logs that were sawed in Saginaw mills came from the Tittabawassee River. Many of them were furnished by two easterners—a lumber tycoon from Vermont by the name of Ammi Willard Wright and his associate, Philip H. Ketcham (a native of Staten Island, New York). In 1881 the two lumbermen established themselves on the banking grounds at Averill (Red Keg). They built a narrow gauge logging railroad, running it north from the rollway on the Tittabawassee—laying down eight miles of track to their big lumber camp called Ketchamville. Eventually, the company laid down almost thirty miles of branches and sidings on which three coal-burning locomotives and seventy-five cars ran to take out logs from ten thousand acres of timberland that the lumbermen owned. Between 1881 and 1884, the Wright and Ketcham line hauled out of the woods over 100 million feet of pine and dumped it on the banks of the Tittabawassee.[5] With logs coming from other camps operating on the F & P M line, the banking grounds at Red Keg became the largest in the world.[6] In the wintertime, as many as five hundred men were employed as loaders to pile up logs as they came in on the cars.

By 1887, sawmillers on the Saginaw River, seeing the pine dwindle from Valley forests, began to look north to the Georgian Bay region in Canada for fresh supplies of logs from rivers like the French, Wahnapitae, Spanish and the Mississagi.[7] Logs were towed 230 miles and more over Lake Huron to Saginaw River sawmills by powerful tugs. Acres of logs were floated loose within circles of boom sticks, chained end to end (called bag-booms).[8] These rafts sometimes covered ten to twenty-five acres of the Lake's surface at

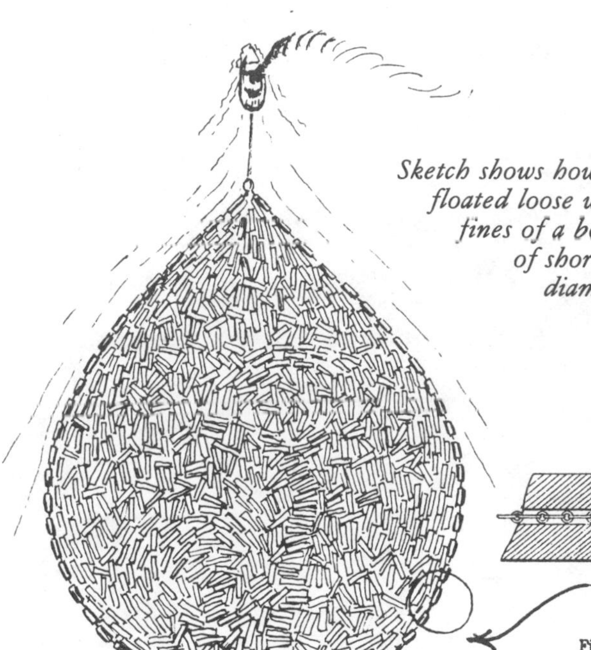

Sketch shows how the logs were floated loose within the confines of a boom, made up of short logs (of large diameter), chained end to end.

359,765. BOOM. BENJAMIN BOUTELL, Bay City, Mich. Filed Nov. 6, 1886. Serial No. 218,191. (No model.)

a time—causing vessel owners, who were already plagued by the hazards of barge tows, to rage about delays and damages caused by logs obstructing canals and narrow passageways.

The disputes and litigations between the vessel owners and the log towers only lasted until 1898, when in retaliation to a United States tariff[9] (which restored an import duty on Canadian white pine), Canada passed a regulation which directed that all logs cut on Crown lands be manufactured within the provinces. The action was the final death blow to the dying pine lumber industry in the Saginaw Valley.

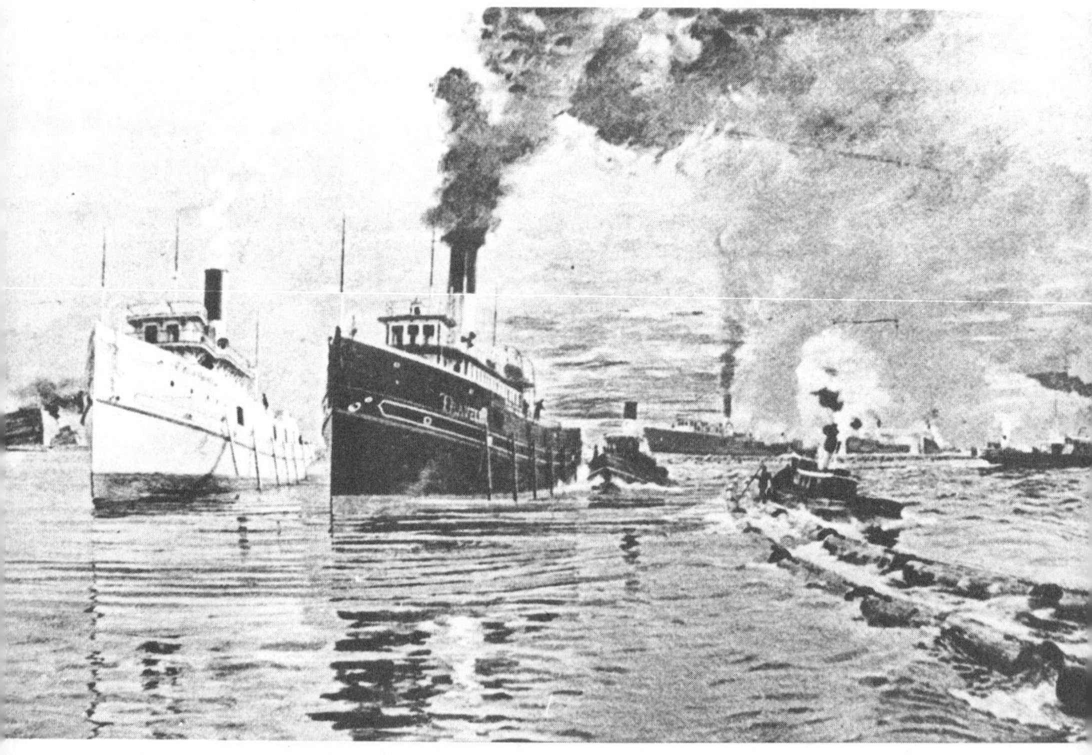

The Boutell Fleet, painted in oils in 1895 by marine artist Howard Freeman Sprague (1871–1899).

The heyday of Saginaw Pine reached its peak in 1882, when Saginaw River mills sawed over one billion, eleven million feet of white pine.[10] From that time on, the pine lumber industry on the Saginaw slowly but steadily declined. In 1894, after thirty years of profitable operation, the Tittabawassee Boom Company closed down, having sorted and rafted logs scaling more than eleven billion feet.[11]

There were still lumber piles on the docks along the Saginaw in the early 1900's, but the lumber in the piles was hardwood and other timber—not pine.[12] Most of the mill owners, the commission merchants and forwarding agents, as well as the raftsmen, woodsmen, rivermen, lakes' sailors and lumber shovers, had left the Valley—the choppers and sawyers continuing their migrations north and west where there was still virgin timber to cut.

Powerful lake tugs like Capt. Ben Boutell's Saginaw Bay Towing Co.'s "Traveler," "Winslow," "Sweepstakes" and "Niagara," towed millions of feet of logs in rafts (bag booms), over the lakes from Canada to feed hungry Saginaw River saws.

The boom shown in the lower right corner of the Sprague painting is the kind used in the 1890's and early 1900's.

ADDENDUM

LOGS SORTED AND RAFTED ON THE TITTABAWASSEE [13] 1856–1894

CHARLES MERRILL & CO. BOOM, 1856–1864	1,700,000,000*
TITTABAWASSE BOOM COMPANY, 1864–1894	11,848,549,293
	13,548,549,293

THE TOTAL AMOUNT of white pine cut in Saginaw River mills will never be known. Records kept between 1851–1891, show the mills manufactured 22,930,757,551 board feet (more than 150 million logs). This figure does not include shingles, laths, barrel shooks, etc. that were cut, or the amount of pine that was manufactured before 1851 and after 1891, which was considerable.

**Board feet*

NOTES

Abbreviations:

MPHC—*Michigan Pioneer and Historical Collections*, published by the Pioneer and Historical Society of the State of Michigan. 39 vols. (1877–1915), Lansing.

MHM—*Michigan History* Magazine, published 1917–1971 by the Michigan Historical Commission, Lansing; from 1972, by the Division of Michigan History, Dept. of State, Lansing.

TBC—Tittabawassee Boom Company.

Introduction

1. Tittabawassee is pronounced Titta-ba-waw-see. The word is a derivation of the Chippewa name *Ta-ta-ba-sink*, which means "running parallel with the bay" (Saginaw Bay, the largest indentation along the west shore of Lake Huron). (Albert Miller, "The Rivers of the Saginaw Valley," *MPHC*, 14 (1889), p. 501). The river is 117 navigable miles long—draining the counties of Ogemaw, Roscommon, Gladwin, Clare, Midland, Gratiot and Saginaw. Tributary to the Tittabawassee are the Chippewa, Pine, Salt, Tobacco, Sugar and Molasses rivers.
2. Northern White Pine (*Pinus Strobus*), also called Eastern White Pine or White Pine, is distinguishable from hard pines in that the cone-bearing tree has soft wood. Its needles grow in clusters of five—its cones are long, drooping and cylindrical in shape. The tree is long-lived; if left undisturbed, it commonly reaches 200 years (maximum age may exceed 450 years). Eastern White Pine is found across southern Canada from southeastern Manitoba to Newfoundland, through our northern and eastern States from Minnesota and northeastern Iowa to the Atlantic coast, and southward along the Appalachians to northern Georgia. Trees exceeding forty inches and 150–200 feet in height were not unusual in the virgin forests of New England, Pennsylvania and Michigan. (*Silvics of Forest Trees of the United States*. Handbook No. 271 (U.S. Dept. of Agriculture, Forest Service, Washington, D.C., 1965), pp. 329–332).
3. For interesting accounts of land cruising see: Chase S. Osborn, *The Iron Hunter* (New York: Macmillan Co., 1919), pp. 82–84, 98–99; John Andrew Russell, *A Quarter of Century of Adventure*: Biography of Joseph Warren Fordney (Boston: The Stratford Co., 1928), pp. 82–103).

   Land lookers worked on a percentage basis—sometimes for as low as one-eighth interest—and infrequently for as high as three-eights interest. (Anita Shafer Goodstein, *Biography of a Businessman: Henry W. Sage* (Ithaca: Cornell University Press, 1962), pp. 159–163.

4. From 1841–1891, lands owned by the U.S. Government or the State of Michigan could be purchased at the government's land sale price, which was $1.25 an acre. During this period, the government, through a confusing number of land disposal laws, put up immense tracts of land for sale. Much of this cheap land was quickly acquired by lumbermen. As soon as these lands went into private hands, however, the price went up to speculative heights. (See: Richard G. Lillard, *The Great Forest* (New York: Alfred A. Knopf, 1947), pp. 188–89; Ralph Hidy, Frank E. Hill, Allen Nevins, *Timber and Men* (New York: The Macmillan Company, 1963), pp. 16–17).
5. Mark Allen, ed. *Michigan Log Marks*. Workers of the Writers Program, W.P.A., comp. (East Lansing, 1941), p. 32.

A is for Ax

1. George S. Thompson, *Up to Date: Or the Life of a Lumberman* (Peterborough, Ontario, 1895), p. 23.
2. *Ibid.*
3. The ax was used to fall trees until about 1875. After that time, the improved crosscut saw was used almost exclusively for the job. (Simmons, *Yesterday and Today*, p. 688).
4. Ax makers David C. and Samuel W. Collins started their business in an old mill in the township of Canton, Hartford County, Connecticut in 1826—improving upon the blunt, rude axes of the period. By 1868, the Collins Co. was operating 23 mills. Besides axes, the company manufactured hatchets, adzes, farming and mining tools, etc. (*Scientific American*, 19 (September 23, 1868), pp. 197–98).
5. Simmons, *Yesterday and Today*, p. 688.
6. When pine was plentiful, lumbermen cut and removed only the best. Later on, when pine became scarce, they went back and cut over the land again and again, until the pine was gone.
7. Cork pines are fully matured, ripe pine trees with very little sap—which makes them even more buoyant (like cork) than less mature pines. (W.J. Beal, "The Names of Pine Trees as Known to Lumbermen," *27th Annual Report*, Michigan State Board of Agriculture (Lansing, 1888), p. 79).

8. A board foot is a standard measure—a fixed quantity accepted by buyers and sellers to determine the contents of a log. It contains 144 cubic inches of sawed lumber, or the equivalent of a board 12″ × 12″ in surface and 1″ thick. 200 feet per log was long used for estimates of value.

B is for Blacksmith

1. Traveling by land, Detroit is about 100 miles from Saginaw. By water, it is 200 miles.
2. An ox, unlike a horse, cannot stand on three legs while being shod. The sling supported his great weight while the blacksmith worked.

C is for Crosscut

1. "Riding the saw" is a term used when a sawyer presses down or pushes the saw away from him on the return stroke. If the saw is sharp, there is no need to push it. At each stroke, one man draws the saw toward himself, while his partner on the other end only guides it. Pushing the saw only upsets the rhythm and makes the job harder for both sawyers.
2. George B. Engberg, "Who were the Lumberjacks," *MHM*, 32 (September, 1948), pp. 238–246.

   For interesting nicknames of woodsmen see: Earl Clifton Beck, *They Knew Paul Bunyan* (Ann Arbor: University of Michigan Press, 1956), pp. 1–3.
3. "The Filer," from the notebook of Marion Ellsworth. Earl Clifton Beck, *Songs of the Michigan Lumberjacks* (Ann Arbor: University of Michigan Press, 1941), pp. 263–64.
4. Later on, iron wedges were used as well as wooden ones, to keep the saw from pinching in the kerf.
5. A hodag was a legendary beast that roamed the woods from Maine to Minnesota—and parts west. He is described by James Stevens (*The Saginaw Paul Bunyan*), (New York: Alfred A. Knopf, 1932), p. 170): "He hungered for the fresh timber that yet fringed Sauger Lake. Paul Bunyan never heard the hodag's hunting howl, but it sounded to the bottom of the lake and brought the dismal sauger up." The mythical animal had calked paws, teeth like saws and horns with prongs like ax bits.
6. Buckwheat pines (*Pinus banksiana*), were pine trees that had large, low tops. They had no value for lumber because of their many branches. (W.J. Beal, "Names of Pine Trees," p. 79).

D for Dough-pounder

1. Fannie Hardy Eckstorm and Mary Windslow Smyth, *Minstrelsy of Maine* (N.Y.: Houghton Mifflin Co., and Boston: The Riverside Press, 1927), p. 156.
2. Before the Civil War, woodsmen slept on the ground or on a long shelf-like bed that was built against the wall of the log shanty. A dozen or more men slept side by side, covered with a big spread made of several blankets sewn together. Turning over presented a problem until someone thought of shouting, "Spoon!" On signal, everyone would turn together.
3. Eckstorm and Smyth, *Minstrelsy,* 152.
4. *Ibid.*, p. 146.

   Sour dough was used in the Maine and Michigan woods long before it was used for cooking in Alaska, where prospectors for gold were called "sour doughs."
5. George S. Thompson, who was a clerk in Canadian lumber camps, wrote: "baking beans (was) done by covering the kettles with hot ashes. Often in taking off the cover or lids, a few pounds of ashes or sand would get into the beans—but a good cook claims that the ashes saves pepper and helps digestion." (Thompson, *Up to Date*, p. 22).
6. Pannikins were pint-size pans or dishes from which the woodsmen drank their tea.
7. Unless a lumber camp was located near a farm (which was unlikely in the early days of pine lumbering), milk, eggs and fresh vegetables were never seen by the woodsmen. And sugar was a luxury. Camp cooks, like Kelly, used instead, a cheap Porto Rican molasses for sweetening. Honey from a honey tree was also used by the camp cooks when they could get it.

E is for Echoes

1. In many lumber camps in the Great Lakes area, card playing was not allowed. Whiskey and rum and talking religion and politics were also on the taboo list.
2. The old woods song *Canada-I-O* was a favorite of woodsmen. There were adaptions of the song—like *Michigan-I-O* and the western song *The Buffalo Skinners*. Several verses of the song as remembered by Mr. Joshua Eldredge of Edinburgh, Maine can be found in Eckstorm and Smyth, *Minstrelsy of Maine,* pp. 21-25. For another version of the song see: Beck, *Paul Bunyan,* pp. 109-115.

F is for Foreman

1. "Haywire outfit" was a term used when a lumber camp had either poor management or poor equipment (or both). Hay was brought into camp baled with wire. The wire was used around the place for everything from clothes lines to mending tools and harness.
2. Franklin B. Hough (Agent of the U.S. Dept. of Agriculture, Forestry Div.), *Report Upon Forestry*, 3 vols. (Washington, D.C. 1882), 3:244.

G is for Gabriel

1. J.J. Heilala, "In an Upper Michigan Lumber Camp," *MHM*, 36 (March, 1952), p. 63.

   In describing the sound made by the cookee's dinner horn, William Hulbert wrote: "—the solo was like a demented bugle call pitched two octaves too low and wandering aimlessly all over the bass clef in the loudest, wildest, weirdest music that ever woke the echoes of the swamp." (William Davenport Hulbert, *White Pine Days on the Tahquamenon* (Lansing: The Historical Society of Michigan, 1949), p. 39.
2. A witness to the old land rush days in Detroit wrote: "The hotels are thronged with an eager, excited crowd of strangers, all rushing about as if afraid Michigan would be bought out ere they had a chance to buy an acre." (George C. Bates, "By-Gones of Detroit," *MPHC*, 22 (1893), p. 386).
3. Silas Farmer, *History of Detroit and Michigan*, 2 vols., 2nd ed. (Detroit, 1889), 1:335.
4. Alvin F. Harlow, *Old Towpaths: The Story of the American Canal Era* (New York and London: D. Appleton and Company, 1926), pp. 123, 338, 343, 345, 362.

   Eric Brunger and Lionel Wyld, "The Grand Canal, New York's First Thruway," *Adventures in Western New York History* series, Buffalo and Erie County Historical Society, 1964), 12:11.

   John F. Hinman, "My First Journey to Michigan, with other Reminiscences," *MPHC*, 14 (1889), pp. 563–571.
5. Pamela Wilson Reeves, "Navigation on Lake Erie, 1825–1860," pt. 4, *Inland Seas* (Great Lakes Hist'l. Soc. quart., Vermilion, Ohio), 16 (Summer, 1960), p. 106.

   J. Harold Stevens, "The Influence of New England in Michigan," *MHM*, 19 (Autumn, 1935), pp. 321–353.

G—is for GABRIEL—continued

6. Alexis de Tocqueville, "A Fortnight in the Wilderness," *Memoir, Letters and Remains of Alexis de Tocqueville*, 2 vols. (Boston: Ticknor and Fields, 1862), 1:147.

7. Bela Hubbard, "The Early Colonization of Detroit," *MPHC*, 1 (1877), p. 359.

8. Farmer, *History of Detroit*, 1:480.

9. Mrs. Bert Barner, "A Notable United States Military Road," *MHM*, 20 (Spring-Summer, 1936, pp. 177–184.

10. Albert Miller, "An Address to the Saginaw Valley Pioneer Society," February 21, 1874," *MPHC*, 4 (1883), p. 360.

11. B. Frank Emery, "Fort Saginaw," *MHM*, 30 (July-Sept. 1946), pp. 476–503.

12. Rev. R.C. Crawford, "Reminiscences of Pioneer Life in Michigan," *MPHC*, 4 (1883), p. 41

13. Fremont J. Tromblé, "The Passenger Pigeon," *Memoirs of Pioneer Days*, (unpub. collection of reminiscences written by members of the Tromblé family), Fremont Tromblé, comp. (Bay City, 1924), Bay County Historical Society, Bay City.

14. William R. McCormick, "Pioneer Life in Saginaw Valley," *MPHC*, 3 (1881), p. 603.

15. George W. Hotchkiss, *History of the Lumber and Forest Industry of the Northwest* (Chicago, 1898), p. 97.
    The engine that powered the Williams Bros. mill was originally in the *Walk-in-the-Water*, the second steamboat to sail on the Great Lakes.

16. *Tromblé Memoirs*, Anecdote by Josephine A. Tromblé-Greening, p. 35. A game played by the Chippewa Indians in the Saginaw Valley was similar to the game of LaCrosse.

17. Mrs. Charles L. Richman, an excerpt from her note book, *MPHC*, 3 (1881), p. 606.

H is for How Red Keg was "born"

1. See: Eugene Thwing, "The Red-Keggers" (New York: Grosset & Dunlap, 1903), p. 2

I is for Ink-slinger

1. In the larger lumber camps men were hired to do nothing but work on the camp records.
2. "Uppers" denoted superior grades of lumber—above "common."
3. Twelve million feet was a good showing for a sawmill in 1867–68.

J is for Jack Driscoll

1. Every lumbering area had its heroes. Silver Jack Driscoll was one who distinguished himself as being something of a fighter in the Saginaw Valley—especially when "in his cups." His exploits were told and retold so many times they became like fiction. Some story tellers say that Jack was a Canadian from Lindsay or Peterborough, Ontario. His death was as legendary as were his deeds in life. E. C. Beck writes: "There are legends about Silver Jack's being shot in an Upper Peninsula train, killed in a Ewen saloon, drowned in the Tahquamenon River; or drifting to an unknown end in the Washington redwoods." (Beck, *They Knew Paul Bunyan*, p. 90).

   The authors have chosen to present their own version of Silver Jack. The physical description of Driscoll was taken from an account in the Bay City *Daily Morning Call*, Nov. 22, 1881.
2. *Elton's Comic All-my-nack*, New York, 1835.
3. Log piracy was considered a part of "shrinkage" by lumbermen. (Other causes were, sunken logs, and logs scattered over the land; logs left behind by high water). Thievery was an ancient art going back to the earliest days in the lumber business. Newspapers faithfully recorded arrests made for stealing logs, but neither jail sentences, fines nor adverse publicity stopped the thieves, as can be seen in the following account: "The timber stealing business is becoming decidedly monotonous of late. The officers find the matter of looking up and making arrests for this misdemeanor a big part of their business and the courts seem to be dealing out justice in liberal doses." (Bay City *Daily Morning Call*, Jan. 26, 1882).

L is for the Loaders

1. White pine logs varied in weight. 6,000 to 8,000 pounds for a thousand board feet of logs, usually considered to be a fair average. Green logs weighed much more. Estimated and actual weighings have produced

L—is for LOADERS—continued

weights which varied from 5,000 to 10,000 pounds for each thousand board feet of lumber in logs. (William G. Rector, *Log Transportation in the Lakes States Lumber Industry, 1840–1918*, (Glendale, Cal., 1953), footnote, p. 192). Using 8,000 pounds for the purpose of computative comparison, a 28″ dia. × 16′ saw log (576 ft. B.M.), would weigh about 2 1/2 tons. A 44″ dia. × 16′ saw log (1600 ft. B.M.), would weigh almost 6 1/2 tons.

2. Three, 44″ dia × 16′ logs would measure 4,800 feet.
3. Lillard, *The Great Forest*, p. 216. Mr. Lillard writes: "It (iced road) was a sort of ice-tracked railroad that lasted like a skinny glacier after the snows alongside had melted away in the spring."
4. Sleigh bunks gradually widened as the lumber industry grew. In the 1880's, they were anywhere from eight, twelve, sixteen to twenty-one feet wide. If the roads were well iced, a team of horses weighing 3200 pounds could pull a load of logs weighing an estimated 180,000 pounds. (*Mississippi Valley Lumberman and Manufacturer*, June 13, 1884). Even though there are many photographs available to show the so-called "brag loads," they were not the usual run-of-the-mill load in the lumber camps. They were for exhibition, or picture taking, only. The famous "World's Fair Load" was one of these. It was exhibited at the Columbian Exposition, Chicago, 1893. The load weighted 145 tons and was comprised of 50 logs (18 feet long) scaling 36,055 feet. This was a single load hauled by horses on a sled (on which it was exhibited) from the forest on the Ontonagon River. The logs were rafted to a point on the Duluth, South Shore and Atlantic Railroad where they were shipped on nine freight cars to Chicago. The logs were loaded by Nester Brothers of Baraga, Michigan. (Souvenir book, *Gems of the World's Fair and Midway Plaisance* (Philadelphia, 1894), p. 11).
5. Sleighs were stoutly built of oak and braced with iron. Although they varied in size, all were built with two pairs of iron-shod runners which were 6 to 7 feet long and set about 5 1/2 to 6 feet apart. The logs rested on two heavy timbers or cross pieces called "bunks." An estimated weight of an empty sleigh and chains was about five tons.
6. John S. Springer, *Forest Life and Forest Trees*, (New York: Harper & Brothers, 1856), pp. 94–98.
7. The following sources were used to describe the loading methods employed at McTaggart's Red Keg: Robert F. Fries;, *Empire in Pine* (Madison: The State Historical Society of Wisconsin, 1951), pp. 31–33; John J. Heilala, "In An Upper Michigan Lumber Camp," *MHM*, 36

L—is for LOADERS—continued

(Mar., 1952), pp. 70–71; Hulbert, *White Pine Days*, pp. 51–67; Stewart Edward White, *The Blazed Trail*, (New York: Grosset & Dunlap, 1902), pp. 13–14, 79.

8. Sources used for road-making and icing roads: Fries, *Empire in Pine*, pp. 26–27; Hulbert, *White Pine Days*, p. 25; Lillard, *Great Forest*, p. 216; Robert E. Pike, *Tall Trees, Tough Men* (New York: W. W. Norton & Co., 1967), p. 77; Rector, *Log Transportation*, pp. 209–213; Springer, *Forest Life*, pp. 83–84; Thompson, *Up to Date*, pp. 21, 82; White, *Blazed Trail*, pp. 8–10.

9. A ten percent drop is a ten-foot drop in a one hundred-foot distance.

M is for Marking Hammer

1. The log rule was also called names like "cheat stick," "robber's cane," and "money robber."

2. For more detailed information on scaling logs, read Donald Bruce and Francis X. Schumacher, *Forest Mensuration* (New York: McGraw-Hill, 1935), pp. 152–170.

3. At one time there were over 50 log rules in use in the U.S. and Canada. None of them gave an accurate reading. The Scribner Rule and the Doyle Rule were the most widely used log rules in the Great Lakes States. Both came from the East—the Scribner Rule being used in Maine as early as 1855. The Scribner rule underestimates the amount of lumber that can be produced, especially in the case of smaller diameter logs. The Doyle Rule gave "ridiculously low value for small logs," while giving too high a value for very large logs. (*Ibid.* p. 159).

   One amusing story about a controversy that arose between a log owner and sawmiller was printed in the *Mississippi Valley Lumberman* (Apr. 1, 1898). The men were arguing about price and scale. Finally, the sawmill operator said he was willing to give in a little if the lumberman would do the same—that it might be well to use the "Golden Rule" and do with others as you would like to be done by.

   "Not by a damn sight," replied the log owner, "I do not go by any Golden Rule—Scribner's Rule is good enough for me to measure my logs by."

4. John Springer wrote: "Frauds are sometimes practiced upon those who purchase logs by driving a knot or piece of a limb of the same tree into the konkus (decayed spot) and hewing it off smoothly so that it has the appearance of a natural knot." (Springer, *Forest Trees*, p. 99). Also see:

M—is for MARKING HAMMER—continued

Hotchkiss, *History of Lumber Industry*, p. 667.

5. Log mark designs were made up of combinations of symbols, initials, lines, triangles, circles and figures. Every camp had its own individual log mark. The marks were changed every season the camp was in operation.
6. Franklin B. Hough, *Report Upon Forestry*, 3 vols. (Washington, D.C., 1878) 1:214–15.
7. Lillard, *The Great Forest*, p. 122.
8. Bark marks consisted of combinations of straight lines that the swamper or chopper cut into the bark of a log with an ax.
9. Allen, *Michigan Log Marks*, p. 8.
10. *Ibid.*, p. 12.
11. Some log owners not only stamped the ends of their logs, but, as an added protection, marked them with the old bark as well. (*Ibid.*, p. 82). Paint marks, dabbed on log ends were sometimes used by lumbermen in Michigan and in Canada to identify logs. (*Ibid.*, pp. 43–45; Thompson, *Up to Date*, p. 96).
12. Laws were passed in Michigan before 1867 to benefit the log owners—but none spelled out log mark regulations until Mar. 27, 1867, when an Act "to protect owners of floating logs and lumber," was approved. The Act was tacked onto an 1861 bill (amended in 1863), entitled, "An Act to provide for the floating of logs and timber in the streams of this State." (Hon. James S. Dewey, comp., *The Compiled Laws of the State of Michigan*, 2 vols. (Lansing, 1872), 1:549–550).
13. Sec. 6 (as amended by Act 100 of the Laws of 1869, approved Apr. 2, 1869): "If any person shall falsely make, forge or counterfeit such mark, and use the same in marking logs or timber, knowing the same to be the mark of another person, and with intent to defraud, shall be deemed guilty of felony, and shall be punished by imprisonment, at hard labor, in the State Prison not to exceed five years, or by fine of not less than one hundred dollars not more than two thousand dollars." (*Ibid.* p. 550).
14. The East Saginaw foundry, A. F. Bartlett & Co. cast hundreds of log mark hammers for Saginaw Valley lumbermen. The stamping irons came in all sizes and shapes.

N is for Needle

1. "Lumberjack's Prayer," is from Earl Clifton Beck, *Lore of the Lumber Camps* (Ann Arbor: University of Michigan Press, 1948), p. 15.

   Sources used to describe Sunday in camp are as follows: John W. Fitzmaurice, *The Shanty Boy*, Cheboygan, 1889, facsimile reprint (Mt. Pleasant: Central Michigan University Press, c 1963), pp. 166–172; Hidy, Hill & Nevins, *Timber and Men*, p. 714; Pike, *Tall Trees*, 96–97; Springer, *Forest Life*, pp. 117–121; Isaac Stephenson, *Recollections of a Long Life*, (Chicago, 1915), p. 86; Thompson, *Up to Date*, pp. 23–25.

2. Arthur Hill, "Life in a Logging Camp," *Scribner's Magazine*, 8 (June, 1893), p. 714.

3. Fitzmaurice, *Shanty Boy*, pp. 112–13; Thompson, *Up to Date*, p. 25.

   In telling about his experiences in Michigan lumber camps, the Rev. William Davis, evangelist of Boston wrote (for the Bay City *Daily Morning Call*, June 12, 1881), "The speaker had been refused permission to hold meetings in but three camps. In one instance, when he asked if he could hold a meeting in camp that evening, the boss replied, 'Yes—which party do you speak for?' And when he understood that it was a religious meeting he desired to hold, he declared he would have none of it; did not allow his men to talk religion, as it made trouble in camp."

   All faiths were represented in lumber camps—and it was almost impossible for one sect to begin talking about religious matters without breeding strife—which was liable to run to excess—trouble that the foreman could well do without. Two notable exceptions to the rule were Preacher Jim Roberts and Holy Bill Poyseor, who visited the camps in the Upper Peninsula in Michigan. Roberts was an Englishman who worked in the iron mines at Ishpeming. Not relishing "being a bloody ground 'og all his life," he turned to preaching. He made the rounds of lumber camps in the U.P., traveling from camp to camp on dog sled and by hand cars. Meetings were held in the cook-camp, "the men giving respectful attention." (Lewis C. Reiman, *When Pine was King* (Ann Arbor: North Woods Publishers, 1952), pp. 135–144).

4. Billy Flynn's "receipt" for hot bran mash was taken from the *Petoskey News-Review*, June 23, 1965. For the care of camp horses and oxen see: Reiman, *When Pine was King*, p. 103; Springer, *Forest Life*, pp. 73–83.

5. Willis C. Ward, "Reminiscences of Michigan's Logging Days," *MHM*, 20 (Autumn, 1936), p. 311; Fitzmaurice, *Shanty Boy*, p. 71.

6. E. A. Grange, "Bone Diseases of Horses," Michigan State Board of

N—is for NEEDLE—continued

Agriculture, *28th Annual Report* (July 1, 1888 to June 30, 1889), pp. 465–474.

7. Epizooty is a kind of distemper that is highly contagious among horses. It is mentioned in connection with lumber camp horses in the *Muskegon Chronicle*, Dec. 11, 1872.
8. *Petoskey News-Review*, June 23, 1965.
9. Fitzmaurice, *The Shanty Boy*, p. 71.

Oh-h-h-h moans the woodsmen

1. Goodstein, *Diary of Businessman*, pp. 268–69.
2. Gerald Carson, *The Old Country Store* (New York: E. P. Dutton & Co., Inc. 1965), p. 257.
3. (*Ibid.*, pp. 240–41).
Philadelphia was once the seat of medical learning. Regular physicians who practiced with a stethoscope and book knowledge were sneeringly called "Philadelphia doctors" by the traveling medicine men who were afraid the regular doctors would hurt their lucrative business.
4. Bay City *Daily Morning Call*, May 25, 1881.
5. The $5 hospital insurance plan was inaugurated by John W. Fitzmaurice, an associate editor of the *East Saginaw Courier*. In 1872, he began agitation in the *Courier* for the establishment of a hospital to care for sick or hurt woodsmen. The operation of the hospital was to be paid for by selling $5 certificates to woodsmen who wished to participate in the plan. Local businessmen quickly responded to the idea and St. Mary's Hospital, organized by the Sisters of Charity in East Saginaw, was the first institution of its kind built to care for injured woodsmen. After that, the $5 hospital plan quickly caught on in other lumbering areas in Michigan as well as in other Great Lakes states. (Fitzmaurice, *The Shanty Boy*, p. 105).
6. Woodsmen purchased their $5 tickets from agents who made the rounds of the lumber camps. An initial payment of nine or ten cents could be made for a certificate—with the idea of paying ten cents a week thereafter, for one year. If the holder of a ticket should become sick or hurt before the five dollars was paid in full, he received hospital care just the same. One of these fortunate individuals was a good Scotsman by the name of Archie MacDonald who hurt his hand in the woods while trying to couple two railroad cars together. His hand became

OH-H-H moans the woodsman—continued

infected and he was sent to a hospital in Port Huron where he spent several months before he was released. In later years he told his children the purchase of the hospital ticket was the best bargain he ever got for one dollar. (Rea MacDonald Trevvett to Irene Hargreaves, July 11, 1967).

R is for the Rivermen

1. "Report of the Superintendent," *Annual Report of the Flint and Pere Marquette Railroad Co.*, for the fiscal year ending Dec. 31, 1869, p. 17.
2. By the early 1870's Michigan Lumbermen were using the railroads to haul logs from the woods. An item in the *Bay City Daily Morning Call* (June 8, 1881) recalls, "only a few years ago such a thing as logging in the summer time was unheard of; but now large quantities of logs are cut in the summer season and brought to the mills by rail. The railroad penetrating the wilderness has almost completely revolutionized logging operations."
3. In spite of the railroads making inroads in the industry, log driving never ceased in the Lakes states. William Rector writes, "most logs still moved down the driving streams in the spring, for in that same year (1887) the twenty-one 'principal booms' in the Lake states actually handled almost twice as many logs as the private logging railroads were capable of handling, even if all of them had operated at full capacity." (William G. Rector, *Log Transportation in the Lakes States Lumber Industry, 1840–1918*, (Glendale, Cal., 1953), p. 205).

   In many instances the log drives worked with the railroads. In 1888, the Flint & Pere Marquette railroad hauled 161 million feet of logs out of the woods to dump about one hundred million feet into the Tittabawassee River at Averill (Red Keg). (*Northwestern Lumberman*, March 17, 1888, p. 24).
4. The "booming grounds" was the place where mixed logs were sorted according to their marks and made into rafts. The Titttabawassee Boom Company (hereafter cited as TBC) activities included running, booming and rafting logs. Their operations extended up river from Green Point, where the Tittabawassee enters the Saginaw. The company operated on the river from 1864–1894. Information about the TBC and its operations was taken from Allen, *Michigan Log Marks*, pp. 29–40; Dustin, *Saginaw History*, pp. 131–135; Ralph W. Stroebel, *Tittabawassee River Log Marks* (Saginaw: Eddy Historical Series, Saginaw Public Libraries, 196); TBC *Minutes*, passim.

R—is for RIVERMEN—continued

5. TBC *Minutes* of monthly board of directors and annul stockholders meetings, 1 vol. (Feb. 8, 1864–Jan. 1, 1881), (Ann Arbor: University of Michigan Historical Collections), p. 117.
6. The description of Bryant's Trip was taken, for the most part, from Fred Dustin, *Saginaw History*, (Saginaw, 1949), pp. 131–134.
7. Mark Allen, Ed. *Michigan Log Marks*, Workers of the Writers Program, W.P.A., Comp. (East Lansing, 1941), p. 31.
8. TBC *Minutes*. "Articles of Association and By Laws," pp. 1–9.
9. Rector, *Log Transportation*, p. 131.
10. Allen, *Michigan Log Marks*, p. 55.
11. TBC *Minutes*. (Mar. 31, 1868), p. 109.
12. Vernon H. Jensen, *Lumber and Labor* (N.Y.: Farrer and Reinhart, 1945), pp. 56, 57.
13. John S. Springer, *Forest Life and Forest Trees*. (New York: Harper & Brothers, 1856), pp. 172–173.
14. Spring peepers are midget frogs. They are so small and well camouflaged, they cannot be seen unless one is looking for them. Only the male sings, making a "peep" each half-second, stringing them together so the result is a trilling sound. Spring peepers appear as early as March in some parts of Southern Michigan and are in full song throughout the state by the end of April. (Gordon Charles, "Midget Frog Bugles Start Off Spring," *Bay City Times*, Apr. 20, 1974).
15. John Kieran, *An Introduction to Nature* (Garden City: Doubleday and Company, Inc., 1955), p. 83. Mr. Kieran: "The lowly Skunk Cabbage (*Symphocarpus foetidus*) deserves honorable mention if for no other reason that in the swamps, low meadows, and wet woods, it is the herald of Spring over a large section of North America, the first wild flowering plant to thrust itself above the ground and burst into bloom."
16. Throughout the TBC *Minutes* there are reports of complaints of damages to shoreline property due to running logs. One record of settlement (five months after the complaint was registered) reads: "Motioned That the Pres. and Sec. be instructed to settle with Wm. Kelly for damage done his farm on Pine river by flooding by paying him the sum of $200, which shall be in full for all demands to July 1, 1867 and to execute a Lease for his Land at $30.00 per year to expire Jan. 1, 1879." (TBC *Minutes*, July 1, 1867, p. 91).

17. Pay on the river drives ran from less than a dollar a day in the 1840's to as high as $2.50-$5.00 in the 1880's and 1890's. (Robert F. Fries, *Empire in Pine* (Madison: The State Historical Society of Wisconsin, 1851), p. 14; An Old Timer, "Around the Campfire," *MHM*, 14 (Autumn, 1930), p. 696).

S is for Sorting Booms and Sawmills

1. In most instances, the rafts of sorted logs were towed to the holding booms at Green Point (located at the river mouth). If the owners neglected to remove their logs from holding areas, the Boom Company reserved the right to remove them at the owner's expense and risk. The Company also leased storage areas on the Saginaw River at a place called Crow Island, which was about half way between Saginaw and the Bay cities.
2. Dustin, *Saginaw History*, pp. 133–34; TBC *Minutes*, pp. 30, 53, 62, 78, 96, 102, 117.
3. Dustin, *Saginaw History*, p. 134; Richard G. Wood, "A History of Lumbering in Maine, 1820–1861," *The Maine Bulletin*, 37:7 (Orono: University Press, Jan. 1935), p. 140.

   The rafting pin (or spike) was used in the Saginaw Valley until about 1874. Because of the enormous amount of rope that was needed to make up rafts using this type of pin, the TBC Board of Directors sought other ways to fasten logs together for delivery to the mills. In the *Minutes* (Mar. 4, 1867), p. 85, the following was noted: "Motioned that the Agent be instructed to purchase of Mr. Raymond a sufficient quantity of his Patent Crotch Rafting pins to test the practicability of using them in Rafting—not to exceed 25,000 at not more than $6.00 per M delivered at the booms where wanted. Adopted."

   The "Crotch Rafting Pin" was finally adopted by the Company but not until after it was well tested, and the chain-and-iron-pin method, as used by the Muskegon Boom Company (Muskegon River), was also tried. (TBC Minutes, Jan. 8, 1867, pp 80–81).

   The crotch pin was used on the Penobscot in Maine as well as on the Tittabawassee in Michigan after the Civil War. The pin was wedge-shaped, about 7 inches long and 2 inches wide. It was tapered to a point at the driving end. To make a raft using the crotched wedge, 1/2" rope was laid down the center of assembled logs. The pin whacker drove the pins over the rope, into the center of each log to "staple" the rope in place. This method eliminated the need to hitch the rope around each

S—is for SORTING BOOMS/SAWMILLS—continued

pin. The D. Hardin & Co. planing mill in Saginaw City manufactured crotched pins for the Boom Company for many years. To the best knowledge of the authors, the only Michigan lumbermen to use this method of rafting logs were those from the Saginaw area. These men also used the method in their operations in the Georgian Bay area in Canada.

4. Albert Miller, "The Rivers of the Saginaw Valley," MPHC, 14 (1889), p. 501.

5. Albert Miller, "*The Saginaw Marshes*," H. R. Page, pub., *History of Bay County, Michigan* (Chicago: H. R. Page & Co., 1883), p. 212.

6. Correspondence between Gen. Frederick Haldimand, (British Commander at Quebec, 1760–65), and others, show that Indian corn grown in the Saginaw region was important to the welfare of the forts at Detroit and Michilimackinac. (The Haldimand Papers on file in the Dominion Archives at Ottawa, Canada. Letters pertaining to Michigan have been copied and printed in the *Michigan Pioneer & Historical Collections*, vols. 9, 10, 11.

7. George W. Hotchkiss, *History of Lumber and Forest Industry of the Northwest* (Chicago, 1898), p. 103.

8. John Gray, "A Small-Town Boy's Big World," *Michigan Natural Resources Magazine* (Lansing, March-April, 1970), p. 11.

9. Lumberman Henry W. Sage wrote to his grandson about the Indians who came to his company store in Wenona (later, West Bay City) to trade goods for "tea, sugar, tobacco, and some bright colored calico." (Henry W. Sage to Harry Sage, July 24, 1877, No. 1557, Dept. of Manuscripts, John M. Olin Library, Cornell University, Ithaca, New York).

10. Henry Crapo was Michigan's "Civil War Governor," (1865–1868). He came to Flint, Michigan (from New Bedford, Mass.), where he made a fortune in the pine lumber industry.

11. Martin Deming Lewis, *Lumberman from Flint: Henry Howland Crapo* (Detroit: Wayne State University, 1958), p. 66.

12. Salt was a by-product of the lumber industry in the Saginaw Valley. Many years before men thought of cutting pine trees for profit, salt brine had been observed by the Indians as it seeped from the ground along the Tittabawassee and Salt rivers. Dr. Douglas Houghton, State Geologist (1837–1845), was the first to realize the commercial possibili-

ties of the brine, which he outlined in a report in 1838. That year a well was sunk in the Valley, but it was not until June, 1860 that a successful operation of salt manufacture was begun in East Saginaw, using cordwood for fuel. Later on, slabs and sawdust, as well as exhaust steam from the mills, were used to heat the brine.

For more information about the salt industry in the Saginaw Valley see: Dustin, *History of Saginaw*, pp. 150–54; *Lumberman's Gazette*, Dec. 29, 1877; Mills, *History of Saginaw*, 1:426–445; William L. Webber, "Salt Making in the Saginaw Region," *MPHC*, 4 (1883), pp. 13–22.

13. The pans used for making salt were made from 1/4" boiler iron, measuring from 90 to 120 feet long. They were 10 or 12 inches deep and 12 or 15 feet wide. (Dustin, *Saginaw History*, p. 150).

14. Vernon H. Jensen, *Lumber and Labor* (New York, Toronto: Farrar & Reinhart, Inc., 1945), pp. 57–58.

15. In 1868, mill workers worked an eleven-hour work day. Traditional hours were from 6:00 to 6:00, with an hour off for lunch.

16. Fries, *Empire in Pine*, p. 106.

17. The circular saw was invented in England sometime before 1777 (Fries, *Empire in Pine*, p. 61), but it didn't come into practical use in the U.S. until about 1850. The early circle saws cut a kerf 1/4" to 1/2" wide, and although they were fast, they were extremely wasteful. H. H. Crapo in Flint, Michigan, operated in his mill an 8-foot-dia. circular to saw cork pine logs. (Hotchkiss, *History of Lumber Industry*, p. 89). For the evolution of sawmill equipment and machinery, see: Hotchkiss, pp. 649–660.

18. Mills, *History of Saginaw*, 1:398.

19. Boards were cut 1-1/16" inches thick and were trimmed 2 inches longer than the called for length, ie: a 12-ft. board measured 12 feet and 2 inches long. (Dustin, *Saginaw History*, p. 142).

20. The various grades of lumber varied from one lumber center to another. Generally, the grades were, 1. Clear or Uppers, 2. Common, 3. Culls.

    In 1873, after years of confusion, rules were adopted for grading lumber by the Chicago Lumberman's Exchange. However, most lumbermen disregarded the rules and continued to grade their lumber as they pleased. Ralph Hidy, Frank Hill and Allen Nevins, *Timber and Men*, p. 175).

S—is for SORTING BOOMS/SAWMILLS—continued

21. The name mulay or muley is derived from the German *mühl säge*, mill saw. (*Appletons' Cyclopaedia of Applied Mechanics*, 2 vols., Park Benjamin, ed. (New York: D. Appleton and Company, 1885), 2:707). For an interesting account of the mulay saw see: Carl Addison Leech, "Sharon Hollow: Story of an Early Mulay Sawmill of Michigan," *MHM*, 17 (Summer-Autumn, 1933), pp. 377–392.
22. Hotchkiss, *History of Lumber Industry*, pp. 94, 103. In the 1880's, circular saws were cutting from 60,000 to 80,000—and in several instances, 200,000 board feet a day.
23. Mills, *History of Saginaw*, 1:398.
24. The band saw—a continuous ribbon of steel no thicker than a knife blade—did not come into general use in the sawmills until almost ten years after it was exhibited at the Centennial Exposition of 1876. The saw measured about 15 inches wide and 50 feet long, with teeth running along one edge. It operated like a pulley belt. The band saw was welcomed by the lumberman because the kerf it made in the saw log was extremely small. By the 1880's, mill men had become very conscious of waste.
25. James J. McCormick, a native of Albany, N.Y., was a pioneer mill operator on the Saginaw River.
26. Hotchkiss, *History of Lumber Industry*, p. 97.
27. Henry D., Edward N. and Charles T. Wickes started business in East Saginaw manufacturing mill equipment in 1860. (Mills, *History of Saginaw* , 2:57–58).
28. Frank Kirby was known, later on, as a marine engineer who designed Great Lakes vessels.
29. Mills, *History of Saginaw*, 2:58.
30. The largest amount of pine lumber ever cut in one season on the Saginaw River was by the Parker and Birdsall mill (formerly the John McGraw mill) in Bay City, when in 1882 it cut 55,260,722 feet). (*Lumberman's Gazette*, Dec. 27, 1882, p. 2). In 1870, the H. W. Sage mill in Wenona cut 34,450,972 feet. (Page, *History of Bay County*, pp. 50–52).

T is for the Towns

1. Water Street in Saginaw City became Niagara Street when East Saginaw and Saginaw City merged "by act of Legislature," in 1889.

T—is for TOWNS—continued

2. Dustin, *Saginaw History*, p. 6. Another Saginaw historian spelled the Indian name *Ke-pay-sho-wink*, which means, "the great camping ground." (Truman B. Fox, *History of Saginaw County from 1819 to Present*) East Saginaw, 1858), p. 9).
3. Dustin, *Saginaw History*, p. 64; Wilbert B. Hinsdale, *Archaeological Atlas of Michigan* (Ann Arbor: University of Michigan Press, 1931), p. 3. The Saginaw Trail began at the southern or lower part of Michigan, and extended through Detroit to the Saginaw River, and on up to Mackinaw—which is about three hundred miles from Detroit.
4. Fox, *History of Saginaw*, p. 9; Dustin, *Saginaw History*, p. 64.
5. Fred Dustin, *The Treaty of Saginaw, 1819*, pamphlet written for the Treaty Centennial celebration, Sept. 19, 1919 (Saginaw, 1919), p. 3; B. Frank Emery, "Fort Saginaw," *MHM*, 30 (July-Sept., 1946), p. 482.
6. Gen. Lewis Cass resigned as territorial governor in Michigan in 1831 to become U.S. Secretary of War. (F. Clever Bald, *Michigan in Four Centuries* (New York: Harper & Row, 1961), pp. 254–55).
7. Emery, *Fort Saginaw*, p. 482.
8. *American State Papers*. II. Indian Affairs, p. 310.
9. Order issued by Major General Macomb by John Biddle, Aide-de-Camp. Emery, *Fort Saginaw*, pp. 483–84.
10. "Treaty with the Chippewa," *American State Papers*. II. Indian Affairs, pp. 194–95; Dustin, *Treaty of Saginaw*, pp. 20–22.
11. *Ibid.* p. 18.
12. For an interesting account of building Fort Saginaw, see: Emery, *Fort Saginaw*, pp. 476–503.
13. General Winfield Scott, Commander, Dept. of the East to Col. Pinkney, Commander, 3rd U.S. Infantry, May 22, 1822. *Ibid.*, pp. 488 89.
14. *Ibid.*
15. "Reminiscences of William L. Webber," Mills, *History of Saginaw*, 1:65.
16. Anson Rudd to Jobish Sawdy, 1854, Mills, 1:183.
17. Tocqueville, *Wilderness*, 1:196.
18. Charles B. Headley, George F. Lewis, *Annual Statement for 1869: A Compilation of Salt and Lumber Statistics for the Saginaw Valley* (East Saginaw, 1870), pp. 6–7.

T—is for TOWNS—continued

19. Holland's *East Saginaw—Saginaw City Directory for 1868–69* (Chicago: Western Pub. Co., 1868), p. 4.
20. *Ibid.*
21. *Ibid.*
22. *Ibid.*
23. W. H. DeLisle, comp., *Directory of Bay City, Portsmouth, Wenona and Bangor for 1868–69* (Bay City, 1868), p. 2.
24. *Ibid.*

    During the decade from 1870 to 1880, while Michigan gained 38% in population, Bay County gained almost 140%. During that period, only two cities in the U.S. exceeded the rate of increase for Bay City. They are Minneapolis and Denver. (George E. Butterfield, *Bay County Past and Present*, Centennial ed. (Bay City, 1957), p. 86.
25. The Flint & Pere Marquette, and the Jackson, Lansing & Saginaw railroads.
26. Jas. M. Thomas and A. B. Galatian, comp. & pub., *East Saginaw, Saginaw City & Bay City Pioneer Directory & Business Advertiser, 1866–67* (East Saginaw, 1866), p. 3.
27. Mills, *History of Saginaw*, 1:240; Dustin, *Saginaw History*, p. 103.
28. Only the main areas of East Saginaw's business district were lighted by gas, but most streets in the Saginaws were still illuminated by oil lamps. (Mills, *History of Saginaw*, 1:193). Bay City was expecting gas fittings to be installed "by July" 1868. (*Bay City Journal*, April 18, 1868).
29. *Ibid.*, Apr. 18, 1868.
30. *Ibid.*, May 30, 1868.
31. *Ibid.*, Aug. 24, 1865.
32. DeLisle, *Bay City Directory, 1868–69*, p. 197.
33. After Saginaw City and East Saginaw were consolidated (1889), Saginaw City's Washington Street became Michigan Avenue.
34. *Bay City Journal*, June 15, 1865.
35. *Ibid.*, July 25, 1868.
36. The business district of East Saginaw is built on slabs and sawdust fill. Every building rests on piles sunk deep in the earth. In 1868, sidewalks were laid over the bayou between Franklin and Jefferson streets on

supports more than six feet high, so people could pass into the shops dry shod.

37. DeLisle, *Bay City Directory*, 1868–69, p. 21.

U is for the Unwinding

1. Stewart Holbrook, *Holy Old Mackinaw* (New York: Macmillan Co., 1938), pp. 67, 195–96.
2. Springer, *Forest Life*, p. 150.
3. *Ibid.*, p. 151.
4. No hard liquor was allowed in most lumber camps operating in the Great Lakes states. It was a different story in the old lumbering days in Maine and in New Brunswick. According to a quotation in John Springer's *Forest Life* (p. 152), " 'The first and most important article,' says Mr. Todd, of St. Stephen's, N.B., 'in all our movements, from the stump in the swamp to the ship's hold, was *Rum*! RUM!' "
5. *Bay City Journal*, June 20, 1868.
6. Fitzmaurice, *Shanty Boy*, p. 177.
7. *Ibid.*, pp. 177–78; R.L. Polk & Co., pub., *Bay City Directory* for 1881–82 (Detroit, 1881), p. 203; *Ibid.*, 1884–85, p. 35.
8. Copyright, 1888, Ira D. Sankey.
9. Copyright, 1888, by the National Temperance Society and Publication House.
10. Allen Eldred, "Temperance Fire Song" (first line second verse), *Youth's Temperance Banner* (New York: National Temperance Society, Mar. 1, 1889), 24:5, p. 20.
11. *Ibid.*, (Apr. 1, 1886), 23:7, p. 28.
12. Weeks & Polk, pub., *Bay City Directory*, for 1873–74 (Detroit, 1873), pp. 380–82, 442.
13. George E. Butterfield, ed., *Bay County, Past and Present* (Bay City, 1918), p. 178. Mr. Butterfield's source was the Bay City Police Dept. *Annual Report*, March, 1887.
14. Holbrook, *Holy Old Mackinaw*, p. 196.
15. A "Michigan bankroll" was newspaper, cut to the length and width of currency, rolled up with a couple of one or two dollar bills wrapped around the outside, to make the fake bundle look like a fat roll of bills.

V is for the Lumber Vessels

1. The steamboat *Evening Star*, 381 g.t., was built by Wiliam Crosthwait in East Saginaw in 1866. The *Star* was still in commission in 1899. (J.B. Mansfield, ed. *History of the Great Lakes*, 2 vols. (Chicago: J.H. Beers & Co., 1899), 1:824.
2. Mills, *History of Saginaw*, 1:716.

   Competition between the Saginaw River passenger-freight boats and tug boat operators was keen—with collisions, bumpings, scrapings and racing almost a daily occurence. The boats not only competed against one another but by 1867, against the railroads as well. Railroad cars ran down river to Bay city from the Saginaws in about thirty minutes—against the two-hour plus ride on the boats. But there were no refreshments (liquid) sold on the trains—and the fare, in 1868, on the Saginaw-Bay City trains cost 40 cents while a boat ticket cost only 25 cents. With such inducements, its no wonder the boats didn't lose their popularity until into the 1890's.
3. The purser and other boat personnel had a lively time keeping their eyes on the lookout for woodsmen or rivermen with calks in their boots. One time, on the side-wheeler *Metropolis* (a luxurious shoreline vessel that ran from Saginaw to Alpena), a woodsman was caught doing a dance on top of the rosewood piano before he was discovered. (Frank Ward Holt, "Side-Wheel Steamer *Metropolis*," *Inland Seas*, 14 (Fall, 1958), pp. 29–230.
4. *Bay City Chronicle*, Aug. 2, 1874.
5. The description of the Saginaw marshes and the wild rice beds was taken from William B. Mershon, *Recollections of My Fifty Years Hunting and Fishing* (Boston: Stratford Co., 1923), pp. 59–87; Albert Miller, "Saginaw Over One Hundred Years Ago," *MPHC*, 17 (1890), pp. 446–449; Iola Fuller, *Loon Feather* (New York: Harcourt Brace & Co., 1940), p. 58.
6. *Bay City Journal*, April 22, 1874.

   The Third Street Bridge report for April 1868 showed that 1,996 steamers, 498 sailing vessels and 106 barges went through the bridge that month. (*Bay City Journal*, May 9, 1868).
7. Water Street in Bay City was called "hell's half mile" during the lumber era.
8. "Yesterdays of Bay City," *Bay City Times Tribune*, Sept. 30, 1923.
9. Sturgeon fish were called "Saginaw beef" in the Saginaw area. The meat

at the fish's head (under the fin) is said to taste much like beef. In New York State, sturgeon was called "Albany beef."

The fishing and shipbuilding industries were only second to the lumber industry in Bay City. Pine barrels, packed with Saginaw Bay fish regularly found their way to eastern markets. In 1866, one Bay City fish dealer shipped about eight tons of frozen fish per week to the New York market (*Bay City Journal*, Feb. 8, 1866).

10. Mansfield, *History of the Great Lakes*, 1:704.

11. *Bay City Journal*, April 11, 1868.

12. The last commercial sailing vessel to be built on the Lakes was the *Cora A.*, built in 1889 at Manitowoc, Wis. She was 146 feet long, 31 feet beam. (Dana Thomas Bowen, *Memories of the Lakes* (Daytona Beach, Florida, 1946), p. 118).

    The largest sailing vessel to be built on the Lakes was the schooner *David Dows*, the world's first five masted vessel and the only one built on the Great Lakes. She was built in Toledo and launched in 1881. Her dimensions were: 265-foot keel, 37-foot beam and 18 feet deep. She had two center boards and carried 70,000 yards of canvas. After operating only a few years as a sailing ship she was made into a tow barge and was lost Thanksgiving night, 1889, off Whiting, Indiana. (George E. Hardy, "The David Dows," *Inland Seas*, 1 (July 1945), p. 54.

13. Howard I. Chapelle, *History of American Sailing Ships* (New York: W. W. Norton & Co., 1935), pp. 268–270.

    The fore and aft schooner rig, which was handier and only required a small crew, was more popular on the Lakes than square rigged vessels, which required more hands and were more difficult to sail in narrow channels. However—many a "laker" carried a yard on her foremast on which was set a great square sail as well as a curious looking triangular "raffee" which was sheeted out on the squaresail yard and hoisted to the foretopmast head—a sight that might have prompted withering remarks from salt water sailors.

    Centerboards—movable keels or "fins," that dropped down through a slot in the bottom of the boat (from a water-tight box in the hull), were built into Great Lakes shoal draft sailing vessels to prevent them from sagging or drifting to leeward when sailing with the wind abeam or when beating windward.

14. Mansfield, *History of the Great Lakes*, 1:510.

15. The last cargo sailing vessel to sail on the Lakes was the *Our Son* (182 feet long, 35 feet wide, 13 feet deep), built in 1875 at Lorain, Ohio. On Sept. 26, 1930 she went down in a gale with distress signals flying, off

V—is for VESSELS—continued

Sheboygan, Wis. Her crew of seven was rescued by the str. *William Nelson*. *Our Son* was carrying pulp wood when she went down. (Bowen, *Memories*, pp. 156–159).

16. The *Empire*, 1200 tons, launched in Cleveland, June 5, 1844, was the first steamboat constructed in the U.S. to measure over 1,000 tons. She was 265 feet long, 36 feet wide (62 feet over the guards) and 15 feet deep. She consumed 600 cords of wood each trip (Albany-Chicago run), which required—as was calculated by one of her passengers—the clearing of over ten acres of well-wooded land. (Mansfield, *History of the Great Lakes*, 1:210).

17. The tug *Reindeer*, 320 tons, was built at East Saginaw in 1863. (Robert C. Laing, "Glimpses of Saginaw's Past," *Saginaw Courier News*, Jan. 21, 1923.

18. Hotchkiss, *History of Lumber Industry*, p. 356.

19. Mansfield, *History of the Great Lakes*, 1:520.

20. Capt. Oscar B. Smith, "The Schooner LaPetite," *Inland Seas*, 26 (Winter, 1970), pp. 275–292.

21. Early tow lines were of 3 and 4 inch dia. hemp. As can be imagined they were very difficult to handle in freezing weather or while riding out a gale. In the late 1880's, when the tow system became more sophisticated, steel hausers replaced the Manila rope. These lines ran on steam-driven winches and capstans which automatically took up or paid out the steel lines between the steamers and their consorts.

    Captain James Davidson, in his West Bay City boat yard, turned out some of the largest wooden steamers and schooner-rigged tow barges that were ever built on the Lakes. When his steamer *City of Glasgow* and consorts *Aberdeen*, *Dundee*, *Grampian* and *Paisley* sailed down the Lakes, the total length of the tow was 3,500 feet. (*Marine Review*, 12 (Cleveland, July 18, 1895), p. 6). His tow barges *Chieftain* and *Montezuma*, built in 1902 and 1903, were each 352 feet long, o.a. Each was strengthened with steel arches and keelsons.

22. To clear the deck load of pine, the fore and aft sails on the lumber vessels had what was known as a "lumber reef" taken in them—the boom jaws were shifted up the mast to clear the load.

23. H.C. Inches and Chester J. Partlow, "Great Lakes Driftwood: Schooner-Scows," *Inland Seas*, 20 (Winter, 1964), pp. 289–294.

24. The charting of all the Lakes was not completed until 1882. The first

V—is for VESSELS—continued

charts were begun in 1841—Lake Erie being the first of the Great Lakes to be charted in 1852. (Pamela Wilson Reeves, "Navigation in Lake Erie, 1825–1860, " 4 parts, *Inland Seas*, 16 (Summer, 1960), pp. 100–110).

25. Mansfield, *History of the Great Lakes*, 1:517.
26. *Ibid.*. 1:707.
27. *Marine Review*, 12 (October 24, 1895). In spite of the *McDougall's* quick launch into the world, she was still "carrying home the bacon" in 1899! (Mansfield, *History of the Great Lakes*, 1:857).
28. *Marine Review*, 9 (Feb. 29, 1894), p. 11, "The method (for caulking) was a simple one. A few bags would be filled with the stuff (manure or sawdust) and after attaching lines to them a man on either side of the vessel would haul them, one at a time, under her bottom. One of the lines was so arranged that a jerk on it would untie the bag, and the suction under the vessel would draw a large part of its contents into the open seams."

Epilogue

1. The Station was known as Red Keg—but its official name was, and still is, Averill—named for the first postmaster, Harrison Averill.
2. Stan Berriman, *Upper Tittabawassee River Boom Towns* (Midland, 1970), p. 152.
3. Paul Wesley Ivey, *The Pere Marquette Railroad Company* (Grand Rapids: Black Letter Press, 1970), pp. 216–217.
4. TBC *Minutes*, p. 103.
5. Berriman, *Upper Tittabawassee*, pp. 154–157.
6. Dustin, *Saginaw History*, p. 130.
7. Robert C. Johnson, "Logs for Saginaw," 2 pts., *Inland Seas*, pt. 2, 5 (Summer, 1949), p. 83.
8. *Ibid.*, p. 84.

   The first successful bag-boom was made by boring longitudinal holes through the centers of 16-ft. logs (3 or 4 feet in dia.) and stringing them on a huge chain like a gigantic necklace. (Pat. No. 359,765, Boom, Benjamin Boutell, Bay City, Mich. Filed Nov. 6, 1886. See: *Official Gazetteer of the U.S. Patent Office*, 38 (Mar. 22, 1887), p.

V—is for VESSELS—continued

1296). The boom was designed to encircle acres of logs which were towed (loose within the floating "fence") by powerful tugs. Another kind of successful boom was made by coupling together sticks of timber (16-ft. logs of large diameter), with 1¼-inch iron chain, which was looped through 6-inch holes bored 2 feet from the ends of each stick—thereby fastening, end to end, the logs that made up the boom. One such bag boom consisted of 690 sticks of pine, which encircled 6 million feet of logs. (Johnson, "Logs for Saginaw," pt. 1 (Spring, 1949), p. 39 and pt. 2 Summer, 1949), p. 83).

9. *Ibid.*, p. 90. The Dingley Tariff, 1897.
10. Allen, *Michigan Log Marks*, p. 39.
11. *Ibid.*
12. Some lumbermen, not willing to terminate their business on the Saginaw, brought rough lumber they manufactured in their mills in Canada to their planing wills on the Saginaw for finishing. Some made wood products like boxes, barrels, shingles and woodenware—others manufactured hardwood flooring. But changing times, conditions and new materials ended these enterprises. The transportation companies that operated fleets of boats to carry logs and lumber from Canada went out of business. By 1914 the Saginaw River was quiet.
13. Allan, *Michigan Log Marks*, p. 39.

Loggers' LINGO

There are hundreds of logging terms—words and phrases commonly employed in work in the woods and on the rivers. Many are unique to a forest region of the United States and Canada. Printed on these pages are words commonly used in the Great Lakes area, brought along with the loggers from the Northeast on their migrations west. For many more terms, please see:

Terms Used in Forestry and Logging. U.S. Dept. of Agriculture, Bureau of Forestry. Bulletin No. 61 (Washington, D.C. 1905). Prepared in cooperation with the Society of American Foresters.

L. G. Sorden and Isabel J. Ebert. *Loggers Words of Yesteryear* L. G. Sorden, pub. (212 N. Allen St., Madison, Wis., 1956).

A

AGUE—Attacks of chills (which causes shaking as with cold) and malarial fever. Caused by the bite of mosquitoes. The following quotation is from a Michigan classic, "Puddleford and its People," written in 1854 by Henry H. Riley:

> "He swore 'that if he ever got well, he'd burn his house, sell his traps, 'bandon his land, pile his family into his cart, hitch on his oxen, and drive 'em, and drive 'em to the north pole, where there warn't no ager, he knew. One minit, 'he was a-freezin', and then he was a-burnin', and then he was a-sweatin' to death, and then he had a well day, and that didn't amount to nothin', for the critter was only gettin' strength to jump on him agin the next.' "

B

BAG-BOOM—White pine logs (16 feet long by 3 or 4 feet in diameter), fastened together, end to end, making a kind of enclosure or "fence." The boom was drawn around loose floating pine logs which were towed by powerful tugs over the lakes from the cutting area to the mill. These gigantic rafts, which sometimes took up as much as ten to twenty-five acres of the lake's surface, resembled the outline of a large bag when floating—hence the name bag-boom.

BANKING GROUNDS—Also called a rollway or decking ground. An inclined way on the river bank where logs were decked or stacked to await the spring log drives.

BANNOCK—Wilderness bread. An old Gaelic word, originally meaning oat or barley-meal cake cooked on a griddle. Woodsmen mixed flour and water, salt and some saleratus (baking powder) into a thick dough. The dough was formed into a loaf and placed into a covered pan which was set in hot coals (with a few coals sprinkled on the lid). Bannock could be cooked in fifteen minutes.

BARK MARK—A log owner's symbol (or mark) chopped into the side of a log to indicate ownership.

BARN-BOSS—The man in charge of the lumber camp stables.

BATEAU—A kind of boat used on river drives to carry tools and other drive paraphernalia. The boats had flat bottoms and sharp, pointed ends. They were operated by oars.

BEAN HOLE—Hole in the ground or fireplace in which beans were cooked.

BINDERS—Springy poles used for tightening binding chains "wrapped" over a sleigh load of logs.

BINDING CHAIN—Chains to hold logs on a sleigh. Also called "wrappers."

BINDLE—A blanket roll. In the early days, a woodsman brought his clothing and other belongings to camp wrapped in his own blanket. Also called "dunnage"—"duffle."

BOARD FEET MEASURE—A standard measure—a fixed quantity accepted by buyers and sellers in determining the contents of a log—containing 144 cubic inches of sawed lumber—or the equivalent of a board 12″ × 12″ in surface and 1″ thick.

BOILING UP—A term the woodsmen used when washing their clothing and blankets. "Boil up day" was usually on Sunday.

BULL-OF-THE-WOODS—The boss or foreman of the lumber camp.

BULL-WHACKER—The man in charge of the oxen.

BUNK(1)—The beds on which the woodsmen slept in the bunk-camp. Nothing more than wide shelves built along one or more walls.
(2) Heavy timbers or cross pieces on logging sleighs on which logs rested when being transported from one place to another.

C

CALKS—Also "corks"—sharp, steel projections in woodsmens' and rivermens' boots to keep them from slipping. Horses wore calked iron shoes.

CAMBOOSE—The word is probably a derivation from the French *kambuis* or English word *caboose* which is a ship's galley or deck room for cooking. In the old fashioned lumber camp, the camboose or caboose was a large fireplace built in the center of the woodsmens' living quarters.

CANNON,To—When loading logs, they would sometimes swing crosswise, or "cannon" instead of rising parallel to the load.

CANT HOOK—A stout, tapered wooden pole, three to five feet long, with a seven to ten inch loose, dangling steel hook near its working end. Used for canting or moving logs, to provide needed leverage. Also call "crooked steel."

CAPERING CLOTHES—Dress up or party clothes.

CAPSTAN—An upright windlass. A drum or cylinder that revolves on a spindle. Used for moving heavy weights by rope or cable passing around the drum.

CAULK—To make a boat or ship watertight by filling the seams with oakum, tar, etc.

CHICKADEE—The men in camp whose task it was to keep the sleigh roads free of horse or ox droppings, tree limbs and other impediments to sleigh runners. Also called "road monkeys."

CHORE BOY—Also "cookee," "crumb boss," "cook's devil," "flunkie," and other unflattering names. The menial around the camp—the cook's helper.

CONK-"KONKUS"—An old expression dating back to the early lumbering days in Maine and in Canada. Used when describing a pine tree that was diseased.

COOK—Also called "stomach-robber, "boiler," "bum-cook," "sizzler," "dough-pounder."

COOKEE—See: Chore Boy.

CORKER—A shipyard worker who works with caulking tools to make vessels watertight.

CORKS—See: Calks.

CROOKED STEEL—See: Cant Hook.

CROSSCUT—A saw, from six to eight feet long, equipped with handles at each end. Before 1876, the saw was used in the lumber woods for cutting up felled trees. After that date, the improved crosscut saws were used to fall the trees as well. Also called "Swedish fiddle," "briar."

CROSS-HAUL—A cleared space where a horse team moved when loading logs at a skidway or landing.

CULL—A log that has been rejected. Also the lowest grade of lumber.

CUTTER—A small lightweight sleigh, usually drawn by one horse.

D

DEACON'S SEAT or BENCH—A bench built along the front of the bunks in the bunk-camp.

DEAL—The word "deal" was used when speaking of boards cut three inches thick. A plank was two inches thick. Anything under two inches was called lumber.

DEATH WARRANTS—A term Great Lakes woodsmen had for hospital tickets or certificates they purchased (after 1873) for $5 a year which entitled them to care in participating hospitals. A kind of hospital insurance.

DINGLE—An alleyway or roofed over space between the cook-camp and bunk-camp. A storage area for things like meat, stove wood and the salt barrel, which was used jointly by the cook and the barn-boss.

DINNER HOLE—Any place where the woodsmen gathered to eat at noon while out on the job.

E

EPIZOOTY—A kind of distemper or the "strangles"—an infectious disease of horses characterized by inflamation of the mucous membrane of the respiratory tract.

EVENER—or EQUALIZER. A wooden device used when two horses are harnessed together as a team—to equalize the pulling power.

F

FID-HOOK—A slender, flat hook used to keep another hook from slipping off a chain.

FLAGGINS—The woodman's noon-time meal, eaten out on the job.

FLY-BOOM—A boom (logs fastened end to end) floating in the midst of free-floating logs. At the cry, "jam below!" the fly-boom was pulled across the stream and fastened to the river banks to stop the flow of logs until the jam was straightened out.

FOREMAN—The boss of the lumber camp. He was also called "king-pin," "bully," "head-rig," the "main-say" and "bull-of-the-woods."

FOUND—Room and board—sometimes in addition to wages.

G

GIN POLE—A single pole with pully and tackle (secured by guy ropes), for loading logs. Also called a "jammer."

GO-BACK ROAD—The return road over which empty sleighs traveled back from the banking grounds for another load of logs.

GRAB HOOK—A hook having a narrow throat, adapted to grasp any link of a chain.

GRAYBACKS—Lice. Also called "blue jackets," "crumbs," "seam squirrels."

GROUSER—(1) The anchor on the wangun. A stick of timber rigged at the stern of the cook-scow—made to drop through a hole in the raft and into the river bottom. The pointed end of the timber held the craft fast. The device was lowered or raised by pegs that were fitted into it.

(2) A piece of wood, a rock—or anything used as a chink to even a load of logs on a sleigh. Also called "dutchman."

H

HARDTACK OUTFIT—A camp that set a poor table. From the hard and cheap Swedish bread (called "hardtack") that was served in some camps.

HOSPITAL TICKET—See: Death Warrant.

I

ICER—A large, horse-drawn wooden tank, mounted on sleigh runners, from which water was distributed on the sleigh roads during freezing weather. Each end of the sprinkler had its own tongue (the pole to which the horses were hitched), so that the contraption could be hauled in either direction without having to turn around. Also called "sprinkler, "water tank," "ice box," "water box."

INK-SLINGER—The lumber camp bookkeeper or clerk—the man who kept the camp records. He sometimes doubled as the scaler.

IRON-BURNER—The camp blacksmith.

J

JAM—A stoppage, or a congestion of logs in a stream due to an obstruction, low water or dangerous bend in the river.

JAMMER—See: Gin Pole.

K

KERF—A gash in a log, made by a saw.

L

LANDING—See: Banking Grounds.

LIGHTER—A shallow draft boat or barge used for unloading or loading vessels not lying at wharves (usually because of low water). Also used for transporting freight around harbors.

LOG MARK—Symbol of ownership. Designs marked or stamped on logs made up of combinations of initials, circles, triangles and figures.

LOG RULE—Also called a log "scale," "cheat stick," "money-robber," "money-maker," "robber's cane." A wooden stick similar in appearance to a common yardstick with which the scaler measured logs.

LUMBER—Boards under two inches thick.

LUMBER SHOVER—A dock worker—also dock "walloper."

M

MARKING HAMMER—A heavy, hammer-like tool that was used to strike a log owner's mark on the ends of his logs. The owner's mark design was raised on the striking surface of the hammer so it would cut into the soft wood of a pine log. Also called "stamping ax," "stamping iron," "branding ax."

MULAY or MULEY SAW—Derived from the German *muhl säge*, a saw with one perpendicular six-foot by ten-inch blade that cut only on the down stroke. This saw was put into retirement by the faster cutting circular saws. Both saws made a quarter-inch kerf in a log.

N

NOSE BAG—(1) A heavy canvas pail-like bag from which horses were fed grain while away from the stables. The bags were held on by straps and fitted over the horses' noses. Also called "feed bags."

(2) A haversack, made out of a grain sack or canvas in

N—continued

which rivermen carried their mid-morning, noon and afternoon lunches. The men wore their "nose-bags" behind them, strapped on or held in place by a piece of rag or rope that they tied over their shoulders or around their waists.

O

OFF OX/NEAR OX—Standing at the rear of a brace of oxen, the right hand ox is the "off ox." The "near ox" is the animal that works on the left side.

P

PACKS or PACS—Winter footwear. Low boots of mocassin construction. Made of soft, oil tanned leather—usually worn with several pairs of wool socks or a felt boot inside.

PEAKER—Top log on a pile or load of logs.

PEAVEY—The riverman's tool, invented by Jospeh Peavey of Stillwater, Maine in 1858. Similar to a cant hook. A stout wooden pole five to seven feet long, fitted with a curved, dangling steel hook at the business end. The tool differs from a cant hook in that the peavey has a pike or sharp point, instead of a toe ring and lip at its end.

PIE-FORK—A three-foot long, hardwood paddle or a two-tined fork the cook used to pull out hot pie pans and bread from the oven. Also used for turning meat in pans, etc.

PIKE POLE—A long pole, from twelve to twenty feet long, with a sharp pike or spike and stationary claw hook on one end. Used by rivermen to maneuver logs.

PIER DAM—Also called a "wing dam." A pier built from the shore, usually slanting downstream, to guide logs past an obstruction or a bend in the river.

PIN WHACKER—Rivermen who worked at the booming grounds. They used a wooden mallet to drive rafting pins into logs being made into rafts. (See: Rafting pins).

POWDER-MONKEY—The man who handled dynamite.

PUNG—A small, hand-sled—usually a rude, oblong box set on poles or runners.

R

RAFTING PIN—(the kind used by the Tittabawassee Boom Company). UP TO 1874, oak spikes (8″–12″ long) and rope were used to fasten logs together to make up rafts. When enough logs were assembled for a raft, rope was laid down the center of the lot. The pin whacker drove a pin about three quarters of the way into the center of each log and then took up the rope and made a double half hitch around each pin—going from pin to pin (log to log) until all the logs were fastened together into a raft. AFTER 1874—a crotched, wedge shape oak pin (2″ × 7″ with 3-1/2" slot) was used to make rafts. The rope was laid down the center of the assembled logs, the pin-whacker going from log to log, driving the crotched pins *over the rope* to staple it in place. Mill men on the Saginaw preferred wooden rafting pins over those made of metal. The saws went through the wooden pins (if they were left in the logs)—unlike metal pins, the kind that was used on the Muskegon.

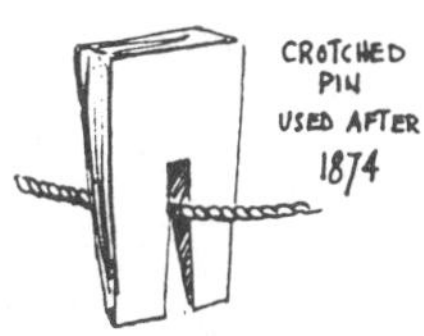

REARS The end of a log drive. Bringing down the "rears," or the "draggers," meant bringing down the last logs on the river drive.

RIVERMEN—The men who worked the logs on the river. Also called "riverhogs," "riverpigs"—and when the water was swift, "white-water men."

ROAD MONKEY—Road workers in the lumber camp. Sometimes called "blue-jays."

ROLLWAY—See: Banking Grounds.

ROSS, To—To trim the bark off the ends of logs so they would slide or drag more easily over the ground.

ROUGH and TUMBLE LANDING—A banking ground, rollway or landing where no attempt was made to arrange logs in tiers—where the pine was dumped like giant-size jack straws on the banks.

RULE—See: Log Rule.

S

SACKERS—Men who retrieved logs left high and dry on the banks or on obstructions in the river. "Shoe-fly" sackers were men who did a poor job.

SAW-LOG JACKS—Old time lumbermen who cut long logs.

SCALER—Man who measured logs in the lumber camp—who determined the number of board feet of lumber the logs would produce at the mill. The scaler's tool was the log rule.

SCANTLINGS—Waste from mill saws.

SEND-UP MEN—The members of the loading crew, who stood with cant hooks at the foot of a log pile (or the side of a sleigh) and guided logs to the top of the load.

SHOOK—A set of staves and "headings" sufficient for one barrel, cask, hogshead.

SKID, To—To haul or drag logs over the ground from the stump, to the skidway, landing, etc.

SKIDDING CHAIN—A chain, usually sixteen feet in length, with a hook at each end. Used for skidding or dragging logs over the ground.

SKIDDING TONGS—Two large curved iron hooks pivoted or hinged together for seizing or grasping the ends of logs for skidding.

SKIDS—Poles or logs, commonly used in pairs, on which logs were piled at the skidways or landings. Also used when logs were rolled up onto sleighs.

SKIDWAY—Way stations along the main sleigh roads where logs were piled as they were brought from the stump.

SNOOSE—Damp snuff—commonly called "Scandihoovian dynamite."

SPRINKLER—See: Icer.

STAGS—A woodsman's boots or pacs with the tops cut off—his "house slippers."

STONE-BOAT—A wooden conveyance with runners that resembles a stout sled on which heavy objects can be transported over the ground.

STUMPAGE—Standing timber. When a lumberman purchased "stumpage," he bought the trees, not the land on which the trees grew. The owner of the land retained his ownership after the trees were cut down.

SWAMP HOOK—A large hook at the end of a chain.

T

TOP-LOADER—The head loader of a loading crew—the man who placed logs as they were sent up on sleighs or log piles.

TOTE—To carry or haul. The tote-beam hauled in camp supplies over the tote-road.

TRAILING DOWN—Skidding from the stump to the skidway.

TRAVOIS—French-Canadian word. Also called a "dray," "go-devil," "bob," "crotch," "drag-sled," "lizard," "scoot," "travoy." A short sled with one set of runners. Used for hauling logs. One end of a log rested on the travois—the other end (which was rossed), dragged behind on the ground. Crotched logs (tree trunk divided into two branches) made good travois'.

TRIP-HOOK—A levered hook fastened at the end of a chain which was attached to the evener of a horse team. When the teamster tripped the lever on the hook, he automatically released the horses from the chain without having to stop the team.

TURKEY—A bag or grain sack used for carrying a woodsman's outfit (his clothing and personal belongings).

V

V-PLOW—A heavy snowplow made of timbers fastened together at the front end and kept apart at the rear by a "spreader." Used in lumber camps for breaking out and smoothing winter roads.

VAN—A small building in the lumber camp where the foreman, the scaler or bookkeeper slept—where a small inventory of store goods (mitts, socks, tobacco, patent medicines, etc.) was kept to sell to the logging crew. The camp office.

W

WALKING BEAM ENGINE—A vertical steam engine used in boats and tugs. A beam, pivoted in the middle and connected to a piston rod at one end and a flywheel at the other which could be

W—continued

observed above the deck. It had a characteristic see-saw motion when under power. (See: Picture of the *Evening Star*).

WALLOPER—See: Lumber Shover.

WANGUN—A raft or rafts (barges) on which the cook-shack and sleeping quarters for rivermen were erected. Also a storage place for drive equipage. Other spellings: "wanigan," "wangan," "wanagun," etc. Also called an "ark."

WHING-DING—A "spree" or a "bender." A good time.

WIDOW-MAKER—A falling tree branch—a potential killer that put many woodsmen out of commission.

WOOD-BUTCHER—Woodsmen who had skill for building. Lumber camp carpenter.

Y

YOU-KNOW—A tin wash basin for taking up collections in camp. Sometimes a coin was soldered to the bottom of the pan to remind the men to give at least that amount.

INDEX

*Illustration.

S

T